D0086515

Construction Safety and the OSHA Standards

Construction Safety and the OSHA Standards

David L. Goetsch

330 Hudson Street, NY NY 10013

Vice President, Portfolio Management: Andrew Gilfillan
Portfolio Manager: Tony Webster
Editorial Assistant: Lara Dimmick
Senior Vice President, Marketing: David Gesell
Field Marketing Manager: Thomas Hayward
Senior Marketing Coordinator: Les Roberts
Marketing Coordinator: Elizabeth MacKenzie-Lamb
Director, Digital Studio and Content Production: Brian Hyland
Managing Producer: Cynthia Zonneveld
Managing Producer: Jennifer Sargunar
Content Producer: Faraz Sharique Ali
Content Producer: Nikhil Rakshit
Manager, Rights Management: Johanna Burke
Operations Specialist: Deidra Smith
Cover Design: Cenveo Publisher Services
Cover Art: Shutterstock
Full-Service Project Management and Composition: R. Sreemeenakshi/SPi Global
Printer/Binder: LSC Communications
Cover Printer: LSC Communications
Text Font: Sabon LT Pro

Library of Congress Cataloging-in-Publication Data

Names: Goetsch, David L.
Title: Construction safety and the OSHA standards / David L. Goetsch.
Description: Second edition. | Boston : Pearson Education, [2018] | Includes
bibliographical references and index.
Identifiers: LCCN 2016042565 | ISBN 9780134420189 (0-13-442018-7)
Subjects: LCSH: Building—Safety measures—Standards—United States.
Classification: LCC TH443 .G575 2018 | DDC 690/.22—dc23 LC record available at https://lccn.loc.gov/2016042565

1 16

ISBN 10: 0-13-442018-7
ISBN 13: 978-0-13-442018-9

Preface

BACKGROUND

The field of construction safety and health has undergone significant change over the past two decades. Some of the more prominent reasons for this change include technological changes that have introduced new hazards at construction sites; proliferation of health and safety legislation and corresponding regulations; increased pressure from regulatory agencies; realization that productivity is increased in a safe and healthy environment; rising health-care and workers' compensation costs; a growing interest in ethics and corporate social responsibility; increased pressure from labor organizations, workers, and the public in general; and an increasing number of incidents of workplace violence.

All of these factors, when taken together, have made the job of the modern construction professional more challenging and more important than ever. These factors have also created the need for an up-to-date book on construction safety and the Occupational Safety and Health Administration (OSHA) standards that contains the latest information needed by people who are responsible for safety and health in one of the most potentially dangerous industries—the construction industry.

WHY WAS THIS BOOK WRITTEN AND FOR WHOM?

This book was written to fulfill the need for an up-to-date practical teaching resource that focuses specifically on the needs of modern construction professionals and on the construction requirements set forth by OSHA and other regulatory agencies. The book is intended for use in universities, colleges, community colleges, technical schools, and corporate training settings where programs, courses, or seminars in construction safety and health are offered. Educators in disciplines such as construction engineering, construction technology, construction management, and the various construction-related trades (e.g., carpentry; heating, ventilation, and air conditioning; electricity; plumbing) will find this book both valuable and easy to use. The direct, straightforward presentation of material focuses on making the theories and principles of construction safety and health practical and useful in a real-world setting. Up-to-date research has been integrated throughout in a down-to-earth manner.

NEW TO THIS EDITION

- Up-to-date photographs that illustrate chapter content have been added throughout the text.
- OSHA standards and related information have been updated throughout the text.
- New section on role of disasters in the safety movement has been added.
- New section on Globally Harmonized System of Classifications and Labelling has been added.
- New section on design for construction safety has been added.
- New section giving examples of the more common types of construction accidents has been added.
- New section on how to find safety solutions through online databases has been added.
- New section on communicating with workers in the aftermath of a violent incident has been added.
- New section on OSHA's Hearing Conservation Program has been added.
- New section on process safety management has been added.
- New section on working near or over water has been added.
- New section explaining the ANSI Z359 Fall Protection Code has been added.
- New section on rescuing a fallen worker has been added.
- New section on OSHA's Fall Protection Directive has been added.
- New sections on 29 CFR 1926 Subparts AA and CC have been added.

ORGANIZATION OF THE BOOK

The text contains 17 chapters, each focusing on a major area of concern for modern construction professionals. The chapters are presented in an order that is compatible with the typical organization of a college-level construction safety and health course. A standard chapter format is used throughout the book. Each chapter begins with a list of major topics covered and ends with a comprehensive summary. After the summary, each chapter contains review questions, key terms and concepts, critical thinking or discussion activities, application activities, and endnotes. Within each chapter there are two or more "Safety Facts & Fines" boxes, which are brief real-world cases that show what may happen when safety regulations and safety practices are ignored.

Fictitious Names

The author occasionally uses fictitious names of both companies and individuals in the "Critical Thinking and Discussion Activities" that appear at the end of each chapter. Such names are also used occasionally in figures within chapters. These fictitious companies and individuals are used to illustrate principles and concepts explained in the chapters in question. Although the situations described are all based on actual events, the names of the companies and individuals are fictitious. Any similarities to actual companies or individuals are purely coincidental.

ABOUT THE AUTHOR

David L. Goetsch is Emeritus Vice President and full professor at Northwest Florida State College. He teaches construction safety and occupational safety in both the traditional and distance learning formats.

ACKNOWLEDGMENTS

The author wishes to acknowledge the invaluable assistance of the following people in developing this book: Savannah Marie King for her help in proofreading and editing the original manuscript, writing numerous "Safety Facts & Fines" boxes, and taking photographs; Caroline McCoy for word processing of the manuscript; and Larry Leiman for reviewing and validating the OSHA standards.

The author also acknowledges the reviewers of this text: Burl George, Bradley University; Eric Marks, University of Alabama; Timothy Piotrowski, Alfred State College; Mark Prenzin, Bowling Green State University; and Dennis Roberts, the University of New Mexico.

Brief Contents

Contents

SAFETY MOVEMENT AND THE CONSTRUCTION INDUSTRY

LEARNING OBJECTIVES

- Provide a brief overview of the construction industry.
- Explain the concept of *liability* as it applies to the construction industry.
- Define the concept of the *competent person*.
- List several major milestones in the safety movement.
- Describe the role organized labor has played in the safety movement.
- Explain the role specific health problems have played in the safety movement.
- Describe the development of accident-prevention programs.
- Describe the development of safety organizations.
- Explain the Globally Harmonized System of Classification and Labeling as it applies to the construction industry.
- Explain how language barriers can affect safety in the construction industry.
- Explain how the concept of *Prevention through Design* applies to construction professionals.

The **safety movement** in the United States has developed steadily since the early 1900s, when workplace accidents were common in this country; for example, in 1907, more than 3,200 people were killed in mining accidents alone. Legislation, precedent, and public opinion all favored management during this period. There were few protections for the safety of workers.

Today, working conditions for employees, including construction employees, have improved significantly. The chance of a worker being killed on the job is less than half of what it was 60 years ago.[1] According to the National Safety Council (NSC), the current death rate from work-related injuries is approximately 4 per 100,000—less than a third of the rate 50 years ago.[2] However, among the various industrial sector jobs, construction continues to be the most dangerous. Construction jobs account for just 5 percent of the workforce in the United States, but account for more than 17 percent of the workplace deaths every year. One construction employee in seven is injured every year. Many of the improvements in safety up to this point have been the result of pressure for legislation to promote safety and health, the steadily increasing costs associated with accidents and injuries, and the emergence of safety management as a profession. Improvements in the future are likely to come as a result of greater awareness of the cost-effectiveness and resultant competitiveness gained from a safe and healthy workforce.

This chapter examines the history of the safety movement in the United States as it relates to the construction industry and how it has developed over the years. Such a perspective helps practicing and prospective construction professionals form a better understanding of the importance of safety and where it fits into their range of responsibilities.

OVERVIEW OF THE CONSTRUCTION INDUSTRY

The term *construction industry* implies the existence of a single uniform entity that is the builder of all the world's projects. In reality, the construction industry is an umbrella concept that encompasses a broad array of specialized crafts, occupations, and professions, (e.g., residential, commercial, industrial, highway, bridge, structural steel, pre-stressed concrete, poured-in-place concrete, carpentry, plumbing, electrical, HVAC, glass, flooring, and demolition contractors to name just a representative sample). Construction is big business in the United States, accounting for more than $600 billion in value annually and employing more than 4.5 million people. Construction companies, both large and small, should be viewed and operated as businesses. One key to success in business is minimizing cost. Providing a safe and healthy workplace is one of the most effective strategies for holding down the cost of doing business.

The construction industry has changed markedly over the years—specialization is just one of the ways, Figure I–1. All of the changes, whether positive or negative, that have occurred in the industry have had an effect on the safety of the construction job site. In fact, some developments have made certain aspects of construction work safer, while also introducing even more dangerous hazards.

FIGURE I–1 Various construction specializations.
Source: Luis Louro/Fotolia

How and Why Construction Has Changed over the Years?

Advances in technology have continually changed the construction industry. In some cases, these advances have made job sites safer, healthier places. In other cases, they have increased the number and seriousness of the hazards that workers confront daily. The following forces have played the most influential roles in changing the nature of the construction industry:

- *Inanimate power.* Electricity, steam, hydraulics, and pneumatics allowed the construction industry to move beyond the limits of human and animal power. Consider the tools and equipment at a modern construction site that would be rendered useless without these various forms of **inanimate power**. On the other hand, consider all of the new hazards these technological marvels have introduced (e.g., electrocution—typically the leading cause of death on the job in the construction industry).
- *Machines.* People historically did much of the work that is now done by machines in the construction industry. For example, forklifts now easily lift, move, and stack materials that, many years ago, would have been handled manually. The forklift has probably reduced back injuries exponentially. However, a forklift operated by a poorly trained or reckless individual is a dangerous piece of equipment.
- *Materials.* Advances in the science of metallurgy, developments in the field of plastics, the introduction of composite materials, and ongoing improvements in the chemistry of concrete have broadened the capabilities of the construction industry immeasurably. On the other hand, materials such as asbestos—once called the *miracle material*—have introduced deadly hazards into the workplace.
- *Work specialization.* Work specialization has created a wide array of focused crafts within the construction industry. Although once a contractor might have performed most or at least many of the tasks associated with constructing a building or another structure, technological advances coupled with increased regulation have forced people to specialize to keep up with all there is to know in a given field. The work of the contractor from an age past is now accomplished by carpenters, lathers, millwrights, painters, glaziers, electricians, masons, roofers, pile drivers, stonemasons, tilesetters, and many other specialists. Specialization typically promotes more thorough knowledge within a field, but it decreases knowledge and understanding across fields. This lack of understanding can increase the potential for accidents when a variety of specialists are working at the same job site.

LIABILITY IN THE CONSTRUCTION INDUSTRY

The United States has become the most litigious of the industrialized countries of the world. Legal liability is always an issue in our lawsuit-prone society. This is true in every industrial sector, but it is especially true in the construction industry because of the hazardous nature of the work. In fact, a construction company can be held at least partially liable for the safety and health of workers, even when it is not the employer of record. For example, a general contractor can be held partially liable when an employee of a subcontractor is injured. The subcontractor and the contractor might, in legal parlance, be assigned *shared liability*.

The concept of shared liability, in turn, tends to promote and encourage lawsuits. It does this by creating greater "targets of opportunity" for those who might want to sue. A target of opportunity is a company that is likely to have the so-called deep pockets to pay a major settlement. Since many subcontractors are small businesses, they might not be viewed as sufficiently able to pay, even if a lawsuit filed against them is clearly viable. Attorneys often shy away from cases involving companies that have insufficient ability to pay. However, when small subcontractors can be coupled with a larger, more financially endowed general contractor—a company perceived to have *deep pockets*—a greater target of opportunity is created.

The concept of shared liability contributes to the following potential problems for construction companies:

- Construction companies can find themselves named in a lawsuit filed by someone they do not even employ when the concept of shared liability is used to couple them with another company.
- Courts often hold general contractors at least partially liable for the actions of their subcontractors, claiming that, if they did not know about unsafe conditions, they should have.
- Construction companies that serve as general contractors are expected to exercise control over all aspects of a construction project and can be held liable.
- The **Occupational Safety and Health Administration (OSHA)** tends to hold all parties accountable (general contractors and subcontractors), even if they are not directly involved in a given violation.

The best defense against lawsuits is to ensure safe and healthy job sites. In addition, the following specific strategies can help protect construction companies that serve as general contractors from both legitimate and malicious lawsuits:

- Make it part of the written contract that subcontractors are expected to comply with all applicable OSHA standards. Make sure that all parties understand their shared responsibilities.
- Make it a common and frequent practice to inspect the work and job sites of subcontractors.
- Require subcontractors to show that they have an established record of working safely.
- Develop a comprehensive safety policy, share it with all employees and subcontractors, and make compliance a part of the written contract with all subcontractors.
- Make sure that all subcontractors carry the appropriate workers' compensation insurance.

COMPETENT PERSON REQUIREMENT

OSHA's construction standard (29 CFR 1926) requires that contractors designate at least one individual to serve as the *competent person* for regularly inspecting the safety and health of job sites. The term *designated* means that the individual is assigned specific responsibilities by the contractor and is expected to carry out those responsibilities. To be considered *competent,* an individual must have subject-matter knowledge of the types of work to be inspected and must be able to demonstrate competence through an appropriate level of education or experience, or both. A competent person must be designated for every project the company has under contract at a given time. More specific requirements relating to the competent person are set forth in various regulations promulgated by OSHA. These requirements are explained in the text when the regulations in question are discussed.

MILESTONES IN THE SAFETY MOVEMENT

The comprehensive safety and health standards (29 CFR 1926; construction standards are found in the grouping numbered 1926) that regulate work in the construction industry are a relatively new phenomenon. There are many historic milestones in the movement that eventually led to passage of the Occupational Safety and Health Act (OSH Act) and establishment of OSHA in 1970.

Just as the United States traces its roots to Great Britain, the safety movement in this country traces its roots to England. During the Industrial Revolution, child labor was common. Work hours were long, the work hard, and the conditions often unhealthy and unsafe. After an outbreak of fever among working children, the people of Manchester, England, began demanding better working conditions. Public pressure eventually forced a government response, and in 1802, the Health and Morals of Apprentices Act was passed. This was a milestone piece of legislation: it marked the beginning of government involvement in workplace safety.

During the early years, hazardous working conditions were common in the United States. Then, after the Civil War, the seeds of the safety movement were sown in this country. Factory inspection was introduced in Massachusetts in 1867. In 1868, the first barrier safeguard was patented. In 1869, the Pennsylvania legislature passed a mine safety law requiring two exits from all mines. The Bureau of Labor Statistics (BLS) was established in 1869 to study workplace accidents and report pertinent information about those accidents.

The following decade saw little new progress in the safety movement until 1877, when the Massachusetts legislature passed a law requiring safeguards for hazardous equipment. The same year also saw passage of the Employer's Liability Law in Massachusetts, establishing the potential for **employer liability** in workplace accidents. In 1892, the first recorded safety program was established in a Joliet, Illinois steel plant in response to a scare caused when a flywheel exploded. Following the explosion, a committee's recommendations were used as the basis for the development of a safety program that is considered to be the first in American industry.

Around 1900, Frederick Taylor began studying efficiency in the workplace. His purpose was to identify the impact of various factors on efficiency, productivity, and profitability. Although safety was not a major focus of his work, Taylor did draw a connection between lost personnel time and management policies and procedures. This connection between safety and management represented a major step toward broad-based safety consciousness.

In 1907, the U.S. Department of the Interior created the Bureau of Mines to investigate accidents, examine health hazards, and make recommendations for improvements. Mining workers definitely welcomed this development because more than 3,200 of their fellow workers were killed in mining accidents in 1907 alone.[3]

One of the most important developments in the history of the safety movement occurred in 1908, when an early form of **workers' compensation** was introduced in the United States. Workers' compensation actually had its beginnings in Germany. The practice soon spread throughout the rest of Europe. Workers' compensation, as a concept, made great strides in the United States when Wisconsin passed the first effective workers' compensation law in 1911. In the same year, New Jersey passed a workers' compensation law that withstood a court challenge.

The common thread among the various early approaches to workers' compensation was that they all provided some amount of compensation for on-the-job injuries regardless of who was at fault. When the workers' compensation concept was first introduced in the United States, it covered a very limited portion of the workforce and provided only minimal benefits. Today, all 50 states have some form of workers' compensation that requires the payment of a wide range of benefits to a broad base of workers. Workers' compensation is examined in more depth in Chapter 5.

The Association of Iron and Steel Electrical Engineers (AISEE), formed in the early 1900s, pressed for a national conference on safety. As a result of AISEE's efforts, the first meeting of the **Cooperative Safety Congress (CSC)** took place in Milwaukee in 1912. What is particularly significant about this meeting is that it planted the seeds for the eventual establishment of the NSC. A year after the initial meeting of the CSC, the **National Council of Industrial Safety (NCIS)** was established in Chicago. In 1915, this organization changed its name to the National Safety Council. It is now an important safety organization in the United States.

From the end of World War I in 1918 through the 1950s, safety awareness grew steadily. During this period, the federal government encouraged contractors to implement and maintain a safe work environment. Also, during this period, industry in the United States arrived at two critical conclusions: (1) there is a definite connection between quality and safety and (2) off-the-job accidents have a negative impact on productivity. The second conclusion became painfully clear during World War II when the call-up and deployment of troops had employers struggling to meet their labor needs. For these employers, the loss of a skilled worker due to an injury, or for any other reason, created an excessive hardship.[4]

The 1960s saw the passage of a flurry of legislation promoting workplace safety. The Service Contract Act of 1965, the Federal Metal and Nonmetallic Mine Safety Act, the Federal Coal Mine and Safety Act, and the Contract Workers and Safety Standards Act all were passed during this decade. As their names indicate, these laws applied to a limited audience of workers.[5]

These were the primary reasons behind passage of the OSH Act of 1970. This federal law represents the most significant legislation to date in the history of the safety movement. Some milestones in the occupational safety movement that have occurred since that time include the following:

- OSHA Training Institute established, OSHA issues its first standard (asbestos), and the first OSHA state plans are approved (South Carolina and Oregon) in 1972
- Federal Mine Safety Act in 1977
- OSHA begins providing training and education grants in 1978
- OSHA coverage is extended to federal workers by executive order of the President in 1980
- OSHA established Voluntary Protection Programs in 1982
- Right-to-Know regulations are passed in 1983
- The "Egregious Violation Enforcement Policy" is established in 1986
- Safety and Health Program Management Guidelines are issued by OSHA in 1989
- California adopts an ergonomics standard in 1997
- The "Global Harmonization System" for hazard communication is established in 2009
- The "Injury and Illness Prevention Program" initiative is established in 2010
- Construction companies show more interest in adopting pro-active safety strategies that will establish a culture of safety in construction 2015-forward

ROLE OF ORGANIZED LABOR

Organized labor has played a significant role in the development of the safety movement in the United States. From the outset of the Industrial Revolution in this country, organized labor has fought for safer working conditions and appropriate compensation for workers injured on the job. Many of the earliest developments in the safety movement were the result of long, hard-fought battles by organized labor.

Although the role of unions in promoting safety is generally acknowledged, one school of thought takes the opposite view. Proponents of this dissenting view hold that union involvement actually slowed the development of the safety movement. Their theory is that unions allowed their demands for safer working conditions to become entangled with their demands for better wages, and, as a result, they met with resistance from management. Regardless of the point of view, there is no question that working conditions in the earliest years of the safety movement often reflected an insensitivity to safety concerns on the part of management.

Among the most important contributions of organized labor to the safety movement was their work to overturn antilabor laws relating to safety in the workplace. These **employer-biased laws** included the fellow servant rule, the statutes defining contributory negligence, and the concept of assumption of risk.[6] The **fellow servant rule** held that employers were not liable for workplace injuries that resulted from the negligence of other employees. For example, if Worker X slipped and fell, breaking his back in the process, because Worker Y spilled oil on the floor and left it there, the employer's liability was removed. In addition, if the actions of employees contributed to their own injuries, the employer was absolved of any liability. This was the doctrine of **contributory negligence**. The concept of **assumption of risk** was based on the theory that people who accept a job assume the risks that go with it; in other words, employees who work voluntarily should accept the consequences of their actions on the job rather than blaming the employer.

Because the overwhelming majority of workplace accidents involve negligence on the part of one or more workers, employers had little to worry about. Therefore, they had little incentive to promote a safe work environment. Organized labor played a crucial role in bringing deplorable working conditions to the attention of the general public. Public awareness and, in some cases, outrage, eventually led to these employer-biased laws being overturned in all states except one. In New Hampshire, the fellow servant rule still applies.

ROLE OF SPECIFIC HEALTH PROBLEMS

Specific health problems that have been tied to workplace hazards have played significant roles in the development of the modern safety and health movement. These health problems contributed to public awareness of dangerous and unhealthy working conditions, which, in turn, led to legislation, regulations, better work procedures, and better working conditions.

Lung disease in coal miners was a major problem in the 1800s, particularly in Great Britain, where much of the Western world's coal was mined at the time. Frequent contact with coal dust led to a widespread outbreak of anthracosis among Great Britain's coal miners. Also known as the *black spit,* this disease persisted from the early 1800s, when it was first identified, until 1875, when it was finally eliminated by such safety and health measures as ventilation and decreased work hours.

In the 1930s, Great Britain saw a resurgence of lung problems among coal miners. By the early 1940s, British scientists were using the term *coal-miner's pneumoconiosis* (CMP) to describe a disease from which many miners suffered. Great Britain designated CMP a separate and compensable disease in 1943. However, the United States did not immediately follow suit, even though numerous outbreaks of the disease had occurred among miners in this country.

The issue was debated in the United States until Congress finally passed the Coal Mine Health and Safety Act in 1969. The events that led up to passage of this act were tragic. An explosion in a coal mine in West Virginia in 1968 killed 78 miners. This tragedy focused attention on mining health and safety, and Congress responded by passing the Coal Mine Health and Safety Act. The act was amended in 1977 and again in 1978 to broaden the scope of its coverage.

Over the years, the diseases suffered by coal miners were typically lung diseases caused by the inhalation of coal dust particulates. However, health problems were not limited to coal miners. Other types of miners developed a variety of diseases—the most common of which was silicosis. Once again, it took a tragic event—the Gauley Bridge disaster—to focus attention on a serious workplace problem.

A company was given a contract to drill a passageway through a mountain near the city of Gauley Bridge, West Virginia. Workers spent as much as 10 hours per day breathing the dust created by drilling and blasting, and this particular mountain had an unusually high silica content. Silicosis is a disease that normally takes 10 to 30 years to show up in exposed workers. At Gauley Bridge, workers began dying within one year. By the time the project was completed, hundreds had died. To make matters even worse, the company often buried employees who died from silica exposure in a nearby field, without even notifying the families. Those who inquired were told that their loved ones left without saying where they were going.

Congress held a series of hearings on the matter in 1936. That same year, representatives from business, industry, and government attended the National Silicosis Conference, convened by the U.S. secretary of labor. Among other outcomes of this conference was a finding that silica dust particulates did, in fact, cause silicosis.

Mercury poisoning is another health problem that has contributed to the evolution of the health and safety movement by focusing public attention on unsafe conditions in the workplace. The disease was first noticed among the citizens of a Japanese fishing village, Minamata, in the early 1930s. A disease with severe symptoms was common in this village, but extremely rare throughout the rest of Japan. After much investigation into the situation, it was determined that a nearby chemical plant periodically dumped methyl mercury into the bay that was the village's primary source of seafood. Consequently, the citizens of this small village ingested hazardous dosages of mercury every time they ate fish from the bay.

Mercury poisoning became an issue in the United States after a study was conducted in the early 1940s that focused on New York City's millinery industry. During that time, many workers in this industry displayed the same types of symptoms as the citizens of Minamata, Japan. Since mercury nitrate was used in the production of hats, enough suspicion was aroused to warrant a study. The study linked the symptoms of workers with the use of mercury nitrate. As a result, the use of this hazardous chemical in making hats was stopped, and a suitable substitute—hydrogen peroxide—was found.

As discussed earlier in this chapter, asbestos was another important substance in the evolution of the modern health and safety movement. By the time it was determined that asbestos is a hazardous material—the fibers of which can cause asbestosis or lung cancer (mesothelioma)—thousands of buildings contained the substance. As these buildings began to age, the asbestos, particularly that which was used to insulate pipes, began to break down. As asbestos breaks down, it releases dangerous microscopic fibers into the air. These fibers are so hazardous that removing asbestos from old buildings has become a highly specialized task requiring equipment and training designed for this particular endeavor.

ROLE OF DISASTERS

Safety disasters in the workplace have played a major role in the safety movement in the United States. Few of the disasters that have helped shape the safety movement have been on construction sites. However, these disasters have been collectively one of the major drivers in the establishment of OSHA's safety standard for construction and other industries. Consequently, construction students and professionals should be aware of them. Some of the worst safety tragedies and disasters in U.S. history are the following:

- *Hawk's Nest Tragedy (1930s).* A construction company was awarded a contract to drill a passageway through a mountain in the Hawk's Nest region of West Virginia near a town named Gauley Bridge (which is why this tragedy is sometimes referred to in the literature as the *Gauley Bridge tragedy*). Workers doing the blasting and drilling spent as much as 10 hours a day breathing what turned out to be deadly silica dust. Hundreds died from what is now known to be silicosis. To make matters worse, the construction company often simply buried those who died in nearby fields without notifying their families.

- *Kepone Chemical Disaster (1975 to 1976).* The pesticide Kepone was being produced at a factory in Hopewell, Virginia. Without proper protections and precautions, 57 workers suffered sterility, tremors, and liver damage. Although this is not a construction example, it led to actions by OSHA that have made chemical use on construction sites less hazardous.
- *Willow Island Disaster (1978).* When the scaffolding on the cooling tower of a power plant collapsed, 51 construction workers plunged to their deaths. Willow Island is still considered one of the worst construction disasters to ever occur in the United States.
- *Phillips 66 Explosion (1989) and BP Refinery Explosion (2005).* An explosion at a Phillips 66 petrochemical plant in Pasadena, Texas killed 23 workers. This disaster led to OSHA issuing the Process Safety Management Standard. An explosion and fire at the BP Refinery in Texas City, Texas killed 15 workers and injured 160 more. Although these were not construction disasters, they did generate support for stricter workplace standards that have affected construction companies and workers.
- *Imperial Sugar Refinery Disaster (2009).* An explosion caused by combustible dust killed 14 workers at the Imperial Sugar Refinery in Port Wentworth, Georgia. This disaster has had an indirect effect on the construction industry because it led to OSHA initiating the rule making process to address fire and explosion hazards relating to combustible dust.

DEVELOPMENT OF ACCIDENT-PREVENTION PROGRAMS

In the modern workplace, there are many different types of accident-prevention programs, ranging from the simple to the complex. Widely used **accident-prevention** techniques include failure minimization, fail-safe designs, isolation, lockouts, screening, personal protective equipment (PPE), redundancy, timed replacements, and many others. These techniques are individual components of broader safety programs that have evolved since the late 1800s.

In the early 1800s, employers had little concern for the safety of workers and little incentive to be concerned. Consequently, organized safety programs were nonexistent—a situation that continued for many years. However, between World War I and World War II, industry discovered the connection between quality and safety. Then, during World War II, troop call-ups and deployments created severe labor shortages. Faced with these shortages, employers could not afford to lose workers to accidents—or for any other reason. This realization created a greater openness toward giving safety the serious consideration that it deserved. For example, around this time industry began to realize the following:

- Improved engineering could prevent accidents.
- Employees were willing to learn and accept safety rules.

- Safety rules could be established and enforced.
- Financial savings from safety improvement could be reaped by savings in compensation and medical bills.

With these realizations came the long-needed incentive for employers to begin playing an active role in creating and maintaining a safe workplace. This, in turn, led to the development of organized safety programs sponsored by management. Early safety programs were based on the "**three Es of safety**": engineering, education, and enforcement. The engineering aspects of a safety program involve making design improvements to processes. By altering the design of a process, it can be simplified and, as a result, made less dangerous. Making the workplace safer for all workers is OSHA's goal, Figure I-2.

The education aspect of a safety program ensures that employees know safe work practices, the importance of following them, and management's expectations regarding adherence to safety regulations, Figure I–3. Safety education typically covers the what, when, where, why, and how of safety.

FIGURE I-2 Construction workers are safer because the OSH Act was passed.
Source: Riccardo Arata/Fotolia

FIGURE I-3 Education is one of the three "Es" of safety.
Source: Monkey Business/Fotolia

The enforcement aspect of a safety program involves making sure that employees abide by safety policies, rules, regulations, practices, and procedures. Supervisors and fellow employees play a key role in the enforcement aspects of modern safety programs.

DEVELOPMENT OF SAFETY ORGANIZATIONS

Today, numerous organizations are devoted in full, or at least in part, to the promotion of safety and health in the workplace. Figure I–4 lists professional organizations with workplace safety as part of their missions. Figure I–5 lists governmental agencies concerned with safety and health. These lists are extensive now, but this has not always been the case. Safety organizations in this country had humble beginnings. The grandfather of them all is the NSC.

Today, the NSC is the largest organization in the United States devoted solely to safety and health practices and procedures. Its purpose is to prevent the losses, both direct and indirect, arising out of accidents or from exposure to unhealthy environments. Although it is

Alliance for American Insurers
American Board of Industrial Hygiene
American Council of Government Industrial Hygienists
American Industrial Hygiene Association
American Insurance Association
American National Standards Institute
American Occupational Medical Association
American Society of Mechanical Engineers
American Society of Safety Engineers
American Society for Testing and Materials
Chemical Transportation Emergency Center
Human Factors Society
National Fire Protection Association
National Safety Council
National Safety Management Society
Society of Automotive Engineers
System Safety Society
Underwriters' Laboratories

FIGURE I–4 Professional organizations concerned with workplace safety.

American Public Health Association
Bureau of Labor Statistics
Bureau of National Affairs
Commerce Clearinghouse
Environmental Protection Agency
National Institute for Standards and Technology (formerly National Bureau of Standards)
National Institute of Occupational Safety and Health
Occupational Safety and Health Administration
Superintendent of Documents, U.S. Government Printing Office
U.S. Consumer Product Safety Commission

FIGURE I–5 Government agencies concerned with workplace safety.

chartered by an act of Congress, the NSC is a nongovernmental, not-for-profit public service organization.

OSHA is the government's administrative arm for the OSH Act. Established in 1970, OSHA sets and revokes safety and health standards, conducts inspections, investigates problems, issues citations, assesses penalties, petitions the courts to take appropriate action against unsafe employers, provides safety training, provides injury-prevention consultation, and maintains a database of health and safety statistics.

Another governmental organization is the **National Institute of Occupational Safety and Health (NIOSH)**. This organization is part of the Centers for Disease Control and Prevention (CDC) of the U.S. Department of Health and Human Services. NIOSH is required to publish annually a comprehensive list of all known toxic substances. NIOSH also provides onsite tests of potentially toxic substances so that companies know what they are handling and what precautions to take.

Integrated Approach to Safety and Health

The integrated approach has become the norm that typifies the health and safety movement of today. OSHA reinforces the integrated approach by requiring companies to have a plan for doing at least the following: (1) providing appropriate medical treatment for injured or ill workers, (2) regularly examining workers who are exposed to toxic substances, (3) providing a qualified first aid person during all working hours, and (4) assigning specific safety-related duties to a competent person.

Smaller companies may choose to contract out the fulfillment of these requirements. Larger companies often maintain a staff of safety and health professionals. The health and safety staff in a modern company may include the following positions:

- *Industrial hygiene chemist or engineer.* Companies that use toxic substances may employ **industrial hygiene chemists** periodically to test the work environment and the people who work in it. In this way, unsafe conditions or hazardous levels of exposure can be identified early, and corrective or preventive measures can be taken. Dust levels, ventilation, and noise levels are also monitored by individuals serving in this capacity.
- *Radiation control specialist.* Companies that use or produce radioactive materials employ **radiation control specialists**, who are typically electrical engineers or physicists. These specialists monitor the radiation levels to which workers may be exposed, test workers for levels of exposure, respond to radiation accidents, develop company-wide plans for handling radiation accidents, and implement decontamination procedures when necessary.
- *Industrial safety engineer or manager.* Individuals serving in this capacity are safety and health generalists with specialized education and training. In larger companies, they may be devoted to safety and health matters. In smaller companies, they may have other duties in addition to safety and health. In either case, **industrial safety engineers** and **industrial safety managers** are responsible for developing and carrying out the company's overall safety and health program, including accident prevention, accident investigation, and education and training.

Other positions that might be included in the team are ergonomist, risk manager, occupational physician, and occupational nurse.

New Materials, New Processes, and New Problems

The field of safety and health is more complex than it has ever been. The materials from which products are made have become increasingly complex and exotic. Engineering metals now include carbon steel, alloy steel, high-strength low-alloy steel, stainless steel, maraging steel, cast steel, cast iron, tungsten, molybdenum, titanium, aluminum, copper, magnesium, lead, tin, zinc, and powdered metals. Each of these metals requires its own specialized process.

Nonmetals are more numerous and have also become more complex. Plastics, plastic alloys and blends, advanced composites, fibrous materials, elastomers, and ceramics also bring their own potential hazards to the workplace.

In addition to the more complex materials being used in the modern construction industry and the new safety and health concerns associated with them, modern construction processes are also becoming more complex. As these processes become more advanced, the potential hazards associated with them often increase.

GLOBALLY HARMONIZED SYSTEM OF CLASSIFICATION AND LABELING (GHS)

Because a variety of chemicals and toxic substances might be used on construction sites, construction students and professionals should know about the *Globally Harmonized System of Classification and Labeling (GHS)*. GHS is a system used to standardize the classification and labeling of chemicals worldwide, Figure I–6. The GHS was developed in response to the need to be able to uniformly communicate the hazards of chemicals and corresponding protective measures anywhere in the world through standardized labels and safety data sheets. In the past, chemicals produced in Asia, for example, and shipped to the United States for use in construction projects might carry labels that could be easily misunderstood. As the world began the trend toward global integration, the potential for classification and labeling problems just increased.

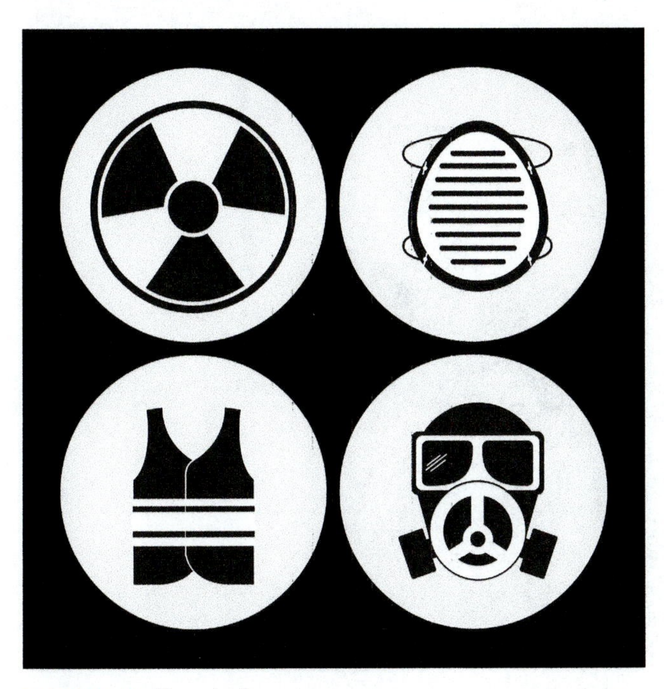

FIGURE I–6 Chemicals are common on construction sites.
Source: Grgroup/Fotolia

The GHS is available through OSHA, but it is not a standard. Rather, it represents a voluntary agreement among nations concerning how to classify and label chemicals. The GHS document is sometimes referred to as the "Purple Book." The Purple Book provides countries with the framework they need for modifying their existing regulatory processes and procedures in ways that ensure *cradle-to-grave* safety with regard to chemicals. In this way, all people who are exposed to a given chemical will have a way of knowing what hazards are associated with it and what safety precautions they should take when handling it.

GHS Classification

Hazard communication begins with classification. Classification is achieved in three steps: (1) identification of relevant data relating to the hazards of a given substance, (2) review of the relevant data to determine specific hazards associated with the substance in question, and (3) determination of whether the substance in question will be classified as hazardous and if so to what degree. The physical hazards the GHS is concerned with include the following:

- Explosives
- Flammable bases
- Flammable aerosols
- Oxidizing gases
- Gases under pressure
- Flammable liquids
- Flammable solids
- Self-reactive substances
- Pyrophoric liquids
- Pyrophoric solids
- Self-heating substances
- Substances that emit flammable gases when in contact with water
- Oxidizing liquids
- Oxidizing solids
- Organic peroxides
- Substances that corrode metals

LANGUAGE BARRIERS IN CONSTRUCTION

Select almost any construction site in the United States and it will be one of the most diverse locations in the community, Figure I–7. Through immigration—both legal and illegal—the United States has become one of the most diverse countries in the world. One of the safety challenges presented by immigration is that the number of people working in construction for whom English is a second language has increased markedly. As the percentage of people for whom English is a second language has increased, so has the number of workplace injuries. This is especially the case in the construction industry where the increase in injuries has actually outpaced the increase in English as a second language or ESL speakers. The trend toward more ESL speakers in the workplace is expected to continue for the foreseeable future.

Although most ESL speakers in the construction industry are Hispanic, those of Asian and Eastern European backgrounds are also increasing. In addition to the language barriers they experience on the job, ESL speakers are more reluctant than their English speaking counterparts to question or speak out concerning the unsafe practices of employers or fellow workers. This reluctance can be attributed to several factors: (1) cultural differences that work against questioning authority, (2) peer pressure from English speaking colleagues, and (3) fear of being discovered as an undocumented worker for those who fall into this category. This last category is especially difficult to overcome because filing a safety complaint could lead to deportation for an undocumented worker.

The problem is exacerbated by the economics of the situation. While it is important, for example, for construction workers to be able to read Safety Data Sheets, many construction companies find it too expensive to provide language training for ESL speakers or to provide bilingual language training. Even those construction companies that do provide bilingual training complain that their highest turnover rates are among ESL speakers. Hence, the construction company invests in providing bilingual training only to have the ESL workers leave.

FIGURE I–7 Construction workers are a diverse group.
Source: Kurhan/Fotolia

OSHA's Response to the Problem

OSHA recognizes the problems associated with language barriers in the construction industry and has taken steps to help alleviate the problems. OSHA does not require that safety training be offered in foreign languages, but has implemented other proactive measures. For example, OSHA established a Spanish language website (go to www.osha.gov and click on "en Espanol"). OSHA also established a toll-free telephone number that puts Spanish speaking workers in touch with OSHA officials who speak Spanish. In addition, OSHA developed Spanish language versions of several safety training courses (including the OSHA 10-hour "Construction Outreach" course). Finally, OSHA has established partnerships with the Hispanic Chamber of Commerce and Hispanic Contractors of America.

One of the more effective of OSHA's responses to language barriers in the workplace has been the awarding of grants to promote the development and provision of bilingual training programs and materials. For example, many of the Susan Harwood Training Grants awarded each year go to non profit organizations that provide bilingual training and/or develop bilingual training materials. Clearly, OSHA recognizes the problems associated with language barriers in the construction industry and is working to do its part to alleviate those problems. Unfortunately, there is a breakdown between OSHA and the individual worker who needs help.

The breakdown that is impeding the dissemination and use of OSHA's foreign language resources is the result of several different factors: (1) it appears that many of the ESL speakers in the construction industry are illiterate not just in English, but in their native language too (hence providing bilingual safety materials does little good), (2) lack of awareness on the part of small and mid-sized construction companies of the types of help OSHA can offer; and (3) inability of small and mid-sized construction companies to compete for grants to pay for bilingual training and materials.

Different organizations are using different methods to deal with the conundrum of language barriers in the construction industry. Some of the more widely used methods are as follows: (1) hiring bilingual supervisors, (2) hiring interpreters, (3) using videos and DVDs for training rather than printed materials, and (4) using symbols rather than words to point out hazards on the job site. All of these methods are useful, but language barriers are likely to continue to be an issue of safety professionals in the construction industry.

DESIGN FOR CONSTRUCTION SAFETY

Prevention through design or PtD is an accident prevention concept. Although it applies more directly to those who design construction projects than those who build them, it is important for construction professionals and students to know about the concept since construction personnel are often called upon to give input into the design process, Figure I–8. PtD involves addressing the safety of workers in the design of the permanent features of a structure. The concept applies directly to architects and engineers and indirectly to safety professionals. PtD is a design concept, not a construction management or job site safety concept. It applies to job site safety, but occurs during the design of a structure.

FIGURE I–8 Construction personnel often give input to designers.
Source: Fotoinfot/Fotolia

The more effectively PtD is applied during the design of a structure, the fewer the number of accidents during its construction. Decreasing the number of accidents has a positive cascading effect on all of the direct and indirect costs associated with accidents (e.g., workers' compensation, medical, lawsuits, wages lost, construction delays, etc.). In essence, PtD encourages designers to think through what will be necessary to construct the permanent features of a structure as it is being designed. What will workers have to actually do to construct a given feature? Will the size, shape, location, or any other factor put construction workers in a hazardous position or expose them to hazardous conditions? If any part of the design creates a potentially hazardous condition, can it be redesigned to eliminate or at least mitigate the hazard? These are the types of questions that designers who apply PtD ask themselves and construction professionals throughout the design process.

Barriers to Acceptance of Design for Construction Safety

Design for Construction Safety or *prevention through design* (PtD) began to gain momentum as a concept in the early 2000s. However, there is still much resistance to the concept. Three types of resistance have slowed the acceptance and application of PtD: (1) professional inertia, (2) process resistance, and (3) education. Professional inertia means that it is always difficult for new concepts to gain a foothold in any profession. People like to continue doing things the way they are accustomed to doing them. PtD represents a whole new mind-set for architects and engineers.

Process resistance means that once a given process is in place and widely used, it can be difficult to change it. The "bid" process that is used for construction products often forces architectural and engineering firms to do everything possible to submit the lowest bid. Some designers resist PtD out of fear that it will increase the size of their bids. Often the effect of increasing the amount of a bid is to cause architects and engineers to lose the bid to a competitor. Finally, many practicing architects and engineers completed their formal education in college before PtD was taught. In fact, even now some colleges or architecture and engineering do not teach the concept. Further, too few design professionals know enough about construction safety to make it a factor in their designs. Designers are not likely to use a concept they have not learned. Consequently, PtD has been slow to catch on.

Advocates continue to encourage its use in spite of the barriers, and progress is being made. Interestingly, the litigious nature of modern society works in favor of PtD. Fear of the expensive, time-consuming lawsuits that can grow out of workplace accidents actually encourages the application of PtD. In the aftermath of a construction accident—especially those that result in death or permanent disability—attorneys look for candidates to sue that are perceived to have "deep pockets." More and more attorneys are including architects and engineers in their lawsuits against construction companies. This fact alone is increasing the interest of design professionals in PtD. Design professionals will have to fight their own battles in coming to grips with PtD. In the meantime, construction professionals should be prepared to offer input and give feedback that will help designers factor construction safety into their designs.

Summary

Construction is big business in the United States, accounting for more than $600 billion annually and employing more than 4,500,000 workers. Over the years, the construction industry has been changed by inanimate power, machines, materials, and work specialization. Legal liability is an important issue for construction companies because construction is a hazardous occupation. Companies must designate a *competent person* to inspect construction sites for safety.

Milestones in the development of the safety movement in the United States include the following: first recorded safety program in 1892, creation of the Bureau of Mines in 1907, passage of the first effective worker's compensation law in the United States in 1911, and establishment of OSHA in 1970.

Organized labor has played a crucial role in the development of the safety movement in the United States. Particularly important was the work of unions to overturn antilabor laws inhibiting safety in the workplace.

Specific health problems associated with the workplace have contributed to the development of the modern safety and health movement. These problems include lung diseases in miners, mercury poisoning, and lung cancer tied to asbestos exposure. Tragedies have changed the face of the safety movement at certain times in the United States. The Gauley Bridge disaster and the **asbestos menace** are examples of such tragedies.

Widely used accident-prevention techniques include failure minimization, fail-safe designs, isolation, lockouts, screening, PPE, redundancy, and time replacements.

The development of the safety movement in the United States has been helped by the parallel development of safety organizations. Prominent among these are the NSC, National Safety Management Society, American Society of Safety Engineers, and American Industrial Hygiene Association.

The safety and health movement today is characterized by professionalization and integration. The safety and health team of a large company may include an industrial chemist or engineer, radiation control specialist, and a safety engineer or manager. New materials and processes are introducing new safety and health problems, making the integrated approach a practical necessity and promoting growth in the profession.

Key Terms and Concepts

Accident prevention
Asbestos menace
Assumption of risk
Contributory negligence
Cooperative Safety Congress (CSC)
Employer-biased laws
Employer liability
Fellow servant rule
Inanimate power
Industrial hygiene chemist
Industrial safety engineer
Industrial safety manager
National Council of Industrial Safety (NCIS)
National Institute of Occupational Safety and Health (NIOSH)
Organized labor
Occupational Safety and Health Administration (OSHA)
Radiation control specialist
Safety movement
The three Es of safety
Workers' compensation

Review Questions

1. Explain the following statement using applicable facts and figures: Construction is the most dangerous industry in the United States.
2. To what cause(s) can improvements in workplace safety be attributed?
3. Describe the circumstances that led to the development of the first organized safety program.
4. What impact did labor shortages during World War II have on the safety movement?
5. Explain how workplace tragedies have affected the safety movement. Give examples.
6. Explain the primary reasons behind the establishment of OSHA.
7. Summarize briefly the role that organized labor has played in the advancement of the safety movement.
8. Explain the three Es of safety.
9. Explain the role tragedies and disasters have played in the safety movement in the United States.
10. Explain how have language barriers affected safety in the construction industry.
11. Explain why it is important for construction professionals to be familiar with the concept of designing for construction safety or "Prevention through Design" (PtD).

Critical Thinking and Discussion Activities

1. Two construction management students are having a debate. The first student says, "I don't know why we have to take a course in construction safety. Employees should have the sense to work safely. That's what they are paid to do." "It's more complicated than that," says the other student. "Construction is a dangerous business. There are so many things that can go wrong. I think everybody who works in construction should be required to take a construction safety course." Join this debate. Who is right? Why?
2. Two construction supervisors are having a friendly discussion over lunch. One says, "Employers in this country are really irresponsible. If it wasn't for the government and OSHA, most companies would eat up workers and spit them out. They don't care about safety." The other supervisor responds, "That is not true. The free market is the main factor driving the emphasis on safety in the workplace. Companies cannot afford to have accidents and injuries. They cost too much." Join this debate. Who is right? Why?
3. "I would never own a construction company," said the job site foreman to his designated competent person. "You're never safe from a lawsuit. This concept of shared liability means you can be sued even if you are not involved in the work when an accident occurs." The competent person responds, "I agree that shared liability is a problem, but as long as you are providing a safe and healthy workplace, you have nothing to worry about." Join this debate. What is your opinion?

Application Activities

1. Research the cost of construction accidents and injuries in the United States last year. What were the costs? How do those costs compare with the costs of accidents and injuries in manufacturing?
2. Find a construction company in your community that will cooperate with you. Ask the company to let you interview one of its designated competent persons. What are this individual's qualifications to serve as a competent person? How did he or she come to earn the designation?

3. Research the role of organized labor in the development of the modern safety and health movement. Write a paper that explains the role and contains specific examples.
4. Select one of the professional organizations listed in Figure I–4 and research its history. How did it get started and when? What is its status now? How many members does it have? What are the requirements for membership? What is its specific mission? How does it go about pursuing that mission?

Endnotes

1. "The Job Safety Law of 1970: It's Passage was Perilous." Retrieved from www.dol.gov/aboutdol/history/osha.htm on September 20, 2015.
2. National Safety Council. *Accident Facts*, 2015.
3. "The Job Safety Law of 1970: It's Passage was Perilous." Retrieved from www.dol.gov/aboutdol/history/osha.htm on September 20, 2015.
4. Ibid.
5. Ibid.
6. Ibid.

PART ONE

THEORIES AND CONCEPTS

CHAPTER **ONE**

COST OF ACCIDENTS: WHY SAFETY IS IMPORTANT?

LEARNING OBJECTIVES

- **Summarize the cost of accidents in the United States on an annual basis.**
- **List the most common causes of accidental deaths in the United States.**
- **Compare and contrast accidental deaths with other causes of death.**
- **Explain the costs and rates of workplace accidents.**
- **Describe how much time is lost on the job annually due to workplace accidents.**
- **Describe how many deaths typically occur on the job annually.**
- **List the most common injuries by type of accident.**
- **List the death rates by industry in order from highest to lowest.**
- **List the most common types of accidents that occur on construction jobs sites.**
- **Explain what is most disturbing about chemical burn injuries in construction.**
- **Explain the special challenge to construction professionals posed by heat burn injuries.**
- **Demonstrate how to estimate the cost of accidents.**
- **Explain how to use *Construction Solutions Database* to improve safety in construction.**
- **Explain how Building Information Modeling can help improve safety in construction.**

Accidents and the corresponding damage they cause to employees, property, equipment, and morale can have a detrimental effect on a construction company's profit and loss statement. Although it can be difficult to measure precisely the economic impact of accidents, the impact is significant.

There is a long history of debate in this country concerning the effect of accidents on industry (the workers and the companies) and the cost of preventing accidents. Historically, the prevailing view was that accident-prevention programs were too costly. The more contemporary view is that the accidents are too costly and that **accident prevention** makes sense economically. As a result, accident prevention, which had been advocated on a moral basis, is now justified also in economic terms.

Accidents are the fourth leading cause of death in this country after heart disease, cancer, and strokes. This ranking is based on all types of accidents, including motor vehicle accidents, **drownings**, fires, **falls**, **natural disasters**, and work-related accidents.

Although deaths from natural disasters tend to be more newsworthy than workplace deaths, their actual impact is substantially less. For example, natural disasters cause fewer than 100 deaths per year on average. **Workplace accidents**, on the other hand, cause more than 10,000 deaths every year in the United States.[1] The following quote from the National Safety Council (NSC) puts workplace accidents and deaths in the proper perspective, notwithstanding their apparent lack of newsworthiness.

> In less time than it takes to read this chapter two people will be killed in America and another 170 will be seriously injured. The costs of these deaths and injuries will approach $3 million. On average accidental deaths in the United States exceed 10 every hour and injuries exceed 1,000 every hour (24 hours a day and seven days a week).[2]

This chapter provides prospective and practicing construction professionals with the information they need to have a full understanding of workplace accidents and their effect on industry in the United States, which helps professionals to play a more effective role in keeping both management and labor focused appropriately on safety and health in the workplace.

COST OF ACCIDENTS

To gain a proper perspective on the economics of workplace accidents, we must view them in the context of all accidents. The overall cost of accidents in the United States is approximately $800 billion annually. This includes **lost wages, medical expenses, insurance administration, fire-related losses, property damage,** and **indirect costs.**

As Figure 1–1 reminds us, workplace accidents are costly. When the costs of the most common accidents are broken down by categories, construction accidents are second only to motor vehicle accidents as the following list shows:

- Motor vehicle accidents (approximately $725 billion annually)
- Workplace accidents (approximately $50 billion annually)

FIGURE 1–1 Workplace accidents are costly.
Source: Photographee.eu/Fotolia

- Home accidents (approximately $20 billion annually)
- Public accidents (approximately $13 billion annually)

As Figure 1–2 illustrates, the costs of accidents and injuries are both direct and indirect. Accident costs in a typical year can be seen in the following list:

- Wages lost (approximately $40 billion annually)
- Medical expenses (approximately $25 billion annually)
- Insurance administration (approximately $30 billion annually)
- Motor vehicle damage (approximately $30 billion annually)
- Fire losses (approximately $10 billion annually)
- Indirect costs (approximately $25 billion annually)

This list shows that the highest cost of accidents and injuries is in wages lost to workers. The category of indirect losses from work accidents consists of costs associated with responding to accidents (i.e., giving first aid, filling out accident reports, handling work slowdowns).

Clearly, accidents on and off the job cost U.S. industry dearly. Every dollar spent responding to accidents is a

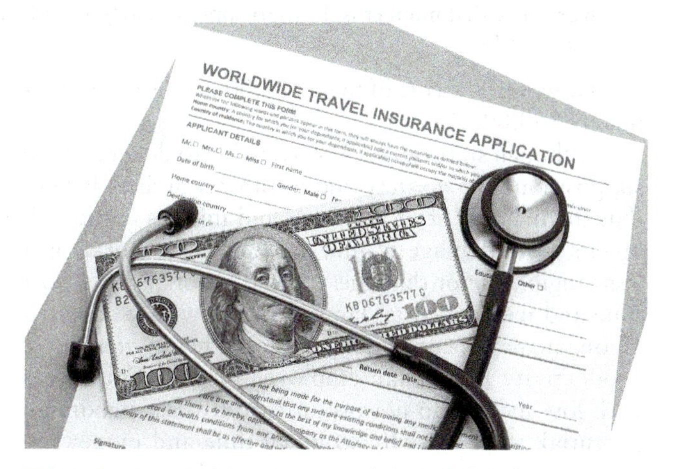

FIGURE 1–2 Accident costs are direct and indirect.
Source: Rukxstockphoto/Fotolia

dollar that could have been reinvested in modernization, employee training, and other competition-enhancing activities.

ACCIDENTAL DEATHS IN THE UNITED STATES

Accidental deaths in the United States result from a variety of causes, including motor vehicle accidents, falls, **poisoning**, drowning, fire-related injuries, **suffocation** (ingested object), firearms, **medical complications**, air transport accidents, injuries from machinery, **mechanical suffocation**, and the impact of falling objects. The NSC periodically computes death totals and **death rates** in each of these categories. The statistics for a typical year are as follows:

- *Motor vehicle accidents.* As the leading cause of accidental deaths in the United States every year, this category includes deaths resulting from accidents involving mechanically or electrically powered vehicles (excluding rail vehicles) that occur on or off the road. In a typical year, there are approximately 47,000 deaths from motor vehicle accidents in the United States.
- *Falls.* This category includes all deaths from falls except those associated with transport vehicles. For example, a person who is killed as a result of falling while boarding a bus or train would not be included in this category. In a typical year, there are approximately 13,000 deaths in the United States from falls.
- *Poisoning.* This category is divided into two subcategories: (1) poisoning by solids and liquids and (2) poisoning by gases and vapors. The first category includes deaths that result from the ingestion of drugs, medicine, recognized solid and liquid poisons, mushrooms, and shellfish; it does not include poisoning from spoiled food or *Salmonella* species. The second category includes deaths caused by incomplete combustion (e.g., gas vapors from an oven or unlit pilot light) or from carbon monoxide (e.g., exhaust fumes from an automobile). In a typical year, there are approximately 6,000 deaths in the first category and 1,000 in the second.
- *Drowning.* This category includes work-related and nonwork-related drowning incidents, but excludes those associated with floods or other natural disasters. In a typical year, there are approximately 5,000 deaths from drowning in the United States.
- *Fire-related injuries.* This category includes deaths from burns, asphyxiation, and falls, as well as from being struck by falling objects in a fire. In a typical year, there are more than 4,000 deaths resulting from fire-related injuries in the United States.

- *Suffocation (ingested object).* This category includes deaths from the ingestion of an object that blocks the air passages. In many such deaths, the ingested object is food. In a typical year, there are approximately 4,000 such suffocation deaths in the United States.
- *Firearms.* This category includes deaths that result when recreational activities or household accidents that involve firearms result in death. For example, a person killed in the home while cleaning a firearm would be included in this category; however, a person killed in combat would not be included. In a typical year, there are approximately 2,000 deaths in this category in the United States.
- *Others.* This category includes deaths resulting from medical complications arising out of mistakes made by health-care professionals, air transport injuries, interaction with machinery, mechanical suffocation, and the impact of falling objects. In a typical year, there are more than 14,000 deaths overall in these subcategories.[3]

ACCIDENTS VERSUS OTHER CAUSES OF DEATH

Although there are more deaths every year from heart disease, cancer, and strokes than from accidents, these causes tend to be concentrated among people at or near retirement age. Among people 37 years of age or younger—prime working years—accidents are the number one cause of death, Figure 1–3. The following list shows deaths by category for people between 25 and 44 years of age—the prime working age group (notice that the leading cause of deaths is accidents):

- Accidents (approximately 30,000 annually)
- Cancer (approximately 21,000 annually)
- Heart disease (approximately 16,000 annually)

FIGURE 1–3 Accidents are a leading cause of disability and death to people of prime working age.
Source: Vladimirfloyd/Fotolia

- Motor vehicle (approximately 17,000 annually)
- Poison (approximately 3,000 annually)
- Drowning (approximately 1,500 annually)
- Falls (approximately 1,200 annually)
- Fire related (approximately 1,000 annually)

Figure 1–3 shows that accidents represent a serious detriment to productivity, quality, and competitiveness in today's workplace. Yet accidents are the one cause of death and injury that companies can most easily control. Although it is true that companies may have some success in decreasing the incidence of heart disease and stroke among their employees through activities such as wellness programs, their impact in this regard is limited. However, employers can have a significant impact on preventing accidents.

WORK ACCIDENT COSTS AND RATES

Workplace accidents cost employers millions every year. For example, one company—the steel-making division of USX—once paid a $3.25 million fine to settle numerous health and safety violation citations.

This example shows the cost of fines only. In addition to fines, employers incur costs for safety corrections, medical treatment, survivor benefits, death and burial costs, and a variety of indirect costs. Clearly, work accidents are expensive. However, the news is not all bad. The trend in the rate of accidents is downward.

Work **accident rates** in this century are evidence of the success of the safety movement in the United States. As the amount of attention given to workplace safety and health has increased, the accident rate has decreased.

> Accident rates have fallen significantly over the years due to advances brought about by the safety movement. For example, in 1912 approximately 21,000 workers were killed on the job in the United States. Now with a workforce that is more than three times the size of the 1912 workforce that number is down to approximately 10,000 annually.[4]

However, the cost of these 10,000 deaths is approximately $50,000.[5]

Although statistics are not available to document the supposition, many safety and health researchers believe that the major cost of accidents and injuries on the job results from damage to morale. Employee morale is a less tangible factor than measurable factors, such as **lost time** and medical costs. However, it is widely accepted among management professionals that few factors affect productivity more than employee morale. Employees with low morale do not produce up to their maximum potential, which is why so much time and money are spent every year to help supervisors and managers learn different ways to help improve employee morale.

Since few things are so detrimental to employee morale as seeing a fellow employee injured, accidents can have a devastating effect. Whenever an employee is injured, his or her colleagues silently think, "that could have been me," in addition to worrying about the employee. Morale is damaged even more if the injured employee is well-liked and other employees know his or her family.

TIME LOST BECAUSE OF WORK INJURIES

An important consideration when assessing the effect of accidents on industry is the amount of time lost due to work injuries.[6] According to the NSC, approximately 35 million hours are lost in a typical year as a result of accidents. This is actual time lost from disabling injuries and does not include additional time lost for medical checkups after the injured employee returns to work. Accidents that occurred in previous years often continue to cause lost time in the current year.

DEATHS IN WORK ACCIDENTS

Deaths on the job have decreased markedly over the years. However, they still occur. For example, in a typical year, there are 10,400 work deaths in the United States. The causes of death in the workplace vary. They include those related to motor vehicles, falls, electric current, drowning, fires, air transport, poison, water transport, machinery, falling objects, rail transports, and mechanical suffocation,[7] Figure 1–4.

FIGURE 1–4 Falls are a leading cause of injuries in construction.
Source: Soja Andrzej/Fotolia

WORK INJURIES BY TYPE OF ACCIDENT

Work injuries can be classified by the type of accident from which they result. The following are the most common causes of work injuries:

- Overexertion
- Impact accidents
- Falls
- Bodily reaction (to chemicals)
- Compression
- Motor vehicle accidents
- Exposure to radiation or caustic chemicals
- Rubbing or abrasions
- Exposure to extreme temperatures

Overexertion, the result of employees working beyond their physical limits, is the leading cause of work injuries. According to the NSC, almost 31 percent of all work injuries are caused by overexertion. The second leading cause, **impact accidents**, involves a worker being struck by or against an object. The next most prominent cause of work injuries is falls.[8] The remaining accidents are distributed fairly equally among the other causes just listed.

DEATH RATES BY INDUSTRY

A variety of agencies and organizations, including the Bureau of Labor Statistics, National Center for Health Statistics, and NSC, collect data on death rates within industrial categories.[9] Such information can be used in a variety of ways, not the least of which is in assigning workers' compensation rates. The most widely used industrial categories are agriculture, including farming, forestry, and fishing; mining and quarrying, including oil and gas drilling and extraction; construction; manufacturing; transportation and public utilities; trade, both wholesale and retail; services, including finance, insurance, and real estate; and federal, state, and local government.

When death rates are computed on the basis of the number of deaths per 100,000 workers in any given year, the industry categories rank as follows (from highest death rate to lowest):

1. Mining and quarrying
2. Agriculture
3. Construction
4. Transportation and public utilities
5. Government
6. Manufacturing
7. Services
8. Trade

The construction industry ranks third in workplace deaths, but first in workplace injuries. The rankings sometimes change slightly from year to year. For example, agriculture and mining and quarrying may exchange the first and second ranking in any given year. This is also true at the lowest end of the rankings, with services and trade. However, generally, the ranking is as shown.

COMMON ACCIDENT EXAMPLES ON CONSTRUCTION SITES

Experience has shown that certain types of accidents are common on construction sites. What follows are representative examples of the types of accidents that occur every year on construction sites:

- Victim fell from a scaffold 60 feet to a concrete surface
- Victim was buried in an excavation
- Victim was crushed between two backhoes
- Victim was working on a circuit breaker when he was electrocuted
- Victim was struck in the head by a crane that was moving
- Victim was installing roofing and fell 40 feet to the ground
- Victim was welding a water tank when the tank collapsed on him
- Victim was backed over by a dump truck
- Victim was working on a platform that collapsed
- Victim was operating a nail gun when his foot was impaled
- Victim was operating a power saw when he cut off a finger
- Victim was hit in the head by a falling object
- Victim sprained his back while attempting to pick up a heavy box
- Victim slipped and fell on a wet spot on a concrete subfloor
- Victim was severely burned when he spilled a barrel containing a toxic chemical

These are just a few examples of the types of workplace accidents that occur on construction sites every year in the United States. The important point to remember is that every one of these accidents could have and can be prevented.

PARTS OF THE BODY INJURED ON THE JOB

To develop and maintain an effective safety and health program, it is necessary to know not only the most common causes of death and injury but also the parts of the body most frequently injured.

Typically, parts of the body prone to injury are as follows (from most frequent to least):

1. Back
2. Legs and fingers
3. Arms
4. Trunk
5. Hands
6. Eyes, head, and feet
7. Neck, toes, and body systems

This ranking shows that one of the most fundamental components of a safety and health program should be instruction on how to lift without hurting the back.[10]

CHEMICAL BURN INJURIES

Chemical burn injuries are a special category with which prospective and practicing construction professionals should be familiar. The greatest incidence of chemical burns occurs in construction and manufacturing.[11]

The chemicals that frequently cause burn injuries include acids and alkalis; soaps, detergents, and cleaning compounds; solvents and degreasers; calcium hydroxide (a chemical used in cement and plaster); potassium hydroxide (an ingredient in drain cleaners and other cleaning solutions); and sulfuric acid (battery acid). Almost 46 percent of all chemical burn injuries occur while workers are cleaning equipment, tools, and vehicles.[12]

What is particularly disturbing about chemical burn injuries is that a high percentage of them occur despite the use of personal protective equipment (PPE), the provision of safety instruction, and the availability of treatment facilities. In some cases, the PPE is faulty or inadequate. In others, it is not properly used, despite instructions.

Preventing chemical burn injuries presents a special challenge to construction professionals. The following strategies are recommended:

- Familiarize yourself, the workers, and their supervisors with the chemicals that will be used and the inherent dangers.
- Secure the proper PPE for each type of chemical that will be used.

SAFETY FACTS & FINES

Failure to properly dispose of hazardous materials can be an expensive mistake for construction companies. It can also subject construction professionals to criminal charges. A construction company in Pensacola, Florida, was fined $100,000 and its site superintendent was sentenced to five years of probation when a woman died from exposure to rodent poison left behind at a job site. The site superintendent had been ordered to remove the rodent poison from the job site, but had failed to do so. The company and the site superintendent were charged with failure to properly dispose of hazardous materials.

- Provide instruction on the proper use of PPE, and then make sure that supervisors confirm that the equipment is used properly every time.
- Monitor workers who are wearing PPE, and replace the equipment when it begins to show wear.

HEAT BURN INJURIES

Heat burn injuries present a special challenge to construction professionals in the modern workplace. The most frequent causes are flame (includes smoke inhalation injuries), molten metal, petroleum asphalt, steam, and water. The most common activities associated with heat burn injuries are welding, cutting with a torch, and handling tar or asphalt—all common activities in construction.[13]

Construction professionals who understand the following negative factors that contribute to heat burn injuries in the workplace are in a better position to prevent heat burn injuries.

- Employer has no health and safety policy regarding heat hazards.
- Employer fails to enforce safety procedures and practices.
- Employees are not familiar with the employer's safety policy and procedures concerning heat hazards.
- Employees fail to use or improperly use PPE.
- Employees have inadequate or worn PPE.
- Employees work in too small a space.
- Employees attempt to work too fast or are pushed to work too fast.
- Employees are careless.
- Employees have poorly maintained tools and equipment.[14]

These factors should be carefully considered by construction professionals when developing accident-prevention programs. Employees should be familiar with the hazards, know the appropriate safety precautions, and have and use the proper PPE. Construction professionals should monitor to ensure that safety rules are being followed and that PPE in good condition is being used correctly.

ESTIMATING THE COST OF ACCIDENTS

Even decision makers who support accident prevention must consider the relative costs of such efforts. Clearly, accidents are expensive. However, to be successful, safety-minded construction professionals must be able to show that accidents are more expensive than their prevention. To do this, they must be able to estimate the cost of accidents.

Cost Estimation Method

To have value, a cost estimate must relate directly to the specific company in question. Applying broad industry cost factors does not suffice. To arrive at company-specific figures, the costs associated with an accident should be divided into *insured* and *uninsured* costs.[15]

Determining the insured costs of accidents is a simple matter of examining accounting records. The next step involves calculating the uninsured costs. Simonds recommends that accidents be divided into the following four classes:

- *Class 1 accidents.* Lost workdays, permanent partial disabilities, and temporary total disabilities
- *Class 2 accidents.* Treatment by a physician outside of the company's facility
- *Class 3 accidents.* Locally provided first aid, property damage of less than $100, or the loss of fewer than eight hours of work time
- *Class 4 accidents.* Minor injuries that do not require the attention of a physician, result in property damage of less than $100, and cause less than eight hours of work to be lost[16]

Average uninsured costs for each class of accident can be determined by pulling the records of all accidents that occurred during a specified period and sorting the records according to class. For each accident in each class, record every cost that was not covered by insurance. Compute the total of these costs by class of accident and divide by the total number of accidents in that class to determine an average uninsured cost for each class, specific to the particular company.

Figure 1–5 is an example of how the average cost of a selected sample of Class 1 accidents can be determined. In this example, there were four Class 1 accidents in the pilot study. These four accidents cost the company a total of $554.23 in uninsured costs or an average of $138.56 per accident. Using this information, accurate cost estimates of an accident and accurate predictions can be calculated.

Other Cost Estimation Methods

The costs associated with workplace accidents, injuries, and incidents fall into broad categories, such as the following:

- Lost work hours
- Medical costs
- Insurance premiums and administration
- Property damage
- Fire losses
- Indirect costs

Class of Accident	Accident Number							
Class 1	**1**	**2**	**3**	**4**	**5**	**6**	**7**	**8**
Cost A	16.00	6.95	15.17	3.26				
Cost B	72.00	103.15	97.06	51.52				
Cost C	26.73	12.62	–	36.94				
Cost D	–	51.36	–	38.76				
Cost E	–	11.17	–	24.95				
Cost F	–	–	–	13.41				
Cost G	–	–	–	–				
Total	114.73	185.25	112.23	142.02				

Grand Total: $554.23

Average Cost per Accident: $138.56 (grand total 4 number of accidents)

Signature: Date:

FIGURE 1–5 Uninsured costs worksheet form.

Calculating the direct costs associated with lost work hours involves compiling the total number of lost hours for the period in question and multiplying the hours times the applicable loaded labor rate. The loaded labor rate is the employee's hourly rate plus benefits. Benefits vary from company to company, but typically inflate the hourly wage by 20–35 percent. A sample of cost-of-lost-hours computation follows:

$$\text{Employee hours lost (fourth quarter)} \times \text{Average loaded labor rate} = \text{Cost}$$

$$386 \times \$13.48 = \$5{,}203.28$$

In this example, the company lost 386 hours due to accidents on the job in the fourth quarter of its fiscal year. The employees who actually missed time at work formed a pool of people with an average loaded labor rate of $13.48 per hour ($10.78 average hourly wage plus 20 percent for benefits). The average loaded labor rate multiplied times the lost hours reveals an unproductive cost of $5,203.28 to this company.

By studying records that are readily available in the company, a construction professional can also determine medical costs, insurance premiums, property damage, and fire losses for the time period in question. All of these costs taken together result in a subtotal cost. This figure is then increased by a standard percentage to cover indirect costs to determine the total cost of accidents for a specific time period. The percentage used to calculate indirect costs can vary from company to company, but 20 percent is a widely used figure.

FINDING SAFETY SOLUTIONS IN ONLINE DATABASES

Construction professionals have an invaluable tool in their safety tool chest for finding potential solutions to problems that are causing accidents and injuries. This invaluable tool is the Internet, Figure 1–6. An example of an excellent source of potential safety solutions is the *Construction Solutions Database*. This particular database contains strategies for preventing accidents and injuries associated with common construction hazards. It may be found at www.cpwrconstructionsolutions.org.

The database is organized so that users pursue solutions in a systematic, well-organized manner. First, the user identifies the "line of work" in question. Next the specific "task" within that line of work is identified. Then the "hazard" related to the task in question is identified. This is followed by a "hazard analysis" for the user. Finally, the user is provided with potential solutions that have the benefit of having been used and tested. The "Line of Work" box where the database

FIGURE 1–6 The Internet gives construction professionals access to safety databases.

Source: Maksym Yemelyanov/Fotolia

begins contains the following categories of construction work:

- Carpentry
- Drywall, glass, and floor coverings
- Electrical
- Excavation and demolition
- General labor
- Heavy equipment
- Insulation and lagging
- Masonry, tile, cement, and plaster
- Paints and coatings
- Pipes and vessels
- Reinforced concrete
- Residential construction
- Roofing
- Sheet metal and HVAC
- Structural steel

This database and other sites on the Internet give construction professionals the benefit of having access to colleagues from around the world who have faced the same or similar situations they are now facing. Using this database and others is the online version of sitting in a room with seasoned professionals and being able to ask their advice on pressing issues relating to construction safety.

BUILDING INFORMATION MODELING

Building Information Modeling is yet another tool for construction professionals. A **Building Information Model** or **BIM** is a digital representation of the physical and functional aspects of a building. A BIM allows construction professionals to view all aspects of a building before it is even built. As such it can be a powerful tool for construction professionals responsible for preventing accidents and injuries as the building is being constructed. Because a BIM allows construction professionals to observe the virtual construction of a building, it provides the advantage of being able to predict hazardous conditions and predetermine solutions to them. The better systems even allow users to introduce possible scenarios and analyze their potential effect on work-site safety. The goal of safety personnel in using BIM is to identify hazardous conditions before construction begins so they can be either eliminated or planned for, Figure 1–7.

Summary

The approximate cost of accidents in the United Sates is $150 billion annually. This includes the direct and indirect costs of accidents that occur on and off the job.

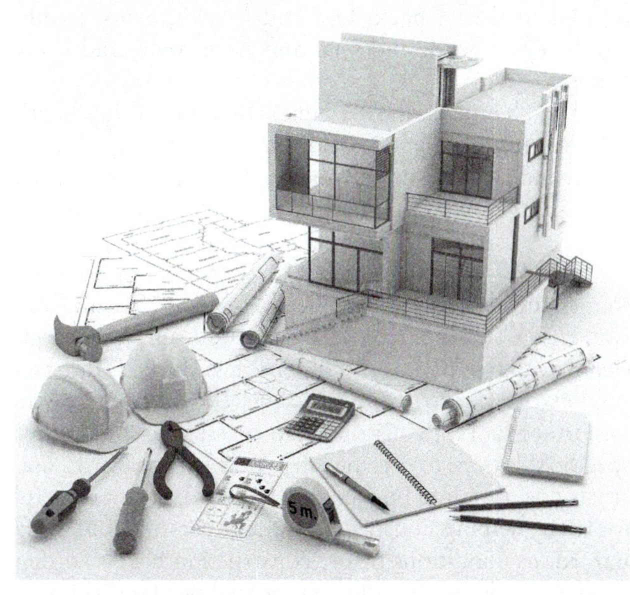

FIGURE 1–7 Building Information Models can be powerful safety tools for construction professionals.
Source: Yulyla/Fotolia

The leading causes of accidental deaths in the United States are motor vehicle accidents, falls, poisoning, drowning, fire-related injuries, suffocation, firearm injuries, medical complications, air transport accidents, machinery-related injuries, mechanical suffocation, and the impact of falling objects.

The leading causes of death in the United States are heart disease, cancer, and stroke. However, these causes are concentrated among people at or near retirement age. Among people aged 37 and younger, accidents are the number one cause of death. Since 1912, the number of accidental work deaths per 100,000 population has declined by 81 percent—from 21 to 4. The leading causes of death in work accidents are motor vehicle-related, falls, electric current, drowning, fire-related and air transport-related injuries, poisoning, and water transport-related injuries.

Approximately 35 million work hours are lost annually as a result of accidents. This is actual time lost from disabling injuries and does not include additional time lost to medical checkups after the injured employee returns to work.

The leading causes of work injuries are overexertion, impact accidents, falls, bodily reaction, compression, motor vehicle accidents, exposure to radiation and caustic chemicals, rubbing or abrasions, and exposure to extreme temperatures.

When death rates are computed on the basis of the number of deaths per 100,000 workers, the industry categories are ranked as follows (from highest death rate to lowest): mining and quarrying, agriculture, construction, transportation and public utilities, government, manufacturing, services, and trade.

Typically, injuries to specific parts of the body are ranked by frequency as follows (from most frequently injured to least): back; legs and fingers; arms; trunk; hands; eyes, head, and feet; and neck, toes, and body systems.

The chemicals most frequently involved in chemical burn injuries include acids and alkalis; soaps, detergents, and cleaning compounds; solvents and degreasers; calcium hydroxide; potassium hydroxide; and sulfuric acid.

The most frequent causes of heat burn injuries are flame, molten metal, petroleum, asphalt, steam, and water. The Internet makes construction safety solutions easily available to construction professionals. Databases such as the *Construction Solutions Database* can be an important tool for construction professionals looking for specific solutions to specific safety problems, hazards, and situations. Building Information Models can be powerful tools for helping construction safety personnel identify hazardous conditions before construction begins so that those hazards can be either eliminated or planned for.

Key Terms and Concepts

Accident prevention
Accident rate
Accidents
Building Information Model
Chemical burn injuries
Death rates
Drowning
Falls
Fire-related losses
Heat burn injuries
Impact accidents
Indirect costs
Insurance administration
Lost time
Lost wages
Mechanical suffocation
Medical complications
Medical expenses
Natural disasters
Overexertion
Poisoning
Property damage
Suffocation
Work injuries
Workplace accidents

Review Questions

1. What are the leading causes of death in the United States?
2. When the overall cost of an accident is calculated, what elements make up the cost?
3. What are the five leading causes of accidental deaths in the United States?
4. What are the leading causes of death in the United States of people between the ages of 25 and 44?
5. Explain how today's rate of accidental work deaths compares with the rate in the early 1900s.
6. What are the five leading causes of work deaths?
7. What are the five leading causes of work injuries by type of accident?
8. When death rates are classified by industry type, what are the three leading industry types?
9. Rank the following body parts according to frequency of injury from highest to lowest: neck, fingers, trunk, back, and eyes.
10. Name three chemicals that frequently cause chemical burns in the workplace.
11. Identify three factors that contribute to heat burn injuries in the workplace.
12. Explain why a construction professional would use a construction-safety related database.
13. Describe how a construction professional might use a *Building Information Model* to enhance safety at a construction site.

Critical Thinking and Discussion Activities

1. "Nobody can prove with hard data that accidents cost the construction industry more than all of these safety and health regulations we have to deal with," said Mike Flint, CEO of Flint Construction Company. As a recent graduate with a degree in construction technology, you are the newest member of Flint's staff. You are worried about what you have seen at the company's various job sites. There are flagrant safety violations occurring at all of Flint's job sites. In your opinion, it is only a matter of time before a tragedy occurs. What should you say to your CEO to try to convince him that establishing a comprehensive safety and health program makes good business sense?
2. You have made your case about developing a comprehensive safety and health program to your CEO, Mike Flint (see Case 1 above). It is two weeks later, and he has called you into his office to discuss your proposal further. Clearly, Flint has been thinking about what you told him earlier. This time he says, "You claim that back and hand injuries are typically the most common injuries in construction. I believe in zeroing in on the heart of the problem. Why not develop a safety program that targets back and hand injuries and leave out all of these other components you say we need?" You are pleased to have made some progress, but are still concerned that a narrowly focused program might leave the company vulnerable in several other critical areas. What should you do? Start small and hope to expand the program over time, or try once again to convince Flint that a more comprehensive program is needed. Explain your reasoning either way.

Application Activities

1. Find a construction company in your community that will work with you and do the following: (1) determine the most common types of accidents that occur in this company each year and (2) determine the amount of time lost annually because of work injuries.
2. Find a construction company in your community that will work with you or conduct a research project in the library to determine how a specific company calculates the annual cost of accidents.
3. Find an insurance company in your community that provides workers' compensation coverage for construction companies. Meet with a representative of the company and determine the following: (1) What are the most frequently reported injuries when construction workers file claims? and (2) What are the most costly types of claims they receive from construction workers?
4. Go to the construction safety database found at http://www.cpwrconstructionsolutions.org and select a "Line of Work" that interests you (e.g., carpentry). Enter a specific task in that line of work (e.g., building roof joists). Go to the "Hazard" step and respond with the hazard you are concerned about (e.g., falling from heights). Summarize the "Hazard Analysis" and potential solutions offered.

Endnotes

1. National Safety Council, *Accident Facts* (Chicago: NSC, 2016), 37.
2. Ibid., 25.
3. Ibid., 4–5.
4. Ibid., 34.
5. Ibid., 35.
6. Ibid.
7. Ibid., 36.
8. Ibid.
9. Ibid., 37.
10. Ibid., 38.
11. Ibid., 39.
12. Ibid., 40.
13. Ibid., 41.
14. Ibid.
15. National Safety Council, *Accident Prevention Manual: Administration and Programs*, 12th ed. (Chicago: NSC, 2001), 158.
16. Ibid.

CHAPTER TWO

ROLES OF CONSTRUCTION PERSONNEL IN SAFETY AND HEALTH

LEARNING OBJECTIVES

- Describe the safety and health team in construction.
- Explain the role of contractors in construction safety.
- Explain the role of managers and other professionals in construction safety.
- Describe the role of supervisors in construction safety.
- Describe the role of employees in construction safety.
- Explain the role of safety and health professionals in construction safety.
- Explain the certification process for safety and health personnel.

SAFETY AND HEALTH TEAM IN CONSTRUCTION

Although one or more people can be given the primary responsibility for coordinating, facilitating, and directing a company's safety and health activities, all construction personnel share the responsibility for safety. At construction sites, safety is and must be a "team sport." The **safety and health team** consists of at least the following players: contractors, managers and other professional personnel, supervisors, and employees or subcontractors. To this group, some companies add one or more **safety and health professionals** (Figure 2–1).

CONTRACTORS AND SAFETY

Contractors, as the owners of the company, bear the ultimate responsibility for safety and health on their job sites. Contractors are responsible for the following: (1) setting a **prosafety tone** for the company, (2) establishing a complete **commitment to safety and health** from the top down, and (3) ensuring that **sufficient resources** are provided to support a comprehensive, company-wide safety effort. By letting their managers and supervisors know that they expect safe and healthy job sites, contractors set a tone for the company that ensures that safety and health will be high priorities. By including safety and health in the company's strategic plan, contractors can establish the necessary commitment on the part of all employees. By providing the resources necessary to support company-wide safety and health programs, contractors ensure that the safety and health program is properly staffed and sufficiently funded.

FIGURE 2–1 Safety on the job is a "team sport."

Source: Ndoeljindoel/Fotolia

MANAGERS AND OTHER PROFESSIONALS AND SAFETY

In companies of sufficient size, contractors have a staff of management personnel and other professional personnel. Management positions may include the following: project managers, financial managers and accountants, marketing representatives, purchasing agents, cost accountants, human resources personnel, and office managers. Other professional personnel may include engineers, designers and drafting technicians, estimators, expeditors, and architects.

Managers and other professionals are responsible for setting a positive example related to safety and health and for translating the commitment of the contractor into everyday practice. This is accomplished by the following types of actions:

1. **Developing job descriptions** that make safety and health part of every employee's job
2. **Developing performance appraisal forms** that contain safety and health criteria
3. **Rewarding safe behavior** on the job by making it an important factor in promotions and pay raises
4. Developing work procedures that emphasize safety and health
5. **Recognizing safe work behavior** as a part of company incentive programs
6. Ensuring that the company has a comprehensive and effective safety and health program
7. Keeping up-to-date with the latest Occupational Safety and Health Administration (OSHA) standards and regulations related to construction
8. Effectively communicating safety and health information to all employees and subcontractors

SUPERVISORS AND SAFETY

Supervisors in the construction industry play a critical role in establishing and maintaining safe and healthy job sites. Supervisors are the first level of management; they interact with employees on job sites more frequently than do higher-level managers, other professional personnel, and contractors. Safety is an everyday, hands-on responsibility for supervisors. Safety- and health-related responsibilities of supervisors fall into the following categories of activity: training, accident prevention, accident investigation, and reporting.

Training Responsibilities of Supervisors

Supervisors are interested primarily in the type of training that is appropriate for their employees. Generally speaking, such training should cover the following subjects:

- **Orientation** to the organization's safety policy and principles
- **General housekeeping procedures**
- Emergency procedures
- Proper use of equipment
- Orientation to hazardous materials present and proper handling of these materials.

The method of instruction used can range from simple one-on-one conversations between supervisor and employee to group discussions and formal instruction. Instruction may be provided directly by supervisors or by individuals designated by supervisors.

Accident-Prevention Responsibilities of Supervisors

Accident prevention requires an ongoing program, consisting of a variety of techniques. Here are some techniques supervisors can use to prevent accidents on their job sites:

- *Involve all employees in an ongoing hazard-identification program.* Employees should be empowered to identify hazards associated with their work and to make recommendations for eliminating them.
- *Involve employees in developing safe job procedures.* Once hazards are identified, supervisors and employees should work together to find productive but safe ways to perform the job in question.
- *Teach employees how to properly use personal protective equipment (PPE), and monitor to make sure they do.* Ensure that employees learn how to use appropriate PPE before beginning a job. Supervisors should also make sure that employees follow through and apply what has been learned. When pressed to meet a deadline, the natural human tendency is to take shortcuts. Shortcuts taken in relation to PPE can cause accidents. Therefore, supervisors must monitor as well as train.
- *Teach employees good housekeeping practices, and require adherence to these practices.* One of the most effective ways to prevent accidents is to maintain a clean, organized, orderly workplace. These things result from good housekeeping. Supervisors should teach good housekeeping, and monitor to ensure it is practiced.
- *Teach employees the fundamentals of safe work practices (e.g., safe lifting, proper dress, use of safety glasses).* General safe work practices are perhaps the most important practices to remember. Some of the most common accidents result from simple mistakes, such as improper bending and lifting. Supervisors should monitor general work practices closely.

Accident Investigation and Reporting

In spite of a supervisor's best prevention efforts, accidents may still occur. It is essential to determine the cause. Why did the accident happen? Answering this question is the purpose of an **accident investigation**. By investigating the cause of an accident, supervisors may be able to prevent a reoccurrence. For this reason, it is also important to conduct the investigation immediately. Time can obscure the facts, the accident scene can change, witnesses can forget what they saw, or unrelated factors can creep in and obscure what really happened.

Figure 2–2 is a checklist that supervisors can use as a guide in conducting on-the-spot accident investigations. This checklist helps supervisors determine if the accident was caused by factors that could have been prevented and what those factors are. Occasionally, the cause is the failure of employees to observe mandatory safety precautions. When this is the case, remember that the purpose of accident investigation is to prevent future accidents, not to assign blame.

The Accident Report. Once the investigation has been completed, an **accident report** should be written.

> An accident report is a comprehensive summary of all the pertinent facts about an accident.

The report format can follow the investigation checklist in Figure 2–2, unless a company uses a standard accident report form. Regardless of the format, supervisors should remember several rules when writing accident reports, including the following:

- Be brief and stick to the facts.
- Be objective and impartial.
- Be comprehensive: Do not leave out facts.
- State clearly what employees and what equipment were involved.
- List any procedures, processes, or precautions that were not being observed at the time of the accident.
- List causal factors and any contributing factors.
- Make brief, clear, concise recommendations for corrective measures.

Checklist for On-the-Spot Accident Investigations

- What was the injured employee doing at the time of the accident?
- Had the injured employee received proper training in the task before being asked to perform it?
- Was the injured employee authorized to use the equipment or perform the task involved in the accident?
- Were other employees present at the time of the accident? If so, who are they, and what were they doing? (Interview them *separately* as soon as possible after the accident.)
- Was the task in question being performed according to properly approved procedures?
- Was proper personal protective equipment being used, and were work procedures being followed at the time of the accident?
- Was the injured employee new to the job in question?
- Was the process, equipment, or system that was involved new? Old? Properly maintained?
- Was the injured person being supervised at the time of the accident?
- Has a similar accident occurred before? If so, were corrective measures recommended? Were they implemented?
- Are there obvious factors that led to the accident or that could have prevented the accident?

FIGURE 2–2 Guidelines for conducting on-the-spot accident investigations.

EMPLOYEES AND SAFETY

The good efforts of contractors, managers, other professionals, and supervisors are to no avail if the individual employee refuses to cooperate. A safety and health policy does not prevent accidents unless employees accept it. Safe and healthy work practices do not prevent accidents unless they are put to use on the job. Ensuring a safe and healthy job site requires the concerted efforts of all members of the team, and employees are critical team members. It is not enough to simply know the rules related to safety; the rules must also be followed. Following the rules related to safety and health on the job site is the responsibility of the individual employee. It is not enough to simply avoid causing accidents on the job site. Employees must also play a positive, proactive role in helping to prevent accidents that might be caused by other employees.

SAFETY FACTS & FINES

Bypassing safety guards and devices on powered machines is as bad as not using them in the first place, and the consequences can be just as negative. In fact, the fines for bypassing can be even higher than those assessed for not having proper machine guards. A company in Birmingham, Alabama, found this out when it was fined $154,000 for allowing employees to bypass safety switches on machines. OSHA inspected the company—on the basis of a complaint filed by union representatives. The company had been cited for the same type of violation in the past.

Paine and Patterson, Inc. (PPI)
Commercial Construction
Vacancy Announcement

Position Title: Safety and Health Manager

Position Description: The Safety and Health Manager is responsible for establishing, implementing, and managing the company's overall safety and health program. The position reports to the CEO. Specific duties include the following:

- Establish and maintain a comprehensive company-wide safety and health program.
- Assess and analyze all jobs, processes, and materials for potential hazards.
- Work with appropriate personnel to develop, implement, monitor, and evaluate accident-prevention and hazard-control strategies.
- Ensure company-wide compliance with all applicable laws, standards, and codes.
- Coordinate the activities of all members of the company's safety and health program.
- Plan, implement, and broker, as appropriate, safety- and health-related training.
- Maintain all required safety- and health-related records and reports.
- Conduct accident investigations as necessary.
- Develop and maintain a company-wide Emergency Action Plan (EAP).
- Establish and maintain an ongoing safety promotion effort.
- Analyze the company's products from the perspectives of safety, health, and liability.

Qualifications Required: The following qualifications have been established by the management team—with input from all levels and all departments.

- *Minimum Education.* Applicants must have an associate of science or applied science degree in construction safety or closely related field.
- *Preferred Education.* Applicants with a baccalaureate degree in any of the following major fields of study will be given priority: construction safety and health, construction technology, construction management, construction engineering technology, or a related field. Degree programs in these fields must include at least one credit course in construction safety and health.

FIGURE 2–3 Safety and health manager job description.

SAFETY AND HEALTH PROFESSIONALS

An important member of the safety and health team is the safety and health professional (safety manager or safety engineer). Companies that are large enough sometimes employ a safety and health manager at an appropriate level in the corporate hierarchy. The manager's position in the hierarchy is an indication of the company's commitment and priorities.

In times past, construction companies with a highly placed safety and health manager were rare. However, passage of the Occupational Safety and Health Act (OSH Act) in 1970 began to change this. The OSH Act, more than any other single factor, put teeth in the job descriptions of safety and health professionals. OSHA standards, onsite inspections, and penalties have encouraged a greater commitment to safety and health. Environmental, liability, and workers' compensation issues and the growing awareness that providing a safe and healthy job site is the right thing to do from both an ethical and a business perspective have also had an impact.

The job of the safety and health manager is complex and diverse. Figure 2–3 is an example of a job description for such a position. The description attests to the diverse nature of the job. Duties include hazard analysis, accident reporting, standards compliance, record keeping, training, and emergency planning.

CERTIFICATION OF SAFETY AND HEALTH PERSONNEL

Professional certification is an excellent way to establish credentials in the construction safety and health field. The most widely pursued accreditations are as follows:

- *Certified Safety Professional (CSP):* Requires a minimum of a Bachelor's degree in any field or an Associate degree in safety, health, or the environment or a closely related field.

SAFETY FACTS & FINES

It is not just companies that can be penalized for poor safety and health conditions. In certain cases, responsible individuals in those companies can also be held accountable—with penalties including prison time. A company in Elk Grove, California, was fined $1.9 million by the Environmental Protection Agency (EPA), and its owner was sentenced to 80 months in prison for improper storage of hazardous chemicals. The company had dumped more than 160,000 gallons of waste oil into the ground. In addition, EPA inspectors found another 1,000 drums of oil on company property, many of them rusted and leaking.

- *Associate Safety Professional (ASP):* Requires a minimum of a Bachelor's degree in any field or an Associate degree in safety, health, or the environment or a closely related field.
- *Graduate Safety Professional (GSP):* This is an alternate route to becoming a CSP. It requires a Bachelor's or Master's degree from an ABET-ASAC or AABI program.
- *Occupational Safety and Health Technician (OHST):* Requires a high school diploma or GED.
- *Construction Health and Safety Technician (CHST):* Requires a high school diploma or GED.
- *Safety Trained Supervisor (STS):* Requires 30 hours of safety and health training.
- *Certified Environmental, Safety, and Health Trainer (CET):* Requires the delivery of 270 hours of training or teaching safety, health, and environmental-related areas.

The requirements summarized here represent the minimums in each case. To determine what the required examinations involve, as well as minimum passing scores, consult the website of the Board of Certified Safety Professionals (www.bcsp.org).

In addition to these safety and health certifications, there are also specialized certifications in the field of ergonomics. These certifications are available from the Board of Certification in Professional Ergonomics (BCPE). There are two levels of certification available from the BCPE: Professional and Associate. Both of these broad areas of certification have three subspecialties for which construction professionals can pursue certification:

- *Professional Certifications.* There are three certifications available in the professional category: **Certified Professional Ergonomist (CPE)**, Certified Human Factors Professional (CHFP), and Certified User Experience Professional (CUXP). All of these certifications require a Bachelor's degree and academic coursework covering BCPE core competencies, minimum of three years of specific work experience, and a passing score on the professional certification examination.
- *Associate Certifications.* There are three certifications available in the Associate category: Associate Ergonomics Professional (AEP), Associate Human Factors Professional (AHFP), and Associate User Experience Professional (AUXP). All of these certifications require a Bachelor's degree and academic coursework covering the BCPE core competencies, less than three years of specific work experience, and transitional status (held a maximum of six years). The application form for these certifications is shortened for graduates of a Human Factors and Ergonomics Society accredited program.

An additional certification—Certified Ergonomics Associate (CEA)—was once available from the BCPE, but has been discontinued. The requirements summarized here represent the minimums in each case. To determine what the required examinations involve as well as minimum passing scores, consult the website of the Board of Certification in Professional Ergonomics (www.bcpe.org).

Summary

The safety and health team in a construction company, depending on the size of the company, may include the following: the contractor, managers, various professional personnel, supervisors, subcontractors, employees, and health and safety professionals. Contractors are responsible for setting a prosafety tone for the company, establishing commitment, and providing sufficient resources to support a comprehensive safety program.

Managers and other professionals are responsible for setting a positive example and for translating the commitment of the contractor into everyday practice. Supervisors are key players on the safety team of a construction company. Their responsibilities include the following: training, accident prevention, and accident investigation and reporting. Employees are responsible for following safety rules and regulations and for playing a proactive role in helping to prevent accidents. Some construction companies are large enough to employ safety and health professionals (managers or engineers). When this is the case, the health and safety professional is the management-level employee who is assigned primary responsibility for establishing, implementing, operating, and evaluating the company's safety and health program.

Key Terms and Concepts

Accident report
Accident investigation
Certified Professional Ergonomist (CPE)
Certified Safety Professional (CSP)
Commitment to safety and health
Developing job descriptions
Developing performance appraisal forms
General housekeeping procedures
Orientation
Prosafety tone
Recognizing safe work behavior
Rewarding safe behavior
Safety and health professionals
Safety and health team
Sufficient resources

Review Questions

1. List the "players" who might be part of the safety and health team for a large construction company.
2. What are the responsibilities of contractors related to safety and health?

3. What are the safety and health responsibilities of managers and other professionals in a construction company?
4. Describe the training responsibilities of supervisors in construction companies.
5. What are the accident-prevention responsibilities of supervisors in construction companies?
6. Define the term ***accident report.***
7. List five rules to follow when writing accident reports.
8. What are the employee's safety and health responsibilities?
9. Explain the responsibilities of safety and health professionals in construction companies.
10. How does one go about becoming a Certified Safety Professional?

Critical Thinking and Discussion Activities

1. "I think we put too much responsibility on the shoulders of management when it comes to maintaining safe job sites," said one student of safety and health. "After all, you can't make an employee work safely. If an employee fails to follow the rules and gets hurt, that's his problem." "It's true that you can't make an employee work safely, but management should still be responsible for establishing, implementing, and enforcing a safety program," responded another student. Join this debate. What is your opinion?
2. "I don't see how you can say that safety is a team sport in those companies that hire safety and health professionals," argued one safety and health student. "If you have a safety manager, everyone is going to expect him or her to be responsible for safety." "That's not true," said another student. "The safety and health manager is just the team captain. Everybody else is still a player and has a job to do." Join this debate. What is your opinion? Will companies have trouble getting all employees to take responsibility for safety if they hire safety managers or engineers?

Application Activities

1. Identify a construction company in your community that will cooperate with you in completing this activity. How is safety handled? Is it a team sport? Do safety-related responsibilities appear in job descriptions? Is safe behavior on the job evaluated? Is it rewarded?
2. Identify a construction company in your community that employs a full-time safety and health manager or engineer. Interview this individual. What are his or her duties? Who does the safety manager or engineer report to? What are the major responsibilities listed in this individual's job description?

CHAPTER THREE

PRO-ACTIVE SAFETY STRATEGIES IN CONSTRUCTION: SAFETY CULTURE, TOTAL SAFETY MANAGEMENT, AND OTHER STRATEGIES

LEARNING OBJECTIVES

- **Define the concept of the safety-first culture as a pro-active strategy.**
- **Explain how to implement Total Safety Management (TSM) as a pro-active strategy.**
- **Describe how to promote pro-active thinking among construction personnel as a pro-active strategy.**
- **Explain the "Must-Do" pro-active safety strategies for construction companies.**
- **Make safety part of the corporate infrastructure for *sustainability*.**

With workplace injuries costing close to $170 billion annually, it is clear that just complying with government regulations is not enough. To remain competitive, construction companies must get out in front of safety hazards and prevent accidents before they can occur. This is not to downplay in any way the importance of OSHA compliance. The costs of failing to comply with OSHA standards are part of the $170 billion mentioned above. Consequently, much of what must be learned about construction safety has to do with OSHA compliance. However, the most effective safety programs are those that get out in front of OSHA compliance by developing and applying pro-active safety strategies. In this way they go beyond just complying and reacting to anticipating and preventing.

Pro-active safety strategies include such things as establishing a safety-first culture, applying the principles of **Total Safety Management (TSM), pro-active thinking** on the part of all construction personnel, and implementing a group of pro-active strategies the author calls the "Must Dos" of construction safety. These and other pro-active safety strategies are explained in this chapter.

ESTABLISHING A SAFETY-FIRST CULTURE IN CONSTRUCTION

An organization's culture consists of the **tacit assumptions, beliefs, values, attitudes, expectations,** and **behaviors** that are widely shared and accepted in the organization. When these things encourage and support a safety-first attitude from top to bottom in an organization, the organization has a safety-first culture, Figure 3–1. Evidence of a safety-first culture can be found in the following factors associated with a given organization:

1. *Its priorities.* Are safety and health top priorities in the organization?
2. *How people in the organization succeed.* Are personnel recognized and rewarded for working safely?
3. *How decisions are made in the organization.* Is safety a major consideration when decisions are made?

FIGURE 3–1 A construction company's culture can determine if its job sites are safe.
Source: Horticulture/Fotolia

4. *Expectations management has of employees.* Do executives and management personnel make it clear that safe behavior is the expected behavior in all cases?
5. *Expectations employees have of management.* Are employees encouraged to make their views known about the quality of the work environment?
6. *Effects of internal peer pressure on safety.* Does peer pressure among workers support or undermine safety?
7. *Unwritten rules that are widely accepted.* Do the organization's unwritten rules support or undermine safety?
8. *How conflict about safety is handled.* When conflicts arise between productivity and safety, are they settled in favor of safety?

These questions make the critical point that the cultural elements of safety and health are part of an organization's larger corporate culture, not some separate and distinct component that stands alone. Safety and health should be so fully integrated into an organization's culture that they are seen to be critical elements not just in OSHA compliance but in the organization's ability to compete in the global marketplace, Figure 3–2.

How an Organization's Culture Is Established?

Many factors contribute to the creation of an organization's corporate culture. The value systems of executive-level decision makers are often reflected in their organization's culture. How managers treat employees and how employees at all levels interact are also factors that contribute to an organization's culture. What management expects of employees and what employees in turn expect of management are factors that contribute to an organization's culture. The stories passed along from employee to employee typically play a major role in the establishment and perpetuation of an organization's culture. All of these factors can either help or hurt an organization.

If supervisors push workers to take shortcuts on safety procedures when management personnel are not looking, it is not likely that there will be a safety-first culture. On the other hand, if supervisors insist on the safe and healthy approach in all cases regardless of who is watching, it is more likely that there will be a safety-first culture. If none of the organization's corporate heroes are people who built a reputation for safety, it is not likely that there will be a safety-first culture. On the other hand, if the stories that are passed down through generations of workers about the organization's corporate heroes include stories about managers, supervisors, or employees who earned a reputation for safety, it is more likely that there will be a safety-first culture.

FIGURE 3–2 In a safety-first culture peer pressure encourages safe work practices.
Source: Lisa F. Young/Fotolia

FIGURE 3–3 Safe work practices must be fully integrated into a construction company's culture.
Source: Madpixblue/Fotolia

Corporate cultures in organizations are established based on what is expected, modeled, passed on during orientation, taught by mentors, included in training, monitored and evaluated, and reinforced through recognition and rewards. If safety is expected by management personnel and individual workers, if it is modeled by people in positions of authority, if it is stressed during the workers' initial orientation to the organization, if it is taught as the right way to do things by mentors, if it is stressed through training, if it is monitored and evaluated by supervisors, and if it is reinforced by management personnel through recognition and rewards, safety will become a fully integrated part of the organization's corporate culture, Figure 3–3.

WHAT A SAFETY-FIRST CULTURE LOOKS LIKE?

Part of the process of establishing a safety-first culture in a construction company is developing an understanding of what one looks like. This is a lot like a person who wants to lose weight taping a picture of a role model to the bathroom mirror. The picture serves not only as a constant reminder of the desired goal, but also as a measurement device that indicates when a goal has been met. If a picture of an organization with a safety-first culture could be taped to a construction company's wall for all employees to see, it would have the following characteristics:

- Widely shared agreement among key decision-makers that providing a safe and healthy work environment is an essential competitive strategy.

- Emphasis on the importance of human resources to the organization and the corresponding need to protect them from hazards.
- Ceremonies to celebrate safety and health-related successes.
- Widely shared agreement that the work environment that is most conducive to peak performance and continual improvement is a safe and healthy work environment.
- Recognition and rewards that are given to high-performing workers and teams include safety and health-related performance on the job.
- Insistence on safety and health as part of supplier relations.
- Effective internal network for communicating safety and health information and expectations.
- Informal rules of behavior that promote safe and healthy work practices.
- Strong pro-safety corporate value system as set forth in the company's strategic plan.
- High expectations and standards of performance relating to safety and health.
- Employee behavior that promotes safe and healthy work practices.

TEN STEPS FOR ESTABLISHING A SAFETY-FIRST CULTURE

The process for establishing a safety-first culture in a construction company consists of 10 broad steps. Those steps are as follows:

1. *Understand* the need for a safety-first corporate culture.
2. *Assess* the current corporate culture as it relates to safety.
3. *Plan* for a safety-first corporate culture.
4. *Expect* appropriate safety-related behaviors and attitudes.
5. *Model* the desired safety-related behaviors and attitudes.
6. *Orient* personnel to the desired safety-first corporate culture.
7. *Mentor* personnel in the desired safety-related behaviors and attitudes.
8. *Train* personnel in the desired safety-related behaviors and attitudes.
9. *Monitor* safety-related behavior and attitudes at all levels.
10. *Reinforce and maintain* the desired safety-first corporate culture.

Understand the Need for a Safety-First Culture

Personnel at all levels need to be shown that providing a safe and healthy work environment is an important responsibility of management, Figure 3–4. Everyone from the CEO of the organization to the newest employee should understand and be able to articulate the following factors that support the need for a safety-first culture in a construction company:

- An organization's corporate culture determines the normal and accepted way things are done in the organization. Consequently, if the normal and accepted way is to be the safe and healthy way, the organization must have a safety-first culture.
- In the same way that the work practices of individuals become habitual, the work practices of organizations become cultural; that is, they become ingrained and codified in the organization's unwritten rules. They become the way things are done when "the boss isn't looking." In order for the way people work when not closely supervised to be the safe and healthy way, an organization must establish and maintain a safety-first culture.

Assess the Current Corporate Culture as It Relates to Safety and Health

Does an organization have a safety-first culture? This is a good question to ask and the answer should never be assumed—either yes or no. The answer to this question, no matter how good or bad the organization's safety and health record might be, should be the result of a thorough assessment. Such an assessment should consider the following factors as a minimum:

- The company's safety record
- Attitudes of key decision makers toward safety and health

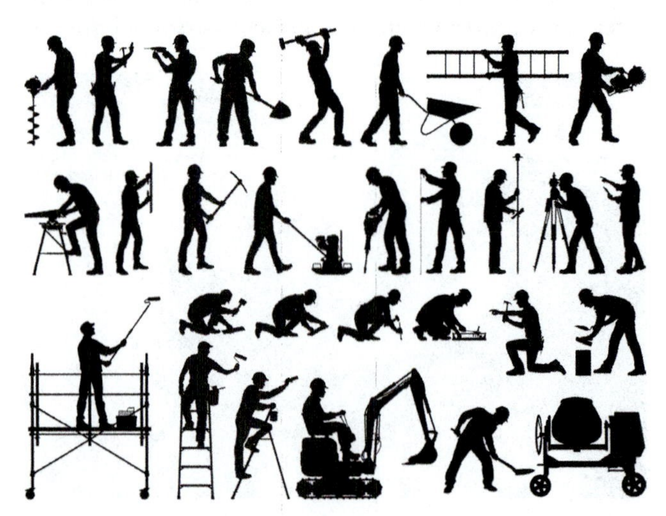

FIGURE 3–4 A safety-first culture requires buy-in from all personnel.
Source: PrintingSociety/Fotolia

- Safety and health concerns are high-priority factors in the decision-making process
- Employees are encouraged to speak out when they have concerns about safety and health issues
- Peer pressure and unwritten rules encourage safety and health
- Employees are recognized and rewarded for working safely and for identifying hazards that might lead to accidents and injuries
- Management personnel are good role models of safe and healthy work practices
- Comprehensive safety and health training are provided on a regular basis
- Working safely is written into all job descriptions and included in performance appraisals

Plan for a Safety-First Culture

Once the results of the assessment process have been tabulated, they should be used as the basis for planning for the establishment of a safety-first culture or for enhancing one that already exists but has weaknesses. For example, assume that the results of the assessment show that employees are not provided regular safety and health-related training. A planning goal to correct this deficiency might read as follows: *Establish a comprehensive training program on all aspects of safety and health for employees and implement it no later than January 1st.*

The plan for establishing a safety-first culture or for enhancing one that has weaknesses should be based on the results of a comprehensive and thorough assessment, Figure 3–5.

Expect Appropriate Safety-Related Behaviors and Attitudes

If you want people to perform well in any capacity, you must first have high expectations of them. It is the same with safety-related behaviors and attitudes. If you want people to work safely, you have to let them know that safety is expected. There are several ways that organizations can let employees know what is expected of them concerning safe and healthy work practices. The most effective of these are job descriptions that specify working safely as an employee responsibility, the examples set by supervisors and managers, safety and health criteria built into performance appraisal documents, including safety and health as criteria for selecting employees for recognition and rewards, mentoring, and daily feedback to employees from supervisors and managers.

FIGURE 3–5 Planning for safety is a critical element of a safety-first culture.

Source: Shotsstudio/Fotolia

Model the Desired Safety-Related Behaviors and Attitudes

One of the worst mistakes supervisors and managers can make is to say to employees either verbally or through actions, "Do as I say, not as I do." Nothing speaks louder to employees than the examples—good or bad—set by supervisors and managers. Consequently, if people in positions of authority want employees to work safely, they must set a consistently positive example of doing so themselves factoring safety and health into their decisions, and showing by their actions that safety and health are high priorities.

Orient Personnel to the Desired Safety-Related Behaviors and Attitudes

Construction companies sometimes miss out on an excellent opportunity to get employees started off on the right foot immediately after they are hired. This opportunity is the organization's orientation for new employees. Too often orientations for new employees are little more than filling out forms, choosing insurance program options, and learning how to navigate the company successfully. This is unfortunate because the only chance a construction company has to make a positive first impression on new employees is during their orientation.

Anything and everything that is relevant to the organization's corporate culture should be introduced and explained during the new employee orientation sessions—including expectations relating to safety and health. Personnel who conduct the orientation sessions should be encouraged to emphasize that in this company the right way is the safe way.

Mentor Personnel in the Desired Safety-Related Behaviors and Attitudes

Once employees have completed a comprehensive orientation, the next step is to assign them an experienced mentor who exemplifies the desired safety-related behaviors and attitudes. Mentors help guide inexperienced personnel until they gain the experience necessary to work safely without assistance, but even more important, they also help them develop a safety-first attitude. Mentors answer questions, make suggestions, and provide guidance, but the most important responsibility of mentors is to set a positive example, an example that includes a safety-first approach.

Train Personnel in the Desired Safety-Related Behaviors and Attitudes

There are two fundamental principles of good management that apply when trying to establish a safety-first culture. The first is that you should never expect employees to do anything they have not been trained to do. The second is that you should never assume that employees know how to do anything without having been trained. If you want employees to work safely, teach them how. If you want employees to have the right attitudes toward safety and health, teach them what such an attitude looks like and the practical applications of such an attitude. Do not assume that employees know how to work safely—teach them how, Figure 3–6.

Monitor and Evaluate Safety-Related Behaviors and Attitudes at All Levels

A principle of effective supervision is "You get the behavior you accept." Supervisors and managers who allow their direct reports to get away with unsafe work practices are saying, "Your unsafe behavior is acceptable to me." Letting unsafe work practices go unchallenged and uncorrected is the same thing as approving them. Consequently, it is critical that supervisors and managers **monitor** employees and correct unsafe work practices immediately. Another reason for monitoring employees and correcting them immediately is that work practices become habitual. Once people become accustomed to doing a task a given way, that way becomes a habit and habits are hard to break. If the habitual way is the unsafe way, the employee in question is heading down a one-way street to disaster.

In addition to monitoring on a daily basis, it is important to make safety and health-related behaviors part of formal performance evaluations. The performance appraisal instrument should contain at least one criterion about safety and health. There is a management principle that says, "If you want performance to improve, measure it." This is why it is important to evaluate the safety and health-related performance of workers. By measuring how safely they work, supervisors and managers have the hard data needed to make specific improvements in the work practices of specific individuals.

Reinforce and Maintain the Safety-First Culture

Just as employees should never stop working safely no matter how good their team's safety record might be, construction companies should never stop doing what is necessary to maintain a safety-first culture. Such a culture is not a goal an organization achieves and then moves on to other matters. It is a state of being that must be reinforced constantly or it will be lost. What follows are some strategies organizations can use to reinforce and maintain their safety-first culture once it has been established:

- Reward safe work behavior by making it an important factor when making promotions, giving wage and salary increases, and awarding performance incentives.
- Recognize safe work behavior by making it an important criterion when singling out workers or teams for recognition awards.
- Recognize employees for pointing out hazards and hazardous conditions before they can cause accidents or injuries.
- Encourage supervisors to verbally and publicly recognize workers who are doing their jobs safely every day as they monitor work performance.

By applying the strategies explained in this section, construction can pro-actively get out in front of OSHA compliance by establishing and maintaining a safety-first culture that will contribute greatly to making them more competitive and more profitable.

FIGURE 3–6 Orientation, mentoring, and training are musts in a safety-first culture.

Source: Rawpixel.com/Fotolia

IMPLEMENTING TOTAL SAFETY MANAGEMENT (TSM)

Another pro-active safety strategy closely related to a safety-first culture is known as Total Safety Management (TSM). TSM is a concept developed by the author back in 1998 to be implemented by organizations that were also implementing the concept of Total Quality Management (TQM). TSM can be defined as: *A performance-oriented approach to safety and health (as opposed to a compliance-oriented approach) that views a safe and healthy work environment as a competitive advantage for organizations. Because they believe there is a strong correlation between job performance and the work environment, TSM practitioners seek to establish a work environment that is conducive to the peak performance of people and processes as well as continual improvement of that performance—including safety and health-related performance. The fundamental elements of TSM include the following:*

- Strategically based and driven by the need to maximize competitiveness (it is understood that accidents and injuries are costly financially, in terms of employee morale, and in terms of the construction company's public image)
- Performance and process oriented (as opposed to compliance oriented)
- Dependent on executive-level commitment (as are all pro-active safety and health strategies)
- Teamwork oriented
- Committed to employee involvement and empowerment
- Based on data-driven decision making
- Committed to continual improvement of the work environment and, in turn, employee performance
- Committed to comprehensive, on-going safety and health training
- Maintained by company-wide unity of purpose

TSM is implemented through the application of a three-phase model: *Planning, identification/assessment,* and *execution/monitoring/adjusting.* Like most pro-active safety and health strategies, TSM requires executive-level buy-in and commitment. However, it can be recommended and initiated by construction professionals below the executive level using the following model (Figure 3–7)

Planning Phase

- Gain executive-level commitment.
- Establish the TSM Steering Committee (this is a safety committee with representatives from all levels of the construction company that oversees the implementation process).
- Mold the TSM Steering Committee into a well-functioning team through teamwork training, including specific safety and health-related training.
- Require the TSM Steering Committee to develop a safety and health vision and related goals for the company that will become part of the company's overall strategic plan.
- Require the TSM Steering Committee to communicate the safety and health vision and goals to all personnel company wide.

FIGURE 3–7 TSM involves all personnel in ensuring safe work practices.
Source: Rawpixel.com/Fotolia

Identification/Assessment Phase

- Identify the organization's weaknesses concerning safety and health.
- Identify safety and health advocates and resisters (those who resist working safely may require special attention or may eventually have to have their employment terminated).
- Identify specific improvements that need to be made as well as larger improvement projects that should be executed.

Execution/Monitoring/Adjusting Phase

- Establish improvement project teams and give them their assignments.
- Execute improvement projects, monitor their effectiveness, and make adjustments as necessary.

PRO-ACTIVE THINKING AND CONSTRUCTION SAFETY

The term "pro-active" means getting out in front of problems rather than just reacting to them and it is important that construction companies do this. Just responding to accidents and injuries after the fact is simply too expensive a proposition. Consequently, the most competitive construction companies encourage their personnel at all levels to think creatively to anticipate safety problems so they can be prevented. Fortunately, thinking pro-actively is a skill that can be taught and learned, Figure 3–8. Construction professionals can encourage pro-active thinking by (1) setting an example of doing the following things, and (2) teaching their personnel to do follow their example and rewarding them for doing so.

- *Think critically.* Be prepared to challenge the way things have always been done and refuse to blindly follow procedures that appear to ignore hazards. Thinking critically means working with one's brains and hands. It means thinking things through before acting rather than simply accepting that everything will be fine as long as one follows procedures. Critical thinking is the opposite of doing things the way they have always been done only because that is how they have always been done. It also ensures that people have their eyes, ears, brains, and experience engaged as they work—that is, they are alert rather than just going through the motions.
- *Turn mistakes into learning opportunities.* When accidents or near misses occur, construction professionals can point the finger of blame to deflect responsibility or they can turn the mistakes that were made into learning opportunities so they do not happen again. The latter course is the better course. This does not mean that construction personnel are not held accountable for working safely. Rather it means that just pointing the finger of blame will do nothing to prevent the same mistakes from occurring in the future. Make sure employees know that company's highest priority is to prevent future mistakes. This is critical because if employees are worried about being disciplined for making mistakes they will simply cover up those mistakes. When this happens accidents and near misses go unreported.
- *Be flexible in seeking safer ways to work.* Following established procedures is important, but this should always be done with an eye to improving those procedures. If there is a safer way to accomplish certain tasks, procedures can be revised accordingly. When discussing proposed revisions to procedures or new procedures, avoid the human tendency to mentally lock onto one proposal and mentally shut out others. One of the other proposals might have merit. It is when considering the broadest spectrum of possibilities that a better way is usually found.
- *Brainstorm with employees and subcontractors.* Construction professionals who are trying to pro-actively anticipate problems and identify potential hazards can multiply their effectiveness using **brainstorming** as a technique. Brainstorming involves bringing together people who work or will work on a given construction project and taking advantage of their eyes, brains, and experience. A facilitator—the construction professional—leads the group through discussions about potential safety problems on the job in question and records their observations on a flip chart or marker board. The idea behind brainstorming is that the observations of one participant will trigger the thinking of the others. Construction professionals who notice that participants are reluctant to speak up can use a

FIGURE 3–8 Pro-active thinking can help construction personnel anticipate and prevent safety problems.
Source: Jerry Sliwowski/Fotolia

modified form of brainstorming called Nominal Group Technique or NGT. With this technique, rather than just ask for verbal input the facilitator hands out 3 × 5 cards and asks participant to write down their concerns and observations—one to a card. The cards are collected and the written observations are recorded on a flip chart or marker board. Allowing participants to write down their concerns and observations rather than giving them verbally in front of the group often increases the level of participation in brainstorming sessions.

MUST-DO PRO-ACTIVE SAFETY STRATEGIES

There are certain strategies that simply must be applied if construction companies are going to enjoy the benefits of their personnel being pro-active. These *must-do* safety strategies include the following:

- *Gain executive commitment to safety.* Without the commitment of the construction company's top executives, it will be difficult to do anything other than comply and react. To get out in front of safety problems construction professionals must invest the time and effort necessary to gain the commitment of top executives because ultimately they decide how resources will be allotted and what the priorities will be in a construction company.
- *Make safety committees cross-functional and give them some teeth.* One of the best ways to get out in front on safety issues is to establish a broadly representative safety committee and give the committee the authority it needs to take whatever actions are necessary to establish and maintain a safe work environment on job sites. The best way to do this is to: (1) Make the safety committee cross-functional (representative of all the different construction trades found on a given job site), and (2) Put construction professionals with authority on the committee. If the committee has only hands-on workers as its members, the committee will lack the teeth (authority) needed to do its job. Consequently, it is important that the safety committee be more than just broadly representative of the workforce (cross-functional), but it also have members who are key decision makers in the company.
- *Reward prevention rather than low accident rates.* When employees are rewarded for maintaining low accident rates, there is a tendency for minor accidents, injuries, and near misses to go unreported. Consequently, it is better to reward people who take the initiative to identify hazards and unsafe conditions before they cause accidents. Reward people who get out in front on safety and health concerns and prevent accidents and injuries.
- *Build safety into every construction project.* Ideally construction professionals will work with architects and engineers during the design phase of construction so that as the structure is planned they can plan how to safely build it. However, if it is not possible for construction personnel to work with designers they can still use the architectural and engineering plans to develop a safety plan for a given project. Considering how to safely build a structure before even clearing the site is an excellent way to get out in front on safety issues.
- *Make safety a consideration in selecting subcontractors.* Construction companies that select subcontractors solely on the basis of low bid may come to regret this approach. In addition to a subcontractor's price, technical competency, and performance on other jobs, construction companies should also consider its safety record. Hiring a subcontractor with a poor safety record—no matter how well it performs in the other aspects of its responsibilities—is buying trouble. Along with submitting a proposed price for its services and references from other jobs, subcontractors should be required to submit their safety records and they should agree in writing to abide by the safety procedures of the construction company, Figure 3–9.
- *Train, train, and train some more.* It is critical that construction companies provide the training their personnel and subcontractors need to work safely under the conditions that exist at any given job site. Never allow a construction worker or subcontractor to perform any task he or she does not know how to perform safely. Also, err on the side of caution when it comes to making decisions about training and make training mandatory. For example, just providing the proper personal protective equipment (PPE) needed for a given job is not sufficient. Construction workers need to be trained in how to properly don, wear, and use PPE, Figure 3–10.

FIGURE 3–9 Make safety a high-priority criterion when selecting subcontractors such as this welding subcontractor.
Source: Maxhalanski/Fotolia

FIGURE 3–10 Always train construction workers in the proper use of PPE.
Source: Yosef19/Fotolia

FIGURE 3–11 Asking questions about different requirements in the company's safety manual is an effective MBWA technique.
Source: Pixelrobot/Fotolia

- *Make predictable safety problems a high priority.* Construction companies keep records of the types of accidents and near misses that occur on their job sites. These records can be a gold mine for preventing future accidents and the corresponding injuries. For example, falls are one of the most common types of accident on construction sites. Consequently, construction professionals should make sure that they plan to prevent falls, train to prevent falls, and talk about preventing falls with construction workers. By focusing first on the predictable types of accidents, construction professionals can eliminate or at least mitigate the biggest threats to job site safety.
- *Talk about safety all the time.* One of the best ways to make sure that construction workers understand that safety is a high priority is for construction professionals to talk about it with them at every chance. In addition to talking, it is important to ask questions, listen to the answers, and encourage construction workers to share their concerns. This is the concept known as MBWA or Management by Walking Around. For example, an effective technique is for safety professionals to gently quiz workers about different requirements in their safety manual or safety guide, Figure 3–11.
- *Be firm in taking on the subject of substance abuse.* Construction ranks second from the top when industries are ranked according to drug use. A drug impaired worker is a dangerous worker—a hazard to himself or herself and others. Consequently, one of the more effective ways to get out in front of safety issues is to establish and maintain a strong "No-Drugs" policy and corresponding program for implementing and enforcing the policy. Drug testing must be a part of the program as well as training and team discussions, Figure 3–12.

FIGURE 3–12 If either of these construction workers is drug impaired, others on the job are in danger.
Source: Andreas Karelias/Fotolia

- *Work with insurance personnel.* Insurance companies have a vested interest in keeping construction sites accident free. Consequently, they can be an invaluable resource for information that will help achieve that goal. Construction companies should partner with their insurance company and work cooperatively to help ensure the safest possible job sites.

SUSTAINABILITY AS A PRO-ACTIVE SAFETY STRATEGY

To ensure that safety is not just a reactive, after-the-fact *add-on* that gets a lot of attention following a serious accident, the various strategies for maintaining a safe work environment must be sustained. To be sustained, a construction company's safety program must become part of its operational infrastructure. In other words, safety must be built into the systems, processes, and procedures of the company so that it can, in turn, become part of the company's corporate culture. The question then becomes, "How can safety be built into a construction company's operational infrastructure?" What follows are methods for making safety part of a construction company's DNA:

- *Make safety part of the company's strategic plan.* A construction company's strategic plan contains a corporate vision, mission, list of core values (sometimes called guiding principles), and broad strategic goals. If safety is going to be sustained and pursued pro-actively, it should be part of the construction company's mission, listed as one of its core values, and have its own strategic goal.
- *Have a corporate safety plan.* Every strategic goal in a company's strategic plan should have a corresponding implementation plan that describes in detail how that specific goal is going to be achieved. Consequently, the strategic goal relating to safety should be assigned to the Safety Committee which then becomes responsible for developing a detailed plan for achieving it or, in other words, establishing and maintaining a safe working environment.
- *Include safety as part of all job descriptions.* Every position from CEO to laborer in a construction company should have a job description and that job description should contain a line that explains the responsibilities of the position in question concerning safety. The point of this exercise is to ensure that safety is part of everyone's responsibility and not something that is delegated to a safety director or foreman or superintendent. All of these positions play important roles in maintaining a safe work environment, but all other positions have their responsibilities concerning safety too.
- *Include safety as part of all performance appraisals.* If construction professionals want their personnel and subcontractors to take safety seriously, it must be included in performance appraisals. The performance appraisal forms used for every level of employee from CEO to laborer should contain a criterion about safety. The wording of the criteria will differ depending on the position in question, but at least one criterion should be part of every performance appraisal.
- *Make training a permanent and on-going enterprise.* Training should become a normal part of the job for construction workers and professionals. Construction companies that want their personnel work safety must commit to teaching them how. Too often training is something that is done reactively in response to a serious accident or incident, but this approach will do little to sustain an effective safety program. To be a sustaining factor, training must be done pro-actively and before the fact. It must be used as a preventive strategy, not as a reactive response, Figure 3–13.

FIGURE 3–13 Training should be pro-active and on-going.
Source: Aeroking/Fotolia

- *Make safety a high-priority criterion when rewarding performance.* When making decisions about performance awards and rewards, safety must be a high-priority factor. Even the best-performing worker should not receive awards or rewards unless his or her outstanding performance was achieved in a safe manner.

Summary

A safety-first culture is one in which the tacit assumptions, beliefs, values, attitudes, expectations, and behaviors that are widely shared and accepted in an organization support the establishment and maintenance of a safe and healthy work environment for all personnel at all levels.

Evidence of a safety-first culture exists in an organization's priorities, how people in the organization succeed, how decisions are made, expectations management has of employees, expectations employees have of management, effects of internal peer pressure, unwritten rules that are widely accepted, and how conflict about safety is handled.

Having a safety-first corporate is important to organizations because it is the right thing to do and because it contributes to more effective regulatory compliance. However, the most fundamental reason for having a safety-first culture is **competition**. The most effective way for an organization to succeed in a competitive marketplace is to consistently provide **superior value** to its customers. This is achieved by consistently providing superior quality, cost, and service. All three of these critical elements of superior value require a safe and healthy work environment and cannot be achieved in the long run without such an environment.

The 10 steps for establishing a safety-first corporate culture are understand, assess, plan, expect, model, orient, mentor, train, monitor, and reinforce and maintain.

Total Safety Management or TSM is a performance-oriented approach to safety and health (as opposed to a compliance-oriented approach) that views a safe and healthy work environment as a competitive advantage for organizations. Because they believe there is a strong correlation between job performance and the work environment, TSM practitioners seek to establish a work environment that is conducive to the peak performance of people and processes as well as continual improvement of that performance—including safety and health-related performance. The fundamental elements of TSM include the following: strategically based, performance oriented, dependent on executive level commitment, teamwork oriented, committed to employee involvement and empowerment, based on data-driven decision making, committed to continual improvement of the work environment, committed to comprehensive, on-going safety training, and maintained by company-wide unity of purpose. TSM is implemented in three phases: planning, identification/assessment, and execution/monitoring/adjustment.

Pro-active thinking means getting out in front of safety problems rather than just reacting to them after the fact. Construction professionals can encourage pro-active thinking in their personnel by setting an example of and teaching the following concepts: critical thinking, turning mistakes into learning opportunities, being flexible in seeking safer ways to work, and brainstorming about job site safety.

Must-do pro-active safety strategies include the following: gain executive commitment, make safety committees cross-functional and give them some teeth, reward prevention rather than low accident rates, build safety into every construction project, make safety a consideration when selecting subcontractors, train, make predictable problems a high priority, talk about safety all the time, be firm in taking on the subject of substance abuse, and work with insurance personnel.

To sustain safety programs rather than allowing them to become reactive after-the-fact add-ons, build safety into the construction company's operational infrastructure by: including safety in the strategic plan, developing a corporate safety plan, including safety in job descriptions and performance appraisals, making training a permanent, on-going enterprise, and making safety a high-priority factor when rewarding performance.

Key Terms and Concepts

Assess
Attitudes
Behaviors
Beliefs
Brainstorming
Competition
Expect
Expectations
Globalization
Mentor
Model
Monitor
Orient
Plan
Pro-active thinking
Reinforce
Safety culture
Safety-first corporate culture
Superior cost
Superior value
Sustainability
Tacit assumptions
Total Safety Management (TSM)
Train
Understand
Values

Review Questions

1. Define the concept of the safety-first corporate culture.
2. Why is the term **safety culture** a misnomer?
3. List at least five ways (evidence) to tell if an organization has a safety-first corporate culture.
4. Why is it important for organizations to have a safety-first corporate culture?
5. How does the concept of **globalization** factor into the need for a safety-first corporate culture?
6. How does the concept of superior value factor into the need for a safety-first corporate culture?
7. Describe how corporate cultures are established.
8. What does a safety-first corporate culture look like?
9. List and explain each of the 10 steps for establishing a safety-first corporate culture.
10. What is Total Safety Management (TSM)?
11. Describe how to implement TSM.
12. What is pro-active thinking as it relates to safety?
13. How can construction professionals encourage pro-active thinking among their personnel?
14. List the *must-do* pro-active safety strategies.
15. Describe how a construction company can sustain its safety program.

Critical Thinking and Discussion Activities

1. A fellow construction professional makes the following comment: "As long as we comply with the applicable OSHA Standards we won't have any safety problems." Another construction

professional responds: "Nonsense. We need to be pro-active and get out in front of safety problems. I don't worry about compliance as long as we are thinking pro-actively about safety." Is either of these construction professionals right? Wrong? Partially so? Take up the discussion. What is your opinion?

2. Two construction professionals are talking about establishing a safety culture in their company. Both are foreman. One comments: "I know all ten steps of the implementation model for establishing a safety-first culture. I don't have the authority to implement some of the steps but I think I can implement enough of the concept to make a difference even without executive level commitment." The other construction professional responds: "I don't think you should do anything until you have gained executive level commitment. Without that you are just wasting your time." Join this discussion. What is your opinion?
3. "I think Total Safety Management is the answer to solving our safety problems in this company. All we ever do is respond after the fact to accidents and injuries. We don't even worry about near misses. If we are going to get out in front on safety we are going to have to stop just reacting and checking off compliance boxes, we are going to have to be pro-active." "I disagree," said another safety professional. The two were discussing their company's poor safety record over lunch. "All we need to do is get serious about OSHA compliance and things will turn around." Join this discussion. What is your opinion? Will stricter OSHA compliance be sufficient or should the company also become more pro-active in its thinking concerning safety?

Application Activities

1. Conduct some Internet research and identify a construction company that prides itself on maintaining a safe and healthy workplace. Determine which of the strategies explained in this chapter the company uses to maintain a safe and healthy work environment.
2. Conduct some Internet research in which you locate the strategic plans for several construction companies. Do any of these strategic plans contain safety-related statements? An alternative approach for completing this activity is to contact local construction companies and ask to see their strategic plans.

CHAPTER FOUR

ETHICS AND SAFETY

LEARNING OBJECTIVES

- Define the term *ethics*.
- Explain how a construction professional can decide what is ethical when making decisions.
- Describe the three personality measures that can influence an individual's ethical behavior.
- Explain the construction professional's role concerning ethics.
- Explain the construction company's role concerning ethics.
- Describe how a construction professional should handle ethical dilemmas.
- List the questions to be asked when making decisions that have ethical ramifications.
- Explain the concept of *whistle-blowing*.

Practically everyone agrees that the business practices of construction companies should be above reproach with regard to ethical standards. Few people are willing to defend unethical behavior. For the most part, industry in the United States operates within the scope of accepted legal and ethical standards.

However, unethical behavior occurs often enough to require that construction professionals be aware of the types of ethical dilemmas they may face and know how to deal with such issues. How to deal successfully with ethics on the job is the subject of this chapter.

AN ETHICAL DILEMMA

Southland Prestressed Concrete (SPC) company has been awarded a contract to build an ultramodern multistory shopping mall. The "University Mall" is the largest contract SPC has ever undertaken. This is the good news. The bad news is that, to win the contract, SPC's vice president of sales and marketing had to agree to a completion date that, if not unreasonable, is at least going to be challenging. In addition, all exposed concrete surfaces on the project must be sprayed with a specified paint that is highly toxic and difficult to apply.

SPC's painting subcontractor has never used this particular type of paint before. Personal protective equipment (PPE) and other engineering controls can minimize the potential hazards, but all precautions must be stringently observed—with absolutely no shortcuts. In addition, the manufacturer of the paint recommends three full days of training for all employees who will work with the paint. But there is a problem: The recommended training cannot be provided soon enough to fit into SPC's expedited schedule for this job.

In a secret meeting, SPC's executive managers decide to purchase the necessary PPE, use the toxic paint as specified, and forgo the recommended training. In addition, the executives decide to withhold from employees all information about the toxicity of the paint. Camillo Rodriguez, safety engineer for SPC, was not invited to the secret meeting. However, the decisions made during the meeting were slipped to him anonymously. Rodriguez now faces an ethical dilemma. What should he do? If he chooses to do nothing, employees of SPC's painting subcontractor might be inappropriately exposed to a highly dangerous substance. If he shares what he knows with the subcontractor, he might be called upon to testify about what he knows—a step that could cost him his job and threaten his career. This is an example of the type of ethical dilemma that construction professionals may face on the job.

ETHICS DEFINED

Any time that ethics is the topic of discussion, terms such as ***conscience***, ***morality***, and ***legality*** are frequently heard. Although these terms are closely associated with ethics, they do not, by themselves, define it. For the purpose of this book, ethics is defined as follows:

> **Ethics** is the study of morality within a context established by cultural and professional values, social norms, and accepted standards of behavior.[1]

Morality refers to the **values** that are subscribed to and fostered by society in general. Ethics attempts to apply reason in determining rules of human conduct that translate morality into everyday behavior. **Ethical behavior** falls within the limits prescribed by morality.

How then does a construction professional know if someone's behavior is ethical? Ethical questions are rarely black and white. They typically fall into a **gray area** between the two extremes of right and wrong. Personal experience, self-interest, point of view, and external pressure often cloud this gray area even further.

Guidelines for Determining Ethical Behavior

Guidelines are needed for construction professionals to use when trying to sort out matters that are not clearly right or wrong. First, however, it is necessary to distinguish between the concepts of *legal* and *ethical*. They are not the same thing. Just because an option is legal, it is not necessarily ethical as well.

In fact, it is not uncommon for people caught in the practice of questionable behavior to use the "I didn't do anything illegal" defense. A person's behavior can be well within the scope of the law and still be unethical. The following guidelines for determining ethical behavior assume that the behavior in question is legal (Figures 4–1 and 4–2).

- Apply the ***morning-after test***. This test asks, "If you make this choice, how will you feel about it tomorrow morning?"
- Apply the ***front-page test***. This test encourages you to make a decision that would not embarrass you if it were a story on the front page of your hometown newspaper.
- Apply the ***mirror test***. This test asks, "If you make this decision, how will you feel about yourself when you look in the mirror?"
- Apply the ***role-reversal test***. This test requires that you trade places with the people affected by your decision and view the decision through their eyes.
- Apply the ***common-sense test***. This test requires that you listen to what your instincts and common sense are telling you. If it feels wrong, it probably is.

Guidelines for Ethical Choices

1. Apply the morning-after test
2. Apply the front-page test
3. Apply the mirror test
4. Apply the role-reversal test
5. Apply the common-sense test

FIGURE 4–1 Guidelines for determining ethical choices.

Checklist for Ethics Models

In addition to the various tests that can be used for determining ethical behavior, there are also numerous models:

- Categorical imperative (black and white)
- Conventionalistic ethic (anything legal is ethical)
- Disclosure rule (explain actions to a wide audience)
- Doctrine of the mean (virtue through moderation)
- Golden rule (do unto others . . .)
- Intuition rule (what is right is known)
- Market ethic (whatever makes a profit is right)
- Means-end ethic (end justifies the means)
- Might-equals-right ethic
- Organization ethic (loyalty to the organization)
- Practical imperative (treat people as ends, not means)
- Equal freedom (full freedom unless it deprives another)
- Proportionality ethic (good outweighs the bad)
- Professional ethic (do only what can be explained to your peers)
- Revelation ethic (answers revealed by prayer)
- Rights ethic (protect rights of others)
- Theory of justice (impartial, evenhanded)

FIGURE 4–2 Models for determining what the ethical choice is when making decisions.

Blanchard and Peale suggest their own test for deciding what the ethical choice is in a given situation.[2] Their test consists of the following three questions:

1. Is it legal?
2. Is it balanced?
3. How will it make me feel about myself?

If a potential course of action is not legal, no further consideration is in order. If an action is not legal, it is also not ethical, because ethical behavior requires that the law be obeyed. If an action is balanced, it is fair to all involved. This means that construction professionals and other personnel have responsibilities that extend well beyond the walls of their unit, organization, and company. If a course of action is in keeping with your own moral structure, it will make you feel good about yourself. Blanchard and Peale also list the following *Five Ps of Ethical Power*:

1. *Purpose*. Individuals see themselves as ethical people who let their conscience be their guide and, in all cases, want to feel good about themselves.
2. *Pride*. Individuals apply internal guidelines and have sufficient self-esteem to make decisions that may not be popular with others.
3. *Patience*. Individuals believe right will prevail in the long run, and they are willing to wait when necessary.
4. *Persistence*. Individuals are willing to stay with an ethical course of action once it has been chosen and to see it through to a positive conclusion.
5. *Perspective*. Individuals take the time to reflect and are guided by their own internal barometer when making ethical decisions.[3]

These tests and guidelines help construction professionals make ethical choices in the workplace. In addition to internalizing the guidelines themselves, construction professionals may want to share these values with all employees with whom they interact.

ETHICAL BEHAVIOR IN ORGANIZATIONS

Research by Trevino suggests that ethical behavior in organizations is influenced by both individual and social factors.[4] Trevino identified three personality measures that can influence an employee's ethical behavior: (1) *ego strength*, (2) *Machiavellianism*, and (3) *locus of control*.

An employee's **ego strength** is his or her ability to undertake self-directed tasks and to cope with tense situations. A measure of a worker's **Machiavellianism** is the extent to which he or she will attempt to deceive and confuse others. **Locus of control** is the perspective of workers concerning who or what controls their behavior. Employees with an internal locus of control feel that they control their own behavior. Employees with an external locus of control feel that their behavior is controlled by external factors (such as rules, regulations, or the supervisor).

Social factors can also influence ethical behavior in organizations. These factors include gender, role differences, religion, age, work experience, nationality, and the influence of other people, who are significant in an individual's life. People learn appropriate behavior by observing the behavior of significant role models (such as parents, teachers, public officials, and supervisors). Because construction professionals represent a significant role model for their employees, it is critical that they exhibit ethical behavior that is beyond reproach in all situations.

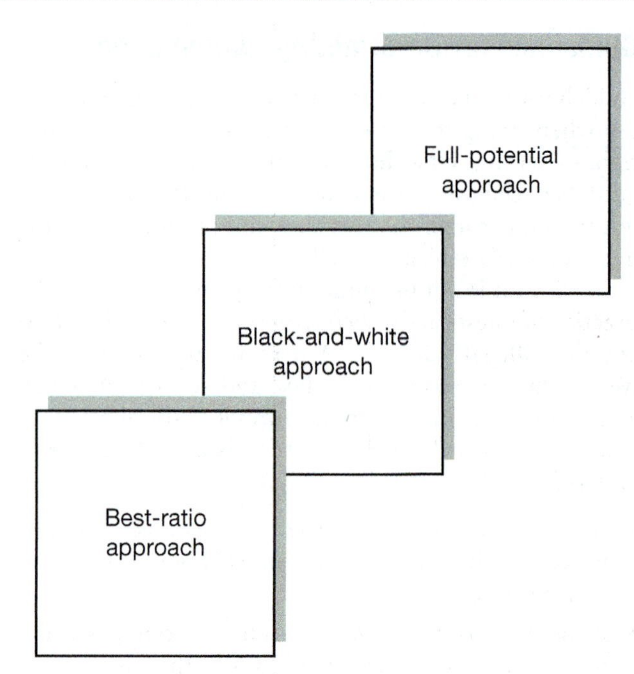

FIGURE 4–3 Three basic approaches to handling ethical problems.

CONSTRUCTION PROFESSIONALS' ROLE IN ETHICS

Using the guidelines set forth in the previous section, construction professionals should be able to make responsible decisions concerning ethical choices. Unfortunately, deciding what is ethical is much easier than actually doing what is ethical. In this regard, trying to practice ethics is like trying to diet. It is not just a matter of *knowing* you should cut down on eating; it is a matter of actually *doing* it.

This fact defines the role of construction professionals with regard to ethics. Their role has three parts. First, they are responsible for setting an example of ethical behavior. Second, they are responsible for helping fellow employees make the right decision when facing ethical questions. Finally, construction professionals are responsible for helping employees follow through and actually undertake the ethical option once the appropriate choice has been identified. In carrying out their roles, construction professionals can adopt one of the following approaches: the best-ratio approach, the black-and-white approach, or the full-potential approach (Figure 4–3).

Best-Ratio Approach

The **best-ratio approach** is the pragmatic option. It assumes that people are basically good and, under the right circumstances, behave ethically. However, under certain conditions, they can be driven to unethical behavior. Therefore, construction professionals should do everything possible to create conditions that promote ethical behavior and try to maintain the highest possible ratio of good choices to bad. When hard decisions must be made, the appropriate choice is the one that does the most good for the most people. This is sometimes referred to as *situational ethics*.

Black-and-White Approach

With the **black-and-white approach**, right is right, wrong is wrong, and circumstances are irrelevant. The construction professional's job is to make ethical decisions and carry them out and, in addition, to help employees choose the ethical route. When difficult decisions must be made, construction professionals should make fair and impartial choices, regardless of the outcome.

Full-Potential Approach

Construction professionals who use the **full-potential approach** make decisions based on how the outcomes affect the ability of those involved to achieve their full potential. The underlying philosophy is that people are responsible for realizing their full potential within the confines of morality. Choices that can achieve this goal without infringing on the rights of others are considered ethical.

Decisions made may differ, depending on the approach selected. For example, consider the ethical dilemma presented at the beginning of this chapter. If the safety engineer, Camillo Rodriguez, applies the best-ratio approach, he may decide to keep quiet, encourage the proper use of PPE, and hope for the best. On the other hand, if he takes the black-and-white approach, he will be compelled to confront SPC's management team with what he knows.

COMPANY'S ROLE IN ETHICS

Construction companies, like all businesses, have a critical role to play in promoting ethical behavior among their employees. Construction professionals cannot set ethical examples alone or expect employees to behave ethically in a vacuum. A company's role in ethics can be summarized as (1) creating an internal environment that promotes, expects, and rewards ethical behavior and (2) setting an example of ethical behavior in all external dealings.

Creating an Ethical Environment

A company creates an **ethical environment** by establishing policies and practices that ensure that all employees are treated ethically and then enforcing these policies. Do employees have the right of due process? Do employees have access to an objective grievance procedure? Are there appropriate safety and health measures to protect employees? Are hiring practices fair and impartial? Are promotion practices fair and objective? Are employees protected from harassment based on race, gender, or other reasons? A company that establishes an environment that promotes, expects, and rewards ethical behavior can answer *yes* to all of these questions.

One effective way to create an ethical environment is to develop an ethics policy and specific, written guidelines for implementing that policy. A sample ethics policy follows:

> J. R. Makin Construction Company will conduct its business in strict compliance with applicable laws, rules, regulations, corporate policies, procedures, and guidelines. We will conduct all business with honesty, integrity, and a strong commitment to the highest standards of ethics. We have a duty to conduct our business with both the letter and the spirit of the law.

This statement would set the tone for all employees at J. R. Makin Construction Company—letting them know that higher management not only supports ethical behavior but also expects it. This approach makes it less difficult for construction professionals who find themselves caught between the pressures of productivity and the maintenance of safe work practices.

Setting an Ethical Example

Companies that take the "do as I say, not as I do" approach to ethics do not succeed. Employees must be able to trust their company leaders to conduct all external and internal dealings in an ethical manner. Companies that do not pay their bills on time, companies that pollute, companies that place short-term profits ahead of employee safety and health, companies that do not live up to advertised quality standards, companies that do not stand behind their guarantees, and companies that are not good neighbors in their communities are not setting a good ethical example. Such companies can expect employees to mimic their unethical behavior.

A final word on the company's role in ethics: In addition to creating an ethical internal environment and handling external dealings in an ethical manner, companies must support employees who make ethically correct decisions. This support must be given not just when such decisions are profitable, but in all cases. For example, in the ethical dilemma presented earlier in this chapter, imagine that Camillo Rodriguez decided that his ethical choice was to confront the management team with his knowledge of the hazards associated with the paint. Management had given the order to withhold critical information. This was obviously the profitable choice in the short run. But was it the ethical choice? If Rodriguez did not think so, would higher management stand behind him? If not, everything else that the company does to promote ethics will fail, Figure 4–5.

HANDLING ETHICAL DILEMMAS

Nobody serves long as a construction professional without confronting an ethical dilemma. How then should one proceed? There are three steps (Figure 4–4):

1. Apply the various guidelines presented earlier in this chapter for determining what is ethical.
2. Select one of the three basic approaches to handling ethical questions.
3. Proceed in accordance with the approach selected, and proceed with consistency.

Steps for Handling Ethical Dilemmas

1. Apply the guidelines.
2. Select the approach.
3. Proceed accordingly and consistently.

FIGURE 4–4 Handling ethical dilemmas.

SAFETY FACTS & FINES

It is not uncommon for the Occupational Safety and Health Administration (OSHA) to inspect a company on the basis of a complaint from an employee or other concerned individual. This is what happened to a company in Houston, Texas. The company was fined $128,000 and cited for unsafe electrical conditions, failure to provide adequate personal protective equipment for employees, and improper electrical grounding of machines. There are no secrets when it comes to unsafe working conditions.

FIGURE 4–5 Construction professionals must set an ethical example at all times.
Source: Antonio Gravante/Fotolia

Step 1: Apply the Guidelines

In Step 1, you should apply as many of the tests set forth in Figure 4–1 as necessary to determine the ethically correct decision. In applying these guidelines, attempt to block out all mitigating circumstances and other factors that tend to cloud the issue. At this point, the goal is only to identify the ethical choice. Deciding whether to implement the ethical choice comes in the next step.

Step 2: Select the Approach

When deciding how to proceed after Step 1, you have three basic approaches, as set forth in Figure 4–3: the best-ratio, black-and-white, and full-potential approaches. These approaches and their ramifications can be debated ad infinitum; however, selecting an approach to ethical questions is a matter of personal choice. Factors that affect the ultimate decision include your personal makeup, the expectations of the company, and the degree of company support.

Step 3: Proceed with the Decision

The approach selected in Step 2 dictates how you should proceed. Two things are important in Step 3. The first is to proceed in strict accordance with the approach selected. The second is to proceed consistently. *Consistency* is critical when handling ethical dilemmas. Fairness is a large part of ethics, and consistency is a large part of fairness. The grapevine will ensure that all employees know how an ethical dilemma is handled. Some will agree and some will disagree, regardless of the decision. Such is the nature of human interaction. However, regardless of their differing perceptions of the problem, employees respect consistency. Conversely, even if the decision is universally popular, you may lose respect if the decision is not consistent with past decisions.

QUESTIONS TO ASK WHEN MAKING DECISIONS

Construction professionals often must make decisions that have ethical dimensions. A constant state of tension exists between meeting contract schedules and maintaining employee safety. Construction professionals will find themselves right in the middle of these issues. The following questions can and should be asked when decisions are being made about issues that have ethical dimensions. Construction professionals should ask these questions themselves, and they should encourage other decision makers within their organizations to do the same.

- Has the issue or problem been thoroughly and accurately defined?
- Have all dimensions of the problem (e.g., productivity, quality, cost, safety, health) been identified?
- Would other stakeholders (employees and customers) agree with your definition of the problem?
- What is your real motivation in making this decision? Meeting a deadline? Outperforming another organizational unit or a competitor? Self-promotion? Getting the job done right? Protecting the safety and health of employees? Some combination of these?
- What is the probable short-term result of your decision? What is the probable long-term result?
- Who will be affected by your decision and in what way? In the short term? In the long term?
- Did you discuss the decision with all stakeholders (or all possible stakeholders) before making it?
- Would your decision withstand the scrutiny of employees, customers, colleagues, and the general public?

Construction professionals should ask themselves these questions, but equally important, they should also insist that other managers do the same. The manager responsible for meeting this month's deadline may be so focused on the date that he or she overlooks safety. The manager who is feeling the pressure to cut costs may make decisions that work in the short term but have disastrous consequences in the long term. Questions such as these can help managers broaden their focus and consider the long-range effects of their decisions, Figure 4–6.

FIGURE 4–6 Construction professionals must ask the difficult questions when facing ethical dilemmas.
Source: Carolyn Franks/Fotolia

ETHICS AND WHISTLE-BLOWING

What can construction professionals do when their employer is violating legal or ethical standards? The first option, of course, should be to bring violations to the attention of appropriate management executives through established channels. In most cases, this is sufficient to stop the illegal or unethical behavior. But what about those occasions when you are ignored or, worse yet, told to "mind your own business"? These are the types of situations that have led to the concept of whistle-blowing, which can be defined as follows:

> **Whistle-blowing** is the act of informing an outside authority or media organ of alleged illegal or unethical acts on the part of an organization or an individual.

Problems with Whistle-Blowing

American society has an interesting attitude toward the concept of whistle-blowing. There seems to be an inherent uneasiness with the concept that is deeply rooted in the American psyche. Even when the illegal or unethical practice in question threatens the safety and health of employees, some people still do not like whistle-blowers. There is often a "don't tell" mentality that causes whistle-blowers to be shunned and viewed as outcasts. As children, we learned not to be tattletales. Many adults still seem to hold to this philosophy.

The "don't be a tattletale" attitude is only one of the problems that works against whistle-blowing. Here are some others.

- *Retribution.* People who blow the whistle on their employer may be subjected to retribution. They may be fired, transferred to an undesirable location, or reassigned to an undesirable job. They may also be shunned. There are numerous ways—legal ways—for an employer to take retribution against a whistle-blower.
- *Damaged relationships and hostility.* Blowing the whistle about an illegal or unethical practice can often damage relationships. Somebody—by commission or omission—is responsible. That person, or those persons, may be disciplined as a result. When this happens, people tend to choose sides, which, in turn, leads to damaged relationships. Damaged relationships are often manifested in the form of hostility directed toward the whistle-blower.
- *Loss of focus.* Whistle-blowers often find that their time, energy, and attention are overtaken by the events surrounding the claim of illegal or unethical behavior. Rather than focusing on doing their jobs, they find themselves dealing with retribution, damaged relationships, and hostility (Figure 4–7).

FIGURE 4–7 Loss of focus by those who blow the whistle on employers that seek retribution can cause accidents.
Source: Lisa F. Young/Fotolia

SAFETY FACTS & FINES

Companies are subject to fines not just from OSHA, but also from organizations such as the Environmental Protection Agency (EPA). The EPA is especially watchful when it comes to how hazardous waste is stored and disposed of. A construction waste processing company in Philadelphia was fined $400,000 by the EPA for violation of waste-storage rules. Specifically, the company was cited for storing more waste than its permits allowed, failing to keep its emergency management plan up to date, and failing to notify the EPA of these violations. In addition to the fine, the company was required to change the way it operates. It was required to scale back the amount of waste it handles by 2,000 tons per day and to hire an environmental manager to oversee its daily operations.

- *Scapegoating.* Negative consequences can certainly occur as a result of whistle-blowing. Because of this, some people may decide to ignore the issue or to raise it to the next level of management and let it drop there. The problem with this approach is the issue of *accountability*. When an employee is injured or the environment is damaged, the actions of construction professionals are certain to be closely scrutinized: "Did you know about the hazardous condition?" "Did you do everything in your power to prevent the accident or incident?" These types of questions are always asked when litigation is brought, as is often the case. An irresponsible organization facing charges of negligence may begin looking for a convenient scapegoat. Managers and supervisors can become targets.

OSHA and Whistle-Blowing

OSHA dealt with the issue of whistle-blowing for certain employees when it adopted regulations governing the *employee protection* provisions of Section 211 (formerly Section 210) of the Energy Reorganization Act (ERA). This statute makes it illegal for an employer covered by the act to discharge an employee or otherwise discriminate against an employee in terms of compensation, conditions, or privileges of employment because the employee or any person acting at the employee's request performs a protected activity.

Employers covered by the ERA include the following:

- Licensees of the Nuclear Regulatory Commission or an agreement state (including applicants for a license)
- A contractor or subcontractor of a licensee or applicant
- A contractor or subcontractor of the Department of Energy

Key provisions of the ERA are summarized as follows:

- *Right to raise a safety concern.* You are engaged in protected activity when you: (1) notify your employer of an alleged violation of the ERA, (2) refuse to engage in any practice made unlawful by the ERA if you have identified the alleged illegality to the employer, (3) testify before Congress or at any federal or state proceeding regarding any provision or proposed provision of the ERA, (4) commence or cause to be commenced a proceeding under the ERA or a proceeding for the administration or enforcement of any requirement imposed under the ERA, (5) testify or are about to testify in any such proceeding, or (6) assist or participate in such a proceeding or in any other action to carry out the purposes of the ERA.
- *Unlawful acts by employers.* It is unlawful for an employer to intimidate, threaten, restrain, coerce, blacklist, discharge, or in any other manner discriminate against any employee because this employee has engaged in a protected activity.
- *Complaint.* An employee or employee representative may file a complaint charging discrimination in violation of the ERA within 180 days of the discriminatory action. A complaint must be in writing and should include a full statement of facts, including the protected activity engaged in by the employee, knowledge by the employer of the protected activity, and the basis for believing that the activity resulted in discrimination against the employee by the employer. A complaint may be filed in person or by mail at the nearest local office of OSHA, U.S. Government, Department of Labor, or with the Office of the Assistant Secretary, OSHA, U.S. Department of Labor, Washington, D.C. 30210.
- *Enforcement.* OSHA reviews the complaint to ensure that it makes an initial showing of discrimination. If not, or if the employer provides clear and convincing evidence that there was no discrimination, there is no investigation. If the required showing is made, OSHA notifies the employer and conducts an investigation to determine whether a violation has occurred. Either the employee or the employer may request a hearing.
- *Relief.* If discrimination is found, the employer is required to provide appropriate relief, including reinstatement (even in the period between the decision and appeal), back wages, or compensation for injuries suffered from the discrimination, and attorney's fees and costs.

Noncovered Whistle-Blowing

The ERA covers a number of different types of organizations, but most employers fall outside of the law's coverage. Consequently, employees in many organizations have no special protection when they blow the whistle on an unethical employer. This is one more reason why construction professionals should encourage their employers to develop comprehensive safety and health policies.

A key element of such a policy should be a mechanism that allows employees to raise questions about safety and health concerns. Such a mechanism should contain provisions that protect employees who have raised the concerns from retribution by the employer.

Summary

Ethics is the study of morality. Morality refers to the values that are subscribed to and fostered by society. Ethics attempts to apply reason in determining rules of human conduct that translate morality into everyday behavior.

Behavior is ethical when it falls within the limits prescribed by morality.

Legal and *ethical* are not the same. If something is illegal, it is also unethical. However, just because something is legal does not mean that it is ethical. An act can be legal but unethical.

To determine if a choice is ethical, you can apply the following tests: *morning after*, *front page*, *mirror*, *role reversal*, and *common sense*.

Safety and health professionals have a three-pronged role with regard to ethics. They are responsible for setting an ethical example, helping employees to make ethical decisions, and helping employees to follow through and actually undertake the ethical option.

Safety and health professionals have three approaches available in handling ethical dilemmas: best ratio, black and white, and full potential.

The company's role in ethics is to create an ethical environment and to set an ethical example. An effective way is to develop a written **philosophy of ethics** and share it with all employees.

Three personality characteristics that can influence an employee's ethical behavior are *ego strength*, *Machiavellianism*, and *locus of control*.

People facing ethical dilemmas should apply the tests for determining what is ethical, select one of the three basic approaches, and proceed consistently.

Whistle-blowing is the act of informing an outside authority or the media of alleged illegal or unethical acts on the part of an organization or individual.

Key Terms and Concepts

Best-ratio approach
Black-and-white approach
Common-sense test
Conscience
Ego strength
Ethical behavior
Ethical environment
Ethics
Front-page test
Full-potential approach
Gray area
Legality
Locus of control
Machiavellianism
Mirror test
Morality
Morning-after test
Philosophy of ethics
Role-reversal test
Values
Whistle-blowing

Review Questions

1. Define the term *morality*.
2. Define the term *ethics*.
3. Briefly explain each of the following ethics tests: *morning after*, *front page*, *mirror*, *role reversal*, and *common sense*.
4. What is the safety and health professional's role with regard to ethics?
5. Briefly explain the following approaches to handling ethical behavior: best ratio, black and white, and full potential.
6. Briefly explain a company's role with regard to ethics.
7. Explain how one should proceed when facing an ethical dilemma.
8. Write a brief philosophy of ethics for a chemical company.
9. List the individual and social factors that may influence an employee's ethical behavior.
10. List and briefly describe the *Five Ps of Ethical Power* as set forth by Blanchard and Peale.
11. What question should safety and health professionals ask when making decisions that have an ethical component?
12. Explain the most common problems associated with whistle-blowing.

Critical Thinking and Discussion activities

1. You are a safety supervisor for Mid-West Construction Company. Part of your job is investigating workers' compensation claims that the company thinks might be overstated or even fraudulent. Your son is a warehouse foreman for a branch of Mid-West that is located in another city. Your son is visiting you while recuperating from a back injury for which he is collecting workers' compensation. You got your son his job at Mid-West. While visiting, he jogs, lifts weights, and plays softball with friends. You realize from this that your son is not really injured. What should you do?
2. The framing supervisor and the union representative have just had a chin-to-chin disagreement about removing the new safety guards from sawing machines used to prefabricate roof trusses. They have asked you—the safety manager for the company—to mediate the dispute. According to the framing supervisor, "We ran these machines for five years without these so-called safety guards and never had an accident. The only reason we put them on was because some OSHA inspector suggested it. They are fine when we aren't in a hurry, but they slow us down when the rush is on, and it's on now." The union representative counters, saying, "These machines are dangerous. He's so worried about making a deadline that management never should have agreed to in the first place that he doesn't care about the safety of my people." What is the right thing to do here? What is your opinion?

3. You are the designated "competent person" responsible for safety at your company's "Bay Bridge construction site." Your company has the contract to build a four-lane bridge across Martin's Bay. Much of the heavy work is done using a crane floated on a barge. The project got off to a good start and was even running ahead of schedule. Then the hurricane season came, and the water was too choppy to use the floating crane for two full weeks. To make matters even worse, the choppy waters banged the barge against a test pile, creating enough damage to put it out of action. Fair weather has returned, and everyone is anxious to get caught up. In fact, the state road department is threatening to enforce the $5,000 per day penalty clause in the contract if the bridge is not completed on time. The pressure to produce is intense. A replacement barge has been located and put in service. This is where your dilemma comes in. The barge is too small. The crane plus the loads it picks up exceed the rated capacity of the barge. Only you and your company's top managers know this. Work has proceeded without incident so far, but you are concerned that it is just a matter of time before tragedy strikes. In your opinion—an opinion supported by the barge's specifications—the barge could actually capsize unless conditions on the water are perfect. They have been so far, but the water conditions change with the weather, and the weather is always unpredictable. What should you do? What is your opinion?

Application Activities

1. Interview management-level personnel from a local construction company. Ask them to share (without naming companies or individuals) ethical dilemmas they have faced or of which they are aware. What were the factors involved? How was the dilemma handled? What were the consequences?
2. Conduct the library research necessary to identify cases in which companies made poor choices when facing ethical dilemmas. What were the circumstances? Who was involved? What happened? What should have been done?
3. Conduct a telephone survey to locate a construction company in your community or region that has a written ethics policy. Request a copy of the policy and analyze its contents. What is your opinion of the policy? Is it weak or strong? How could it be improved?

Endnotes

1. Goetsch, D. L., *Effective Supervision* (Columbus, OH: Prentice Hall, 2001), 81.
2. Blanchard, K., and N. V. Peale, *The Power of Ethical Management* (New York: Ballantine Books, 1988), 10–17.
3. Ibid., 79.
4. Trevino, L. K, "Ethical Decision Making in Organizations: A Person-Situation Interactionist Model," *Academy of Management Review* 11, no. 3 (1996): 601–17.

CHAPTER FIVE

WORKERS' COMPENSATION AND OTHER KINDS OF CONSTRUCTION INSURANCE

LEARNING OBJECTIVES

- Provide a summary overview of workers' compensation.
- Explain the historical perspective of workers' compensation.
- Summarize significant workers' compensation legislation.
- Explain the state of modern workers' compensation.
- Explain the types of workers' compensation insurance.
- Describe how workers' compensation disputes are settled.
- Explain the types of injuries that require workers' compensation.
- Explain the types of disabilities that are compensable from workers' compensation.
- Describe the various monetary benefits of workers' compensation.
- Summarize the medical treatment and rehabilitation aspects of workers' compensation.
- Describe how workers' compensation claims are managed.
- Explain the concept of cost allocation in workers' compensation.
- Summarize the most problematic aspects of workers' compensation.
- Explain how to spot fraud and abuse in workers' compensation.
- Summarize the future of workers' compensation.
- Explain the most common cost-reduction strategies for workers' compensation.
- Summarize the other types of insurance construction professionals should know about.

OVERVIEW OF WORKERS' COMPENSATION

The concept of **workers' compensation** developed as a way to allow injured employees to be compensated appropriately without having to take their employer to court. The underlying rationale for workers' compensation has two aspects: (1) fairness to injured employees, especially those without the resources to undertake often lengthy and expensive legal actions, and (2) reduction of costs associated with workplace injuries (e.g., legal, image, and morale costs). Workers' compensation is intended to be a no-fault approach to resolving workplace accidents by rehabilitating injured employees and minimizing the personal losses that result because of their reduced ability to perform and compete in the labor market.[1] Since its inception as a concept, workers' compensation has evolved into a system that pays out approximately $70 million in benefits and medical costs annually. Some of the highest workers' compensation insurance rates are in construction fields (e.g., roofers).

Workers' compensation represents a compromise between the needs of employees and the needs of employers. Employees give up their right to seek unlimited compensation for pain and suffering through legal action. Employers award the prescribed compensation (typically through insurance premiums) regardless of the employee's negligence. The theory is that, in the long run, both employees and employers benefit more than either would through legal action. As you will see later in this chapter, although workers' compensation has reduced the amount of legal action arising out of workplace accidents, it has not eliminated legal actions.

Objectives of Workers' Compensation

Workers' compensation laws are not uniform across states. In fact, there are extreme variations. However, regardless of the language contained in the enabling legislation in a given state, workers' compensation has several widely accepted objectives:

- Replacement of income for injured employees
- Rehabilitation of the injured employee
- Prevention of accidents
- Allocation of cost[2]

The basic premises underlying these objectives are described in the following paragraphs.

Replacement of Income. Employees injured on the job lose income if they are unable to work. For this reason, workers' compensation is intended to replace the lost income adequately and promptly. Adequate

income replacement is viewed as replacement of current and future income (minus taxes) at a level of two-thirds of actual income (in most states). Workers' compensation benefits are required to continue, even if the employer goes out of business.

Rehabilitation of the Injured Employee. A basic premise of workers' compensation is that the injured worker will return to work in every case possible, although not necessarily in the same job or career field. For this reason, a major objective of workers' compensation is to rehabilitate the injured **employee**. The **rehabilitation** program provides the necessary medical care at no cost to the injured employee until he or she is pronounced fit to return to work. The program also provides vocational training or retraining as needed. Both components seek to motivate the employee to return to the labor force as soon as possible.

Prevention of Accidents. Preventing future accidents is a major objective of workers' compensation. The theory underlying this objective of **accident prevention** is that employers will invest in accident-prevention programs to hold down compensation costs. The payoff to employers comes in the form of lower insurance premiums, which result from fewer accidents (theoretically).

Allocation of Cost. The potential risks associated with different occupations vary. For example, working as a roofer is generally considered more hazardous than working as an architect. The underlying principle of **cost allocation** is to spread the cost of workers' compensation appropriately and proportionately among industries ranging from the most to the least hazardous. The costs of accidents should be allocated in accordance with the accident history of the industry, so that high-risk industries pay higher workers' compensation insurance premiums than do low-risk industries.[3] Construction is a high-risk industry.

Who Is Covered by Workers' Compensation?

Workers' compensation laws are written at the state level, and there are many variations among these laws. As a result, it is difficult to make generalizations. Complicating the issue further is the fact that workers' compensation laws are constantly being amended, revised, and rewritten. In addition, some states make participation in a workers' compensation program voluntary; others excuse employers with fewer than a specified number of employees.

In spite of the differences among workers' compensation laws in the various states, approximately 80 percent of the employees in the United States are covered by workers' compensation. Those employees who are not covered or whose coverage is limited vary as the laws vary. However, employees who are not can be categorized in general terms as follows[4]:

- Agricultural employees
- Domestic employees
- Casual employees
- Hazardous work employees
- Charitable or religious employees
- Employees of small organizations
- Railroad and maritime employees
- Contractors and subcontractors
- Minors
- Extraterritorial employees

Coverage in these types of employment, to the extent there is coverage, varies from state to state as follows:

- *Agricultural employees* have limited coverage in 38 states, Puerto Rico, and the Virgin Islands. In 15 states, workers' compensation coverage for agricultural employees is voluntary. In these states, employers are allowed to provide coverage if they wish, but are not required to do so.
- *Domestic employees* have coverage available in all 50 states and Puerto Rico. However, coverage tends to be limited and is subject to minimum requirements regarding hours worked and earnings.
- *Casual employees* are employed in positions in which the work is occasional, incidental, and scattered at irregular intervals. Such employees are not typically afforded workers' compensation coverage.
- *Hazardous employment* is the only type afforded workers' compensation coverage in some states. To qualify, a particular type of employment must be on an approved list of hazardous or especially hazardous jobs. However, the trend in these states is to broaden the list of approved jobs.
- *Charitable* or *religious employees* are not afforded workers' compensation in most states when this work is irregular, temporary, or short term.
- *Small organizations* that employ fewer than a specified number of employees do not fall under the umbrella of workers' compensation in 26 states.
- *Railroad* and *maritime workers* are not typically covered by workers' compensation. However, in most cases, they are covered by the Federal Employer's Liability Act. This act disallows the use of common law defenses by employers if sued by an employee for negligence.
- *Contractors* and *subcontractors* are those who agree to perform a job or service for an agreed amount of money in a nondirected, nonsupervised format. In essence, contract and subcontract employees are viewed as being self-employed. For

this reason, they are not covered by workers' compensation. Most states build in safeguards to prevent employers from setting up employees they supervise as subcontractors as a way to avoid paying workers' compensation premiums.

- *Minors* are afforded regular workers' compensation coverage as long as they are legally employed. In some states, the cost of coverage is significantly higher for minors who are working illegally.
- *Extraterritorial employees* are those who work in one state but live in another. In these cases, the employee is usually on temporary duty. Such employees are typically afforded the workers' compensation coverage in their home state.

HISTORICAL PERSPECTIVE

Before workers' compensation laws were enacted in the United States, injured employees had no way to obtain compensation for their injuries except to take their employer to court. Although common law did require employers to provide a safe and healthy work environment, injured employees had to prove that negligence in the form of unsafe conditions contributed to their injuries. Before passage of workers' compensation, employees often had to sue their employer to receive compensation for injuries resulting from a **workplace accident** or **occupational disease**.

Proving that an injury was the result of employer negligence was typically too costly, too difficult, and too time consuming to be a realistic avenue of redress for most injured employees. According to Somers and Somers, a New York commission determined that it took from six months to six years for an injured worker's case to work its way through the legal system.[5] Typically, injured workers, having lost their ability to generate income, could barely afford to get by, much less pay medical expenses, legal fees, and court costs. Another inhibitor was the **fear factor**. Injured employees who hoped to return to work after recovering were often afraid to file suit because they feared retribution by their employer. Employers not only might refuse to give them their jobs back, but also might **blackball** them with other employers. Add to this that fellow employees were often afraid to testify to the negligence of the employer, and it is easy to see why few injured workers elected to take their employers to court.

Even with all of these inhibitors, some injured employees still chose to seek redress through the courts in the days before workers' compensation. Those who did faced a difficult challenge, because the laws at that time made it easy for employers to defend themselves successfully. All an employer had to do to win a decision denying the injured plaintiff compensation was to show that at least one of the following conditions existed at the time of the accident:

1. *Contributory negligence.* **Contributory negligence** meant that the injured worker's own negligence contributed to the accident. Even if the employee's negligence was a very minor factor, it was usually enough to deny redress in the days before workers' compensation.
2. *Negligence on the part of a fellow worker.* As with contributory negligence, negligence by a fellow employee, no matter how minor a contributing factor it was, could be sufficient to deny compensation.
3. *Assumption of risk on the part of the injured employee.* If an employee knew that the job involved risk, he or she could not expect to be compensated when the risks resulted in accidents and injuries.[6]

Since the majority of workplace accidents involve at least some degree of negligence on the part of the injured worker or fellow employees, employers typically won these cases. Because it required little more than a verbal warning by the employer to establish grounds for **assumption of risk**, the odds against an injured employee being awarded compensation were great.

In his book *American Social Science*, Gagliardo gives an example of a case that illustrates how difficult it was to win redress in the days before workers' compensation.[7] He relates the case of an employee who contracted tuberculosis while working under clearly hazardous conditions. She worked in a wet, drafty basement that admitted no sunlight. Dead rats floated in the overflow of a septic tank that covered the basement floor, and a powerful stench permeated the workplace. Clearly, these were conditions that could contribute to the employee contracting tuberculosis. However, she lost the case and was denied compensation. The ruling judge justified the verdict as follows:

> We think that the plaintiff, as a matter of law, assumed the risk attendant upon her remaining in the employment (217 N.Y. Supp. 173).

SAFETY FACTS & FINES

Exposing workers to dust without taking the necessary precautions is dangerous and can be costly. A renovation contractor in West Seneca, New York, was fined $176,620 for failing to provide proper respiratory devices and for the improper use of scaffolding. Workers were assigned to repoint brick on the outside of a building. Not only did they have no respiratory protection, the scaffold they were using was enclosed in plastic, thereby increasing the concentration of silica to which the workers were exposed.

Situations such as this eventually led to the enactment of workers' compensation laws in the United States.

WORKERS' COMPENSATION LEGISLATION

Today, all 50 states, the District of Columbia, Guam, and Puerto Rico have workers' compensation laws. However, these laws did not exist before 1948. Considering that Prussia passed a workers' compensation law in 1838, the United States was obviously slow to adopt the concept. In fact, the first workers' compensation law enacted in the United States did not pass until 1908, and it applied only to federal employees working in especially hazardous jobs. The driving force behind passage of this law was President Theodore Roosevelt, who, as governor of New York, had seen the results of workplace accidents firsthand. Montana was the first state to pass a compulsory workers' compensation law. However, it was short lived. Ruling that the law was unconstitutional, the Montana courts overturned it.

In 1911, the New York Court of Appeals dealt proponents of workers' compensation a serious blow. The New York State legislature had passed a compulsory workers' compensation law in 1910. However, in the case of *Ives v. South Buffalo Railway Company* (201 N.Y. 271, 1911), the New York Court of Appeals declared the law unconstitutional based on the contention that it violated the due process clause in the Fourteenth Amendment to the U.S. Constitution.[8]

This ruling had a far-reaching impact. According to *Occupational Hazards* magazine, "The prestige of the New York court influenced legislation in many of the other states to believe that any compulsory law also would be held unconstitutional."[9] However, even with such precedent-setting cases as *Ives* on the books, pressure for adequate workers' compensation grew as unsafe working conditions continued to result in injuries, diseases, and deaths. In fact, shortly after the New York Court of Appeals released its due process ruling, tragedy struck in a New York City textile factory.

On March 25, 1911, the building that housed the Triangle Shirtwaist Factory on its eighth floor caught fire and burned.[10] As a result of the fire, 149 of the company's 600 workers were dead and another 70 were injured. Although the cause of the accident could not be determined, it was clear to investigators and survivors alike that unsafe conditions created by the management of the company prevented those who died or were injured from escaping the fire. Exit passageways on each floor of the building were unusually narrow (20 inches wide), which made it difficult for employees to carry out bolts of material. A wider exit on each floor was kept locked to force employees to use the narrow exit. The two elevators were slow and able to accommodate only small groups at a time.

As the fire quickly spread, employees jammed into the narrow passageways, crushing each other against the walls and underfoot. With all exits blocked, panic-stricken employees began to jump out of windows and down elevator shafts. When the pandemonium subsided and the fire was finally brought under control, the harsh realization of why so many had been trapped by the deadly smoke and flames quickly set in.

The owners were brought into court on charges of manslaughter. Although they were not convicted, the tragedy did focus nationwide attention on the need for a safe workplace and adequate workers' compensation. As a result, new, stricter fire codes were adopted in New York, and in spite of the state court's ruling in *Ives*, the state legislature passed a workers' compensation law.

The next several years saw a flurry of legislation related to workers' compensation in other states. In response to demands from workers and the general public, several states passed limited or noncompulsory workers' compensation laws. Many such states held back out of fear of being overturned by the courts. Others, particularly Washington, publicly disagreed with the New York Court of Appeals and passed compulsory laws. The constitutionality debate continued until 1917, when the U.S. Supreme Court ruled that workers' compensation laws were acceptable.

MODERN WORKERS' COMPENSATION

Since 1948, all states have had workers' compensation laws. However, the controversy surrounding workers' compensation has not died. As medical costs and insurance premiums have skyrocketed, many small businesses have found it difficult to pay the premiums. Unrealistic workers' compensation rates are being cited more and more frequently as contributing to the demise of small business in America.

The problem has even become an economic development issue. Businesses are closing their doors in those states with the highest workers' compensation rates and moving to states with lower rates. States with lower rates are using this as part of their recruiting package to attract new businesses. Where low-rate states border high-rate states, businesses are beginning to move their offices across the border to the low-rate state while still doing business in the high-rate state.

Critics are now saying that workers' compensation has gotten out of hand and is no longer fulfilling its intended purpose. To understand whether this is the case, one must begin with an examination of the purpose of workers' compensation. The U.S. Chamber of Commerce identified the following basic objectives of workers' compensation:

1. To provide an appropriate level of income and medical benefits to injured workers or to provide income to the workers' dependents, regardless of fault

2. To reduce the amount of personal injury **litigation** in the court system
3. To relieve public and private charities from the financial strain created by workplace injuries that go uncompensated
4. To eliminate time-consuming and expensive trials and appeals
5. To promote employer interest and involvement in maintaining a safe work environment through the application of an experience-rating system
6. To prevent accidents by encouraging frank, objective, and open investigations of the cause of accidents[11]

Early proponents of workers' compensation envisioned a system in which both injured workers and their employers would win. Injured workers would receive prompt compensation, adequate medical benefits, and appropriate rehabilitation to allow them to reenter the workforce and be productive again. Employers would avoid time-consuming, expensive trials and appeals and improve relations with employees and the public in general.

Proponents of workers' compensation did not anticipate the following factors:

1. Employees who see workers' compensation as a way to ensure themselves a lifelong income without the necessity of working
2. Enormous increases in the costs of medical care with corresponding increases in workers' compensation insurance premiums
3. The radical differences among workers' compensation laws passed by the various states

Not all employees abide by the spirit of workers' compensation (e.g., rehabilitation in a reasonable amount of time). Attempted abuse of the system is perhaps inevitable. Unfortunately, such attempts result in a return to what workers' compensation was enacted to eliminate: time-consuming, drawn-out, expensive legal battles and the inevitable appeals.

Medical costs have skyrocketed in the United States since the 1960s, for many reasons. During this same period, the costs associated with other basic human needs, including food, clothing, transportation, shelter, and education, have also increased markedly. Increases in medical costs can be explained, at least partially, as the normal cost of living increases experienced in other sectors of the economy. However, the costs associated with medical care have increased much faster and much more than the costs in these other areas. The unprecedented increases can be attributed to two factors:

1. Technological developments that have resulted in extraordinary but costly advances in medical care
2. A proliferation of litigation that has driven the cost of **malpractice** insurance steadily up

Each of these factors has contributed to higher medical costs. For example, X-ray machines that cost thousands of dollars have been replaced by magnetic resonance imaging systems that may cost millions. Malpractice suits that once might not even have gone to court now result in multimillion-dollar settlements. Such costs are, of course, passed on to whomever pays the medical bill—in this case, employers who must carry workers' compensation insurance.

Early supporters of the concept did not anticipate the radical differences among workers' compensation laws in the various states. The laws themselves differ, as do their interpretations. The differences are primarily in the areas of benefits, penalties, and workers covered. These differences translate into differences in the rates charged for workers' compensation insurance. As a result, the same injury incurred under the same circumstances, but in different states, can yield radically different benefits for the employee.

The potential for abuse, steadily increasing medical costs that lead to higher insurance premiums, and differences among workers' compensation laws all contribute to the controversy that still surrounds this issue. As business and industry continue to protest that workers' compensation has gotten out of hand, workers' compensation will continue to be a heated issue in state legislatures as states try to strike the proper balance between meeting the needs of the workforce and simultaneously maintaining a positive environment for doing business.

WORKERS' COMPENSATION INSURANCE

The costs associated with workers' compensation must be borne by employers as part of their overhead. In addition, employers must ensure that the costs are paid even if they go out of business. The answer for most employers is workers' compensation insurance.

In most states, workers' compensation insurance is compulsory. Exceptions to this are New Jersey, South Carolina, Texas, and Wyoming. New Jersey allows 10 or more employers to form a group and self-insure. Texas requires workers' compensation for *carriers only*—as defined in Title 25, Article 911-A, Section II of the Texas state statutes. Wyoming requires workers' compensation only for employers involved in specifically identified *extrahazardous occupations*.

A common thread woven through all of the various compensation laws is the requirement that employers carry workers' compensation insurance. There are three types: **state funds, private insurance**, and **self-insurance**. Figure 5–1 summarizes the methods of insurance coverage chosen; rates can vary greatly from company to company and state to state. Rates

State	State Fund	Private Insurer	Individual Employer Self-Insurance	Group of Employers Self-Insurance
Alabama	No	Yes	Yes	Yes
Arkansas	No	Yes	Yes	Yes
California	Competitive	Yes	Yes	Yes
Florida	No	Yes	Yes	Yes
Indiana	No	Yes	Yes	No
Kansas	No	Yes	Yes	Yes
Maryland	Competitive	Yes	Yes	Yes
Montana	Competitive	Yes	Yes	Yes
Nebraska	No	Yes	Yes	No
New Jersey	No	Yes	Yes	No*
New Mexico	Competitive	Yes	Yes	Yes
Vermont	No	Yes	Yes	No
Washington	Exclusive	No	Yes	Yes
Wisconsin	No	Yes	Yes	No

FIGURE 5–1 Workers' compensation coverage methods that are allowed for selected states.

*Permits 10 or more employees licensed by the state to self-insure as a group.

Source: From Bureau of Labor Statistics. Published by U.S. Department of Labor.

are affected by a number of factors, including the following:

- Number of employees
- Risk involved in types of work performed
- Accident history of the employer
- Potential future losses
- Overhead and profits of the employer
- Quality of the employer's safety program
- Estimates by actuaries

Insurance companies use one of the following methods in determining the premium rates of employers[12]:

1. *Schedule rating.* Insurance companies establish baseline safety conditions and evaluate the employer's conditions against the baselines. This is known as **schedule rating**. Credits are awarded for conditions that are better than the baseline, and debits are assessed for conditions that are worse. Insurance rates are adjusted accordingly.
2. *Manual rating.* A manual of rates is developed that establishes rates for various occupations. This is known as **manual rating**. Each occupation may have a different rate based on its perceived level of hazard. The overall rate for the employer is a prorated combination of all the individual rates.
3. *Experience rating.* Employers are classified by type. Premium rates are assigned based on predications of average losses for a given type of employer. Rates are then adjusted either up or down according to the employer's actual experience over the past three years.
4. *Retrospective rating.* Employees pay an established rate for a set period. This is known as **retrospective rating**. At the end of the period, the actual experience is assessed and an appropriate monetary adjustment is made.
5. *Premium discounting.* Large employers receive discounts on their premiums based on their size. This is known as **premium discounting**. The theory behind this method is that it takes the same amount of time to service a small company's account as it does a large company's, but the large company produces significantly more income for the insurer. Premium discounts reward the larger company for its size.
6. *Combination method.* The insurer combines two or more of the other methods to arrive at premium rates.

The trend nationwide for the past decade has been for premiums to increase markedly. For example, over the past 10 years, some states experienced increases of more than 60 percent. This trend ensures that workers' compensation will remain a controversial issue in the state legislatures.

RESOLUTION OF WORKERS' COMPENSATION DISPUTES

One of the fundamental objectives of workers' compensation is to avoid costly, time-consuming litigation. Whether this objective is being accomplished is questionable. When an injured employee and the employer's insurance company disagree on some aspect of the compensation owed (e.g., weekly pay, length of benefits, degree of disability), the disagreement must be resolved. Most states have an arbitration board for this purpose. Neither the insurance company nor the injured employee is required to hire an attorney. However, many employees do. There are a number of reasons for this. Some don't feel they can adequately represent themselves. Others are fearful of the "big business running over the little guy" syndrome. In any case, workers' compensation litigation is still very common and expensive.

In most states, allowable attorney fees are set by statute, administrative rule, or policy. In some states, attorney fees can be added to the injured employee's award. In others, the fee is a percentage of the award.

INJURIES AND WORKERS' COMPENSATION

The original workers' compensation concept envisioned compensation for workers who were injured in on-the-job accidents. What constituted an accident varied from state to state. However, all original definitions had in common the characteristics of being *sudden* and *unexpected.* Over the years, the definition of an accident has undergone continual change. The major change has been a trend toward the elimination of the *sudden* characteristic. In many states, the gradual onset of a disease as result of prolonged exposure to harmful substances or a **harmful environment** can now be considered an accident for workers' compensation purposes.

The harmfulness of an environment does not have to be limited to its physical components. Psychological factors (such as stress) can also be considered. In fact, the highest rate of growth in workers' compensation claims over the past two decades has been in the area of stress-related injuries.

The National Safety Council maintains statistical records of the numbers and types of injuries suffered in various industries in the United States. Industries are divided into the following categories: agriculture, mining, construction, manufacturing, transportation and public utilities, trade, services, and the public sector. Injuries in these industrial sectors are classified according to the type of accident that caused them. Accident types include overexertion, being struck by or against an object, falls, bodily reactions, being caught in or between objects, motor vehicle accident, coming in contact with radiation or other caustics, being rubbed or abraded, and coming in contact with temperature extremes.

When all industry categories are viewed in composite, more than 30 percent of total disabling work injuries are the result of overexertion. The next most frequent cause of injuries is being struck by or against objects (24 percent). Falls account for just over 17 percent. The remainder are fairly evenly distributed among the other accident types.[13]

AOE and COE Injuries

Workers' compensation benefits are owed only when the injury **arises out of employment (AOE)** or occurs in the **course of employment (COE)**. When employees are injured during work prescribed in their job description, work assigned by a supervisor, or work normally expected of employees, they fall into the AOE category. Sometimes, however, different circumstances determine whether the same type of accident is considered to be AOE. For example, say a welder burns her hand while joining two steel beams at her construction job. This injury would be classified as AOE. Now suppose the same technician burns her hand while welding a personal project at home. This injury would not be covered, because the accident did not arise from her employment. Determining whether an injury should be classified as AOE or COE is often a point of contention in workers' compensation litigation.

Who Is an Employee?

Another point of contention in workers' compensation cases is the definition of the term *employee.* This is an important definition because it is used to determine AOE and COE status. A person who is on the company's payroll, receives benefits, and has a supervisor is clearly an employee. However, a person who accepts a service contract to perform a specific task or set of tasks and is not directly supervised by the company is not considered an employee. Although definitions vary from state to state, there are common characteristics. In all definitions, the workers must receive some form of remuneration for work done, and the employer must benefit from this work. Also, the employer must supervise and direct the work—both process and result. These factors—supervision and direction—are what set **independent contractors** apart from employees and exclude them from coverage. Employers who use independent contractors sometimes require the contractors to show proof of having their own workers' compensation insurance.

DISABILITIES AND WORKERS' COMPENSATION

Injuries that are compensable typically fall into one of four categories (Figure 5–2):

1. Temporary partial disability
2. Temporary total disability
3. Permanent partial disability
4. Permanent total disability

Determining the extent of disability is often a contentious issue. In fact, it accounts for more workers' compensation litigation than any other issue. Further, when a disability question is litigated, the case tends to be

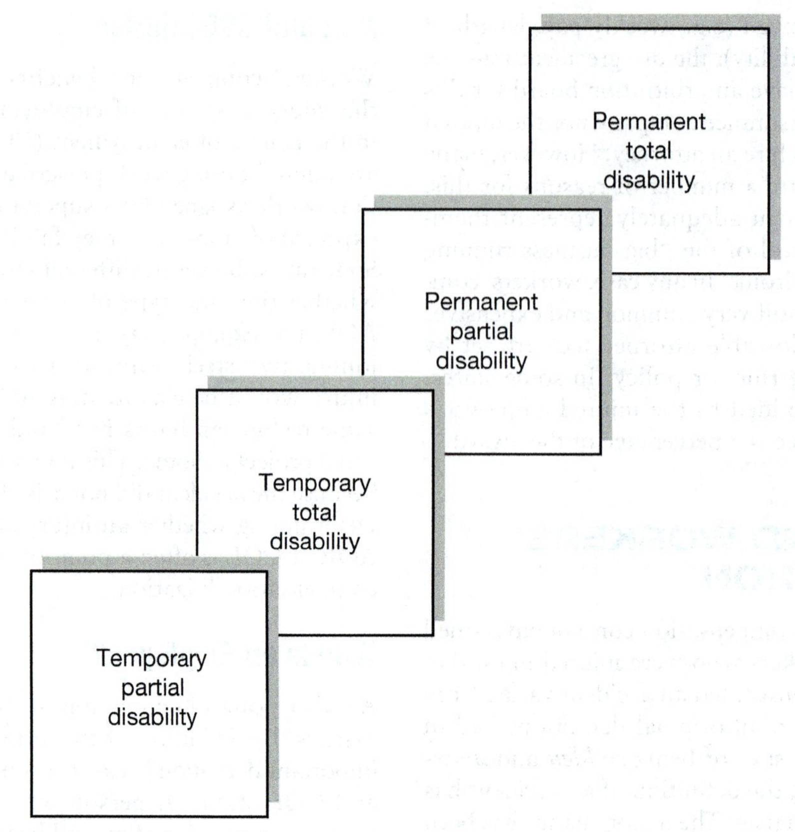

FIGURE 5–2 Types of Disabilities.

complicated because the evidence is typically subjective, and it requires hearing officers, judges, or juries to determine the future.

Temporary Disability

Temporary disability is the state that exists when it is probable that an injured worker, who is currently unable to work, will be able to resume gainful employment with no or only partial disability. Temporary disability assumes that the employee's condition will substantially improve. Determining whether an employee is temporarily disabled is not normally difficult. Competent professionals can usually determine the extent of the employee's injuries, prescribe the appropriate treatment, and establish a timeline for recovery. They can then determine if the employee will be able to return to work and when the return might take place.

There is an important point to remember when considering a temporary disability case. The ability to return to work relates only to work with the company that employed the worker at the time of the accident.

Temporary disability can be classified as either **temporary total disability** or **temporary partial disability**. A temporary total disability classification means the injured worker is incapable of any work for a period of time, but is expected to recover fully. Most workers' compensation cases fall into this classification. A temporary partial disability means the injured worker is capable of light or part-time duties. Depending on the extent of the injury, temporary partial disabilities sometimes go unreported. This practice is allowable in some states. It helps employers hold down the cost of their workers' compensation premium. This is similar to not reporting a minor fender-bender to your automobile insurance agent.

Most states prescribe in law the benefits owed in temporary total disability cases. Factors prescribed typically include a set percentage of an employee's wage that must be paid and a maximum period during which benefits can be collected. Figure 5–3 shows this information for a geographically distributed selection of states. Since workers' compensation legislation changes continually, this figure is provided only as an illustration of how benefits are prescribed in the laws of the various states. Actual rates are subject to change.

Permanent Partial Disability

Permanent partial disability is the condition that exists when an injured employee is not expected to recover fully. In such cases, the employee will be able to work again, but not at full capacity. Often employees who are partially disabled must be retrained for another occupation.

State	Percentage of Employee's Wage	Maximum Period
Alabama	66⅔	Duration of disability
Arkansas	66⅔	450 weeks
California	66⅔	Duration of disability
Florida	66⅔	104 weeks
Indiana	66⅔	500 weeks
Kansas	66⅔	Duration of disability
Maryland	66⅔	Duration of disability
Montana	66⅔	Duration of disability*
Nebraska	66⅔	Duration of disability
New Jersey	70⅔	400 weeks
New Mexico	66⅔	Duration of disability
Vermont	66⅔	Duration of disability
Washington	60–75	Duration of temporary disability
Wisconsin	60–75	Duration of disability

FIGURE 5–3 Temporary total disability benefits for selected states.

*Or until worker is released to preinjury job or similar employment.
Source: From Bureau of Labor Statistics. Published by U.S. Department of Labor.

State	Percentage of Employee's Wage	Maximum Period
Alabama	66⅔	300 weeks
Arkansas	66⅔	450 weeks
California	66⅔	619.25 weeks
Florida*	—	364 weeks
Indiana	66⅔	500 weeks
Kansas	66⅔	415 weeks
Maryland	66⅔	Duration of disability
Montana	66⅔	350 weeks
Nebraska	66⅔	300 weeks
New Jersey	70	600 weeks
New Mexico	66⅔	500 weeks
Vermont	66⅔	330 weeks
Washington	—	—
Wisconsin	66⅔	1,000 weeks

FIGURE 5–4 Permanent partial disability benefits for selected states.

*Shall not exceed 66.67 percent.
Source: From Bureau of Labor Statistics. Published by U.S. Department of Labor.

Permanent partial disabilities can be classified as *schedule* or *nonschedule* disabilities. ***Schedule disabilities*** are typically the result of nonambiguous injuries, such as the loss of a critical but duplicated body part (e.g., arm, ear, hand, finger, or toe). Because such injuries are relatively straightforward, the amount of compensation that they generate and the period of time that it will be paid can be set forth in a standard schedule. A compilation of information from such schedules for a geographically distributed list of states is shown in Figure 5–4. Since workers' compensation legislation changes continually, this figure is provided only as an example. Actual rates are subject to continual change.

Nonschedule injuries are less straightforward and must be dealt with on a case-by-case basis. Disabilities in this category tend to be the result of head injuries—the effects of which can be more difficult to determine. The amount of compensation awarded and the period over which it is awarded must be determined by studying the evidence. Awards are typically made on the basis of a determination of percentage of disability. For example, if it is determined that an employee has a 25 percent disability, the employee might be entitled to 25 percent of the income he or she could have earned before the injury, with normal career progression taken into account.

Four approaches to handling permanent partial disability cases have evolved. Three of these are based on specific theories, and the fourth is based on a combination of two or more of these theories. The three theories are as follows:

1. Whole-person theory
2. Wage-loss theory
3. Loss of wage-earning capacity theory

Whole-Person Theory. The **whole-person theory** is the simplest and most straightforward of the theories for dealing with permanent partial disability cases. Once it has been determined that an injured worker's capabilities have been permanently impaired to some extent, this theory is applied like a subtraction problem. What the worker can do after recuperating from the injury is determined and subtracted from what he or she could do before the accident. Factors such as age, education, and occupation are not considered.

Wage-Loss Theory. The **wage-loss theory** requires a determination of how much the employee could have earned had the injury not occurred. The wages actually being earned are subtracted from what could have been earned, and the employee is awarded a percentage of the difference. No consideration is given to the extent or degree of disability. The only consideration is loss of actual wages.

Loss of Wage-Earning Capacity Theory. The most complex of the theories for handling permanent partial disability cases is the **loss of wage-earning capacity theory**, because it is based not just on what the employee earned at the time of the accident, but also on what he or she might have earned in the future. Making such a determination is obviously a subjective undertaking. Factors considered include past job performance, education, age, gender, and advancement potential at the time of the accident, among others. Once future earning capacity has been determined, the extent to which it has been impaired is estimated, and the employee is awarded a percentage of the difference. Some states prescribe maximum amounts of compensation and maximum periods within which it can be collected. For example, in Figure 5–4, Alabama sets 300 weeks as the maximum period for collecting on a nonschedule injury. Maryland, on the other hand, awards compensation for the duration of the disability. Disabilities on the schedule are typically compensated for the duration of the disability in all states.

The use of schedules has reduced the amount of litigation and controversy surrounding permanent partial disability cases. This is the good news aspect of schedules; the bad news aspect is that they may be inherently unfair. For example, a surgeon who loses his hand would receive the same compensation as a laborer with the same injury, if the loss of a hand is scheduled.

Permanent Total Disability

A **permanent total disability** exists when an injured employee's disability means that he or she cannot compete in the job market. This does not necessarily mean that the employee is helpless. Rather, it means an inability to compete reasonably. Handling permanent total disability cases is similar to handling permanent partial disability cases except that certain injuries simplify the process. In most states, permanent total disability can be assumed if certain specified injuries have been sustained (e.g., loss of eyes or arms). In some states, compensation is awarded for life. In others, a time period is specified. Figure 5–5 shows the maximum period over which compensation can be collected for a geographically distributed list of states. Note that the time periods range from 401 weeks (in Texas) to life (in California and Wisconsin).

State	Maximum Paid
Alabama	Duration of disability
Arkansas	Duration of disability
California	Life
Florida	Duration of disability
Indiana	500 weeks
Kansas	Duration of disability
Maryland	Duration of disability
Montana	Duration of disability
Nebraska	Duration of disability
New Jersey	450 weeks (life in some cases)
New Mexico	Life
Vermont	Duration of disability*
Washington	Life
Wisconsin	Life

FIGURE 5–5 Permanent total disability benefits for selected states.

*Minimum of 330 weeks

Source: From Bureau of Labor Statistics. Published by U.S. Department of Labor.

MONETARY BENEFITS OF WORKERS' COMPENSATION

The **monetary benefits** accruing from workers' compensation vary markedly from state to state. The actual amounts are of less importance than the differences among them. Of course, the amounts set forth in schedules change frequently. However, for the purpose of comparison, consider that at one time the loss of a hand in Pennsylvania resulted in an award of $116,245; the same injury in Colorado brought only $8,736.

When trying to determine a scheduled award for a specific injury, it is best to locate the latest schedule for the state in question. One way to do this is to contact the following agency:

U.S. Department of Labor
Employment Standards Administration
Office of State Liaison and Legislative Analysis
Division of State Workers' Compensation Programs
200 Constitution Avenue, N.W.
Washington, D.C. 20210
http://www.dol.gov

Death and Burial Benefits

Workers' compensation benefits accrue to the families and dependents of workers who are fatally injured. Typically, the spouse receives benefits for life or until remarriage. However, in some cases, a time period is specified. Dependents typically receive benefits until they reach the legal age of maturity—unless they have a condition or circumstance that makes them unable to support themselves even after attaining that age. Figure 5–6 contains the death benefits accruing to surviving spouses and children for a geographically distributed list of states. Since the actual amounts of benefits are subject to change, these are provided for the purpose of illustration and comparison only.

Further expenses are provided in addition to death benefits in all states except Oklahoma. As is the case with all types of workers' compensation, the amount of burial benefits varies from state to state and is subject to change.

MEDICAL TREATMENT AND REHABILITATION

All workers' compensation laws provide for payment of the medical costs associated with injuries. Most states provide full coverage, but some limit the amount and duration of coverage. For example, in Arkansas, employer liability ceases six months after an injury occurs in cases in which the employee is able to continue working, or six months after he or she returns to work in cases

State	Percentage of Employee's Wage (Spouse Only)	(Spouse and Children)	Maximum Period
Alabama	50	66⅔	500 weeks
Arkansas	35	66⅔	Widow/widowerhood Children until 18 or married
California	66⅔	66⅔	N/A
Florida	50	66⅔	Widow/widowerhood Children until 18
Indiana	66⅔	66⅔	500 weeks
Kansas	66⅔	66⅔	Widow/widowerhood; Children until 18
Maryland	66⅔	66⅔	Widow/widowerhood; Children until 18
Montana	66⅔	66⅔	Surviving spouse—10 years; Children until 18
Nebraska	66⅔	75	Widow/widowerhood; Children until 18
New Jersey	50	70	Widow/widowerhood; Children until 18
New Mexico	66⅔	66⅔	700 weeks
Vermont	66⅔	66⅔	Widow/widowerhood until 62; Children until 18
Washington	60	70	Widow/widowerhood; Children until 18
Wisconsin	66⅔	N/A	300 weeks

FIGURE 5–6 Death benefits for surviving spouses and children for selected states.

Source: From Bureau of Labor Statistics. Published by U.S. Department of Labor.

in which there is a period of recuperation. In either case, the employer's maximum financial liability is $10,000.[14] In Ohio, medical benefits for silicosis, asbestosis, and coal miner's pneumoconiosis are paid only in the cases of temporary total or permanent total disability.[15]

The laws also specify who is allowed or required to select a physician for the injured employee. The options can be summarized as follows:

- *Employee selects physician of choice.* This option is available in Alaska, Arizona, Delaware, Hawaii, Illinois, Kentucky, Louisiana, Maine, Massachusetts, Mississippi, Nebraska, New Hampshire, North Dakota, Ohio, Oklahoma, Oregon, Rhode Island, Texas, the Virgin Islands, Washington, West Virginia, Wisconsin, and Wyoming.
- *Employee selects physician from a list provided by the state agency.* This option applies in Connecticut, Nevada, New York, and the District of Columbia.
- *Employee selects physician from a list provided by the employer.* This option applies in Georgia, Tennessee, and Virginia.
- *Employer selects the physician.* This option applies in Alabama, Florida, Idaho, Indiana, Iowa, Maryland, Montana, New Jersey, New Mexico, North Carolina, South Carolina, and South Dakota.
- *Employer selects the physician, but the selection may be changed by the state agency.* This option applies in Arkansas, Colorado, Kansas, Minnesota, Missouri, Utah, and Vermont.
- *Employer selects the physician, but after a specified period of time, the employee may choose another.* This option applies only in Puerto Rico.

Rehabilitation and Workers' Compensation

Occasionally an injured worker needs rehabilitation before he or she can return to work. There are two types of rehabilitation: medical and vocational. Both are available to workers whose ability to make a living is inhibited by physical or mental work-related problems.

Medical rehabilitation consists of providing whatever treatment is required to restore—to the extent possible—any lost ability to function normally. This may include such services as physical therapy or the provision of prosthetic devices. **Vocational rehabilitation** involves providing the education and training needed to prepare the worker for a new occupation. Whether the rehabilitation services are medical or vocational in nature or both, the goal is to restore the injured workers' capabilities to the level that existed before the accident.

MEDICAL MANAGEMENT OF WORKPLACE INJURIES

Out-of-control workers' compensation cases led to the concept of **medical management of workplace injuries.**[16] Through better management of workers' compensation claims, more than 30 states have merged the concepts of workers' compensation and managed care. The goals of these state-level efforts are to

1. speed up the processing of workers' compensation claims,
2. reduce costs,
3. reduce fraud and abuse, and
4. improve medical management of workplace injuries.

Workers' compensation and managed care have been merged through the creation of Health Partnership Programs (HPP). HPPs are tie-ups between employers and their state's Bureau of Workers' Compensation (BWC). Employers who choose to participate (some states mandate participation) are required to have a managed care organization (MCO) that provides medical management of workplace injuries and illnesses.

The HPPs are effective because an MCO coordinates the paperwork generated by the injured employee, the employer, health-care providers, and the BWC.[17] The results of this coordination are clear. The average time required to process an injury report before implementation of the HPP was more than 66 days. The HPP reduced this to approximately 33 days—effectively cutting reporting time in half. Whether managed medical care can continue this positive trend remains to be seen, but the states that have implemented HPPs have learned the following lessons:

- It is better to mandate HPPs than to make them optional.
- Cost containment is only part of the goal. Managed care programs must also include criteria such as lost wages, ability to return to work, and administrative costs to the employer.
- Employees want choice in selecting health-care providers.
- Smart return-to-work programs are critical.

ADMINISTRATION AND CASE MANAGEMENT

Even though the Occupational Safety and Health Administration (OSHA) specifies what constitutes a recordable accident, it is not uncommon for minor injuries to go unreported. The employee may be given the rest of the day off or treated with first aid and returned to work. This is done to avoid time-consuming paperwork and to hold down the cost of workers' compensation insurance. However, if an accident results in a serious injury, several agencies must be notified. What constitutes a serious injury—like many workers' compensation issues—can differ from state to state. However, as a rule, an injury is serious if it requires more than 24 hours of active medical treatment (this does not include passive treatment, such as observation). Of course, a fatality, a major disfigurement, or the loss of a limb or digit is also considered serious and must be reported.

At a minimum, the company's insurer, the state agency, and the state's federal counterpart must be notified. Individual states may require that additional agencies be notified. All establish a time frame within which notification must be made. Once the notice of injury has been filed, there is typically a short period before the victim or dependents can begin to receive compensation unless inpatient hospital care is required. However, when payments do begin to flow, they are typically retroactive to the date of the injury.

State statutes also provide a maximum time period that can elapse before a compensation claim is filed. The notice of injury does not satisfy the requirement of filing a **claim notice**. The two are separate processes. The statute of limitations on claim notices varies from state to state. However, most limit the period to no more than a year except in cases of work-related diseases in which the exact date of onset cannot be determined.

All such activities—filing injury notices, filing claim notices, arriving at settlements, and handling disputes—fall under the collective heading of administration and case management. Most states have a designated agency that is responsible for administration and case management. In addition, some states have independent boards that conduct hearings or hear appeals when disputes arise.

Once a workers' compensation claim is filed, an appropriate settlement must be reached. Three approaches can be used to settle a claim:

1. Direct settlement
2. Agreement settlement
3. Public hearing

The first two are used in uncontested cases; the third is used in contested cases.

1. *Direct settlement.* The employer or its insurance company begins making what it thinks are the prescribed payments. The insurer also sets the period over which payments will be made. This is known as a **direct settlement**. Both factors are subject to review by the designated state agency. This approach is used in Arkansas, Michigan, Mississippi, New Hampshire, Wisconsin, and the District of Columbia.
2. *Agreement settlement.* The injured employee and the employer or its insurance company work out an agreement on how much compensation will be paid and for how long. Such an agreement must be reached before compensation payments begin. Typically, the agreement is reviewed by the designated state administrative agency. In cases in which the agency disapproves of the agreement, the worker continues to collect compensation at the agreed-on rate until a new agreement can be reached. If this is not possible, the case becomes a contested case. This is known as an **agreement settlement.**
3. *Public hearing.* If an injured worker feels he or she has been inadequately compensated or unfairly treated, a hearing can be requested. Such cases are known as *contested cases*. The hearing commission reviews the facts surrounding the case and renders a judgment concerning the amount and duration of compensation. This is known as a **public hearing**. Should an employee disagree with the decision rendered, civil action through the courts is an option.

COST ALLOCATION

Workers' compensation is a costly concept. From the outset, one of the basic principles has been cost allocation. Cost allocation is the process of spreading the cost of workers' compensation across an industry so that no individual company is overly burdened. The cost of workers' compensation includes the cost of premiums, benefits, and administration. These costs have risen steadily over the years.

When workers' compensation costs by industry are examined, there are significant differences. For example, the cost of workers' compensation for a bank is less than 0.5 percent of gross payroll. For a construction company, the percentage might be as high as 3 or 4 percent.

Cost allocation is based on the **experience rating** of the industry. In addition to being the fairest method (theoretically) of allocating costs, this approach is also supposed to give employers an incentive to initiate safety programs. Opinions vary as to the fairness and effectiveness of this approach. Arguments against it include the following:

1. Small firms do not have a sufficient number of employees to produce a reliable and accurate picture of the experience rating.
2. Firms that are too small to produce an experience rating are rated by class of industry, thereby negating the incentive figure.
3. Premium rates are more directly sensitive to experience levels in larger firms, but are less so in smaller companies.
4. To hold down experience ratings, employers may put their efforts into fighting claims rather than preventing accidents.

Not much hard research has been conducted to determine the real effects of cost allocation. Such research is badly needed to determine if the theoretical construct of workers' compensation is, in reality, valid.

PROBLEMS WITH WORKERS' COMPENSATION

There are serious problems with workers' compensation in the United States. On the one hand, there is evidence of abuse of the system. On the other hand, many injured workers who are legitimately collecting benefits suffer a substantial loss of income. Complaints about workers' compensation are common from all parties involved with the system (employers, employees, and insurance companies).

Consider the actual example of the overweight deputy sheriff who applied for benefits because he was distraught over the breakup of his extramarital love affair: a poor performance evaluation caused him stress. This individual is just one of thousands who are claiming that job stress has disabled them to the point that workers' compensation is justified. In 1980, there were so few **stress claims** that they were not even recorded as a separate category. Today, they represent a major and costly category.

Stress claims are more burdensome than physical claims because they are typically reviewed in an adversarial environment. This leads to the involvement of expert medical witnesses and attorneys. As a result, even though the benefits awarded for stress-related injuries are typically less than those awarded for physical injuries, the cost of stress claims is often higher because of litigation.

Although the cost of workers' compensation is increasing steadily, the amount of compensation going to injured workers is often disturbingly low. In a given year, if workers' compensation payments in the United States amount to around $27 billion (which is typical), $17 billion of this goes to benefits. Almost $10 billion is taken up by medical costs.[18] The amount of wages paid to injured workers in most states is 66 percent. This phenomenon is not new.

The most fundamental problem with workers' compensation is that it is not fulfilling its objectives: Lost income is not being adequately replaced, the number of accidents has not decreased, and the effectiveness of cost allocation is questionable. Clearly, the final chapter on workers' compensation has not yet been written.[19]

SPOTTING WORKERS' COMPENSATION FRAUD OR ABUSE

There is evidence of waste, fraud, and abuse of the system in all states that have passed workers' compensation laws. However, the public outcry against fraudulent claims is making states much less tolerant of and much more attentive to abuse. For example, the Ohio legislature passed a statute that allows criminal charges to be brought against employees, physicians, and lawyers who give false information in a workers' compensation case. This is a positive trend. However, even these measures will not completely eliminate abuse.

For this reason, companies must know how to spot employees who are trying to abuse the system by filing fraudulent workers' compensation claims. The following are some factors that should cause employers to view claims with suspicion. However, just because one or more of these factors are present does not mean that an employee is attempting to abuse the system. Rather, these are simply factors that should raise cautionary flags.[20]

- The person filing the claim is never home or available by telephone or has an unlisted telephone number.
- The injury in question coincides with a layoff or termination.
- The person filing the claim is active in sports.
- The person filing the claim has another job.

- The person filing the claim is in line for early retirement.
- The rehabilitation report contains evidence that the person filing the claim is maintaining an active lifestyle.
- No organic basis exists for disability. The person filing the claim appears to have made a full recovery.
- The person filing the claim receives all mail at a post office box and will not divulge a home address.
- The person filing the claim is known to have skills, such as carpentry, plumbing, or electrical skills, that could be used to work on a cash basis while feigning a disability.
- There are no witnesses to the accident in question.
- The person filing the claim has relocated out of state or out of the country.
- Demands for compensation are excessive.
- The person filing the claim has a history of filing.
- Doctor's reports are contradictory in nature.
- A soft-tissue injury is claimed to have produced a long-term disability.
- The injury in question occurred during hunting season.

If one or more of these factors are present, employers should investigate the claim carefully before proceeding. These factors can help organizations spot employees who may be trying to abuse the workers' compensation system. Legitimate users of the system are hurt, as are their employers, by abusers of the system.

FUTURE OF WORKERS' COMPENSATION

The future of workers' compensation can be summarized in one word: **reform**. The key to reforming workers' compensation is finding a way to allocate more of the cost to benefits and medical treatment and less to administration and litigation. California, which is the trendsetting state for workers' compensation, is the leader in the reform movement with its Workers' Compensation Improvement Act. Key elements of the act included the following:

- Stabilizing workers' compensation costs over the long term
- Streamlining administration of the system
- Reducing the costs associated with the resolution of medical issues
- Limiting stress-related claims
- Limiting vocational rehabilitation benefits
- Increasing benefits paid for temporary and permanent disabilities
- Reducing the amount that insurers may charge for overhead
- Providing more public input into the setting of rates

At the time the act was proposed, California's workers' compensation costs had a history of increasing at a rate of more than 11 percent per year. California paid out more in claims every year than any other state, but ranked among the lowest nationwide in benefits paid to injured workers. California's problems with workers' compensation paint a picture in microcosm of the problems facing all states.

COST-REDUCTION STRATEGIES

Construction professionals are responsible for helping their companies hold down workers' compensation costs. Of course, the best way to accomplish this goal outside of the legislation process is to maintain a safe and healthy workplace, thereby preventing the injuries that drive the costs up. This section presents numerous other strategies that have proven effective in reducing workers' compensation costs after injuries have occurred.

General Strategies

Regardless of the type of organization, there are several rules of thumb that can help reduce workers' compensation claims. These general strategies are as follows:

1. *Stay in touch with the injured employee.* Let injured employees know that they have not been forgotten and that they are not isolated. Answer all of their questions, and try to maintain their loyalty to the organization.
2. *Have a return-to-work program and use it.* The sooner an injured employee returns to work, even with a reduced workload, the lower workers' compensation costs will be. Reduced costs can, in turn, lower the organization's insurance premium. When using the return-to-work strategy, be cautious. Communicate with the employee and his or her medical treatment team. Have a clear understanding of the tasks that can be done and those that should be avoided, such as how much weight the employee can safely lift.
3. *Determine the cause of the accident.* The key to preventing future accidents and incidents is determining the cause of the accident in question. And the key to holding down workers' compensation costs is preventing accidents. Eliminating the root cause of every accident is fundamental to any cost-containment effort.

Colledge and Johnson recommend using the SPICE model for improving the effectiveness of return-to-work programs.[21] It consists of the following components:

- Simplicity
- Proximity
- Immediacy
- Centrality
- Expectancy

Simplicity means that the medical professionals who treat injured employees should work closely with employers to prevent system-induced complications. Such complications occur when employees become convinced their injuries are more serious than they really are because of ominous-sounding diagnostic terminology and complicated tests and treatments. Medical professionals and safety personnel should work together to keep the terminology simple and to explain tests and treatments in easily understood lay terms.

Proximity means keeping the injured employee as close to the job as possible. Employees who are physically separated from their place of employment and their fellow employees also become mentally separated. Within a short time, what used to be "us" can become "them." Giving as much injury care as possible at the work site, providing light-duty assignments, and communicating regularly with employees whose injuries preclude onsite assignments or treatment keeps employees connected and maintains the advantages of proximity.

Immediacy means that the faster an employee's injury claims can be handled, the less likely he or she is to develop psychosocial issues that can complicate the recovery process. The longer it takes to process a claim, conduct a diagnosis, and begin treatments, the more likely the employee is going to begin to worry about the injury and to become accustomed to being off from work. Immediate diagnosis, processing, and treatment can decrease the amount of time that elapses before the employee can begin a return-to-work program.

Centrality means getting the employee, his or her family, the medical professionals handling the case, insurance personnel, and the employer to agree on a common vision for successfully returning the injured party to work as soon as possible. Injured employees and their families must know that everyone involved has the same goal and is working in good faith to achieve that goal.

Expectancy means creating the expectation that getting the employee well and back to work is the goal of all parties involved. It is achieved by communicating this message clearly to all parties and reinforcing it by establishing short-term goals and timelines for actually being back on the job. Achievement of each respective short-term goal should move the employee one step closer to recovery and return to work.

Specific Strategies

In addition to the general strategies just presented, there are numerous specific cost-containment strategies that have proven to be effective. These specific strategies are presented in this section.

1. *Make safety part of the culture.* The following steps can be useful for making safety part of the organizational culture and reducing workers' compensation costs:
   - Ensure the visible, active leadership; involvement; and commitment of senior management.
   - Involve employees at all levels in the safety program, and recognize them for their efforts.
   - Provide comprehensive medical care, part of which is a return-to-work program.
   - Ensure effective communication throughout the organization.
   - Coordinate all safety and health processes.
   - Provide orientation and training for all employees.
   - Have written safety practices and procedures, and distribute them to all employees.
   - Have a comprehensive written safety policy.
   - Keep comprehensive safety records, and analyze the data contained in those records.

SAFETY FACTS & FINES

It pays to follow safety regulations at all times. You never know when someone might be watching. A construction company in Lithia, Florida, found this out the hard way when an OSHA inspector just happened to be driving by a job site and noticed a safety violation. The inspector stopped and looked things over. By the time he was done, the construction company had been cited for numerous violations that eventually led to a fine of $229,250. The violations cited were as follows:

- Unstable scaffolding that lacked cross-bracing
- Using loose masonry blocks stacked up to reach scaffolding
- No guardrails on scaffolds
- Workers were exposed to impalement hazards
- No fall protection provided

2. *Have a systematic cost-reduction program.* To reduce costs, an organization should have a systematic program that can be applied continually and consistently. The following strategies are recommended:
   - Insert safety notes and reminders in employee's paycheck envelopes.
   - Call injured employees at home to reassure them that they will have a job when they return.
   - Keep supervisors trained on all applicable safety and health issues, procedures, and rules.
   - Hold monthly meetings to review safety procedures, strategies, and techniques.
   - Reward employees who give suggestions for making the workplace safer.
   - Make safety part of every employee's job description and performance appraisal.
3. *Use integrated managed care.* Managed care is credited by many with reducing workers' compensation costs nationwide. Others claim that cost reduction has occurred because managed care dangerously restricts the types and amount of health care provided to injured employees.

OTHER KINDS OF CONSTRUCTION INSURANCE

Workers' compensation is critical for contractors and builders, but there are other kinds of insurance that also apply. These other kinds of construction-related insurance include the following:

- *Health insurance* for employees to cover all or a portion of their normal medical bills (should the employer opt to provide it)
- *Life insurance* for employees (should the builder opt to provide it)
- *Disability insurance* to cover the overhead costs of builders in the event they are unable to work
- *Crime insurance* to protect against the risk of theft, vandalism, and other crimes at the job site or to the builder's facilities and property
- *Stop loss insurance* to cover the costs of losses that exceed normal policy limits
- *Builder's risk insurance* to protect against the risks of physical loss or damage on the job site during construction

There are also several different types of policies that protect against the specific types of risk associated with building projects in foreign countries, including terrorist, kidnapping, and political risk insurance policies. Of these various kinds of insurance, the most pertinent to builders is builder's risk insurance.

Builder's Risk Insurance

Much can go wrong during a construction project. Fire, hurricanes, tornados, floods, theft, vandalism, lightning, and even sabotage are just a few of the possible threats faced by those who undertake building projects. Builder's risk insurance is purchased to mitigate the potential financial losses of such risks. It may be purchased by the owner of the building being constructed or the builder depending on the conditions set forth in the construction contract. For this reason, it is important for builders to ensure that there is builder's risk insurance on their projects and to know who is going to purchase it.

Summary

Workers' compensation was developed to allow injured employees to be compensated without the need for litigation. It has four main objectives: income replacement, rehabilitation, accident prevention, and cost allocation.

Before the enactment of workers' compensation laws, the employee's only recourse when injured was through the courts, and the prevailing laws favored employers. Early workers' compensation laws were ruled unconstitutional. The constitutional debate continued until 1917, when the U.S. Supreme Court ruled that workers' compensation laws were acceptable.

All 50 states have workers' compensation laws, but they vary markedly. All laws are enacted to provide benefits, pay medical costs, provide for rehabilitation when necessary, decrease litigation, and encourage accident prevention. There are three types of workers' compensation insurance: state funds, private insurers, and self-insurance. Six methods are used for determining insurance premium rates for employers: schedule rating, manual rating, experience rating, retrospective rating, premium discounting, and a combination of these.

Although an often-stated objective of workers' compensation is the reduction of costly litigation, many cases still go to court. This is particularly true of stress-related cases.

Workers' compensation applies when an injury can be categorized as arising out of employment (AOE) or occurring in the course of employment (COE).

The definition of an employee can vary from state to state. However, a key concept in distinguishing between an employee and an independent contractor is direction (supervision). Employees are provided direction by the employer; contractors are not.

Injuries that are compensable through workers' compensation fall into one of four categories: (1) temporary partial disability, (2) temporary total disability, (3) permanent partial disability, and (4) permanent total disability.

Three theoretical approaches to handling permanent partial disability cases are (1) the whole-person theory, (2) the wage-loss theory, and (3) the loss of wage-earning capacity theory.

Workers' compensation benefits accrue to families and dependents of workers who are fatally injured.

Typically, the spouse receives benefits for life or until remarriage. Dependents typically receive benefits until they reach the legal age of maturity.

All workers' compensation laws provide for payment of medical expenses within specified time periods, but there are differences concerning how a physician may be selected. The options are: (1) employee selects physician of choice, (2) employee selects physician from a list provided by the state agency; (3) employee selects physician from a list provided by the employer; (4) employer selects the physician; (5) employer selects the physician, but the selection may be changed by the state agency; or (6) employer selects the physician, but after a specified period of time the employee may choose another.

Workers' compensation claims can be settled in one of three ways. The first two are for uncontested claims; the third is for contested ones. The approaches are: (1) direct settlement, (2) agreement settlement, or (3) public hearing.

The definition of a serious injury can vary from state to state. However, as a rule an injury is serious if it requires more than 24 hours of active medical treatment. This does not include passive treatment such as observation. The goals of medical management of workplace injuries are to: (1) speed up the processing of claims, (2) reduce costs, (3) reduce fraud and abuse, and (4) improve medical treatment.

Cost allocation is the process of attempting to spread the cost of workers' compensation across an industry so that no individual company is overly burdened. The cost of workers' compensation includes the cost of premiums, benefits, and administration. Cost-reduction strategies include staying in touch with injured employees and determining the causes of accidents.

The problems being experienced with workers' compensation can be summarized as follows: Workers' compensation is not achieving its intended objectives. It has not succeeded in taking litigation out of the process. The cost of the system rises steadily, but the benefits to injured workers have decreased in real terms. There is substantial evidence of abuse. Therefore, the future of workers' compensation can be summarized in one word: reform.

Key Terms and Concepts

Accident prevention
Agreement settlement
Arises out of employment (AOE)
Assumption of risk
Blackball
Claim notice
Contributory negligence
Cost allocation
Course of employment (COE)
Direct settlement
Employee
Experience rating
Extraterritorial employees
Fear factor
Harmful environment
Income replacement
Independent contractor
Litigation
Loss of wage-earning capacity theory
Malpractice
Manual rating
Medical management of workplace injuries
Medical rehabilitation
Monetary benefits
Occupational disease
Permanent partial disability
Permanent total disability
Premium discounting
Private insurance
Public hearing
Reform
Rehabilitation
Retrospective rating
Schedule disabilities
Schedule rating
Self-insurance
State funds
Stress claims
Temporary partial disability
Temporary total disability
Vocational rehabilitation
Wage-loss theory
Whole-person theory
Workers' compensation
Workplace accident

Review Questions

1. Explain the underlying rationale of workers' compensation as a concept.
2. List four objectives of workers' compensation.
3. List five types of employees who may not be covered by workers' compensation.
4. What is meant by the term *contributory negligence*?
5. What is meant by the term *assumption of risk*?
6. Explain the reasons for the unprecedented increases in medical costs in the United States.
7. What are the three types of workers' compensation insurance?
8. Insurance companies use one of six methods for determining the premium rates of employers. Select three and explain them.
9. How can one determine if an injury should be considered serious?
10. Explain the concepts of AOE and COE.
11. Distinguish between an employee and an independent contractor.
12. Define the following terms: *temporary disability* and *permanent disability*.
13. Explain the following theories of handling permanent partial disability cases: *whole-person, wage-loss*, and *loss of wage-earning capacity*.
14. Distinguish between medical and vocational rehabilitation.
15. What are the three approaches for settling workers' compensation claims?
16. Explain the concept of medical management of workplace injuries.
17. Explain the theory of cost allocation.
18. Summarize briefly the problems most widely associated with workers' compensation.
19. What types of actions are workers' compensation reform movements likely to recommend in the future?
20. Explain the most common workers' compensation cost-reduction strategies.

Critical Thinking and Discussion Activities

1. In California, medical care costs for back injuries are 43 percent higher when part of a workers' compensation claim than when part of a group medical insurance plan. Critics of workers' compensation claim that this difference can be attributed to abuse of the system by both injured employees and their physicians. Proponents of workers' compensation argue that the higher costs indicate exactly the opposite: that is, under workers' compensation, injured employees are receiving more and better (hence more costly) medical care. They claim that group medical plans are shortchanging employees by allowing insurance companies to make medical decisions. Join the debate. What is your opinion?
2. "If workers' compensation is going to work, we are going to have to cut down on fraud and abuse of the system," says Mark Baker, a supervisor for Middletown Construction Company. "You'll never do that," says his colleague Josh Menton. "Some people are just dishonest, and nothing we can do will change that." "What about the company in Oklahoma that reduced its workers' compensation costs by 150 percent in just two years by running its employees through a character training program? I just read about it in the *Journal of Construction Safety*. The program focuses on character traits such as honesty, attentiveness, gratitude, and dependability." "Sorry," said Menton. "I don't think people change. If they are honest, they are honest, and if they are not, they are not." The case in Oklahoma is real, as are the results. Join this debate. What is your opinion?
3. "Most of the abuse of workers' compensation is in the area of mental and psychological trauma caused by stress. The problem is, who can really know if someone is suffering psychological problems that were caused by work. If a person is suffering from stress, how do we know it's not his or her marriage, stock market account, or a child that is causing problems? How do we know he or she isn't suffering over a lost loved one or a relationship that has fallen apart? I think workers' compensation should cover physical problems only. There is just too much potential for abuse with mental and psychological problems." Assume one of your classmates just made this statement. Is this classmate right or wrong? What is your opinion?

Application Activities

1. Get a copy of the workers' compensation legislation for your state. Does it favor injured workers or their employers? Has it been successful in satisfying the goals of workers' compensation set forth in this chapter? Who is covered and who is excluded in your state? What type of coverage is allowed in your state: state fund, private insurer, individual employer self-insurance, or group of employers self-insurance?
2. Interview a manager or human resources specialist from a construction company in your community. What problems do they have with workers' compensation? What do they like about it? What would they change if they could?
3. Contact the website of the U.S. Department of Labor, and print out the monetary benefits for workers' compensation for your state (http://www.dol.gov).

Endnotes

1. "Workers' Comp Update," *Occupational Hazards* 61, no.10: 51.
2. Ibid.
3. Ibid.
4. Ibid.
5. Somers, H. M., and A. R. Somers, *Workmen's Compensation* (New York: John Wiley & Sons, 1945), 29.
6. Ibid.
7. Gagliardo, D., *American Social Science* (New York: Harper & Row, 1949), 149.
8. "Workers' Comp Update," 51.
9. Ibid.
10. Ibid.
11. U.S. Chamber of Commerce. "2016 Analysis of Workers Compensation Laws," U.S. Chamber of Commerce, Washington, D.C., 2016, 2–11, January 2016.
12. "Workers' Comp Update," *Occupational Hazards* 60, no. 7: 93.
13. Ibid.
14. U.S. Department of Labor, *State Workers' Compensation Laws* (Washington, D.C.: U.S. Department of Labor, 2008), 16.
15. Ibid., 20.
16. Strazewski, Len. "Managed Care: A Prescription for Worker's Compensation." Retrieved from http://businessfinancemag.com/hr/managed-care-prescription-workers-compensation on December 1, 2015.
17. Ibid., 68.
18. U.S. Department of Labor, *Federal Worker 2000* (Washington, D.C.: U.S. Department of Labor, 2000), 2.
19. Ibid., 20.
20. Goetsch, D., "Workers' Comp Fraud Detection Checklist (Revised)," report no. 2016–3, The Development Institute (January 2006), 8.
21. Colledge, A., and H. Johnson, "The S.P.I.C.E. Model for Return to Work," *Occupational Health and Safety* 69, no. 2: 64–69.

CHAPTER SIX

OSHA COMPLIANCE

LEARNING OBJECTIVES

- Explain the rationale for the OSH Act.
- Summarize OSHA's mission/purpose.
- Explain OSH Act coverage.
- Summarize how OSHA standards are developed, adopted, amended, and revoked.
- List OSHA's requirements for record keeping and reporting.
- Explain OSHA's requirements for keeping employees informed.
- Describe the process for workplace inspections by OSHA.
- Summarize the various citations and penalties OSHA can assess.
- Describe how OSHA citations and penalties can be appealed.
- Explain the concept of **state-level OSHA programs**.
- Summarize the services available from OSHA.
- List the rights and responsibilities of employers concerning OSHA.
- List the rights and responsibilities of employees concerning OSHA.
- Describe how a construction professional can keep up to date concerning OSHA standards.
- Explain some of the most pressing problems with OSHA.
- List the other federal agencies and organizations with an interest in workplace safety.
- Distinguish between a standard and a code.
- Summarize the body of law pertaining to workplace safety.

Since the early 1970s, the amount of legislation passed—and the number of subsequent regulations—concerning workplace safety and health have increased markedly. Of all the legislation, by far the most significant has been the **Occupational Safety and Health Act (OSH Act)**. Prospective and practicing construction professionals must be knowledgeable about the OSH Act and the agency established by it—the **Occupational Safety and Health Administration (OSHA)**. This chapter provides students with the information they need about the OSH Act, OSHA, and other pertinent federal legislation and agencies.

RATIONALE FOR THE OSH ACT

Perhaps the most debilitating experience one can have on the job is to be involved in or exposed to a work-related accident or illness. Such an occurrence can be physically and psychologically incapacitating for the victim, psychologically stressful for the victim's co-workers, and extraordinarily expensive for the victim's employer. In spite of this, until 1970, laws governing workplace safety were limited and sporadic. Finally, in 1970, Congress passed the OSH Act with this stated purpose: "To assure so far as possible every working man and woman in the nation safe and healthful working conditions and to preserve our human resources."[1]

According to the U.S. Department of Labor, in developing this comprehensive and far-reaching piece of legislation, Congress considered the following statistics:

- Every year, an average of 14,000 deaths were caused by **workplace accidents**.
- Every year, 2.5 million workers were disabled in workplace accidents.
- Every year, approximately 300,000 new cases of occupational diseases were reported.[2]

Clearly, a comprehensive, uniform law was needed to help reduce the incidence of work-related injuries, illnesses, and deaths. The OSH Act addressed this need; it is contained in Title 29 of the Code of Federal Regulations (CFR), Parts 1900 through 1910. The act also established OSHA, which is part of the U.S. Department of Labor and is responsible for administering the OSH Act.

OSH'S MISSION AND PURPOSE

According to the Department of Labor, the mission and purpose of OSHA can be summarized as follows[3]:

- Encourage employers and employees to reduce workplace hazards.
- Implement new safety and health programs.

- Improve existing safety and health programs.
- Encourage research that leads to innovative ways of dealing with workplace safety and health problems.
- Establish the rights of employers regarding the improvement of workplace safety and health.
- Establish the rights of employees regarding the improvement of workplace safety and health.
- Monitor job-related illnesses and injuries through a system of reporting and record keeping.
- Establish training programs to increase the number of safety and health professionals and to improve their competence continually.
- Establish mandatory workplace safety and health standards, and enforce those standards.
- Provide for the development and approval of state-level workplace safety and health programs.
- Monitor, analyze, and evaluate state-level safety and health programs.

OSH ACT COVERAGE

The OSH Act applies to most employers. If an organization has even one employee, it is considered an employer and must comply with applicable sections of the act. This includes all types of employers from manufacturing and construction to retail and service organizations. There is no exemption for small businesses, although organizations with 10 or fewer employees are exempted from OSHA inspections and the requirement to maintain injury/illness records.

Although the OSH Act is the most comprehensive and far-reaching piece of safety and health legislation ever passed in this country, it does not cover all employers. In general, the OSH Act covers employers in all 50 states, the District of Columbia, Puerto Rico, and all other territories that fall under the jurisdiction of the U.S. government. Exempted employers are as follows:

- Persons who are self-employed
- Family farms that employ only immediate family members
- Federal agencies covered by other federal statutes (in cases where these other federal statutes do not cover working conditions in a specific area or areas, OSHA standards apply)
- State and local governments (except to gain OSHA's approval of a state-level safety and health plan, states must provide a program for state and local governments)
- Coal miners (coal mines are regulated by mining-specific laws)

Federal agencies are required to adhere to safety and health standards that are comparable to and consistent with OSHA standards for private sector employees. OSHA evaluates the safety and health programs of federal agencies. However, OSHA cannot assess fines or monetary damages against other federal agencies as it can against private sector employers.

There are many OSHA requirements to which employers must adhere. Some apply to all employers—except those exempted—whereas others apply only to specific types of employers. These requirements cover such areas of concern as the following:

- Fire protection
- Electrical safety
- Sanitation
- Air quality
- Machine use, maintenance, and repair
- Posting of notices and warnings
- Reporting of accidents and illnesses
- Maintaining written compliance programs
- Employee training

In addition to these, other more important and widely applicable requirements are explained later in this chapter.

OSHA STANDARDS

The following statement by the U.S. Department of Commerce summarizes OSHA's responsibilities related to standards:

> In carrying out its duties, OSHA is responsible for promulgating legally enforceable standards. OSHA standards may require conditions, or the adoption or use of one or more practices, means, methods, or processes reasonably necessary and appropriate to protect workers on the job. It is the responsibility of employers to become familiar with standards applicable to their establishments and to ensure that employees have and use personal protective equipment when required for safety.[4]

The general duty clause of the OSH Act requires that employers provide a workplace that is free from hazards that are likely to harm employees. This is important because the general duty clause applies when there is no specific OSHA standard for a given situation. Where OSHA standards do exist, employers are required to comply with them as written.

How Standards Are Developed?

OSHA develops standards on the basis of its perception of need and at the request of other federal agencies, state and local governments, other agencies that set standards, labor organizations, and even individual private citizens. OSHA uses the committee approach for developing standards—both standing committees within OSHA and special ad hoc committees. Ad hoc committees are appointed to deal with issues that are beyond the scope of the standing committees.

OSHA's standing committees are the National Advisory Committee on Occupational Safety and Health (NACOSH) and the Advisory Committee on Construction Safety and Health. NACOSH makes recommendations on standards to the U.S. secretary of health and human services and the secretary of labor. The Advisory Committee on Construction Safety and Health advises the secretary of labor on standards and regulations related specifically to the construction industry.

The **National Institute for Occupational Safety and Health (NIOSH)** was established by the OSH Act. Whereas OSHA is part of the Department of Labor, NIOSH is part of the Department of Health and Human Services. NIOSH has an education and research orientation. The results of this agency's research are often used to assist OSHA in developing standards.

OSHA Standards versus OSHA Regulations

OSHA issues both standards and regulations. Construction professionals need to know the difference between the two. OSHA standards address specific hazards, such as working in confined spaces, handling hazardous waste, or working with dangerous chemicals. Regulations are more generic in some cases than standards and more specific in others. However, even when they are specific, regulations do not apply to specific hazards. Regulations do not require the rigorous review process that standards go through. This process is explained in the next section.

How Standards Are Adopted, Amended, or Revoked?

OSHA can adopt, amend, or revise standards. Before any of these actions can be undertaken, OSHA must publish its intentions in the *Federal Register* in either a notice of proposed rule making or an advance notice of proposed rule making. The **notice of proposed rule making** must explain the terms of the new rule, delineate proposed changes to existing rules, or list rules that are to be revoked. The advance notice of proposed rule making may be used instead of the regular notice when it is necessary to solicit input before drafting a rule.

After publishing the notice, OSHA must conduct a public hearing if one is requested. Any interested party may ask for a public hearing on a proposed rule or rule change. When this happens, OSHA must schedule the hearing and announce the time and place in the *Federal Register*.

The final step, according to the Department of Labor, is as follows:

> After the close of the comment period and public hearing, if one is held, OSHA must publish in the Federal Register the full, final text of any standard amended or adopted and the date it becomes effective, along with an explanation of the standard and the reasons for implementing it. OSHA may also publish a determination that no standard or amendment needs to be issued.[5]

How to Read an OSHA Standard?

OSHA standards are typically long and complex and are written in the language of lawyers and bureaucrats, making them difficult to read. However, reading OSHA standards can be simplified somewhat if one understands the system.

OSHA standards are part of the CFR published by the Office of the Federal Register. The regulations of all federal agencies are published in the CFR. Title 29 contains all of the standards assigned to OSHA; it is divided into several parts, each carrying a four-numeral designator (such as Part 1910 or Part 1926). These parts are divided into sections, each carrying a numerical designation. For example, 29 CFR 1926.1 means *Title 29, Part 1926, Section 1, Code of Federal Regulations.*

These sections are divided into four different levels of subsections, each with a particular type of designator as follows:

First Level: Alphabetical, using lowercase letters: (a) (b) (c) (d)

Second Level: Numerical, using numerals: (1) (2) (3) (4)

Third Level: Numerical, using roman numerals: (i) (ii) (iii) (iv)

Fourth Level: Alphabetical, using uppercase letters: (A) (B) (C) (D)

Occasionally, the standards go beyond the fourth level of subsection. In these cases, the sequence just described is repeated with the designator shown in parentheses underlined. For example: (a), (1), (i), and (A).

Understanding the system used for designating sections and subsections of OSHA standards can guide readers more quickly to the specific information needed. This helps to reduce the amount of cumbersome reading needed to determine what is necessary to comply with the standards.

Temporary Emergency Standards

The procedures described in the previous section apply in all cases. However, OSHA is empowered to pass temporary standards on an emergency basis without undergoing normal adoption procedures. Such standards remain in effect only until permanent standards can be developed.

To justify passing **temporary emergency** standards, OSHA must determine that workers are in imminent danger from exposure to a hazard not covered by existing standards. Once a temporary standard has been developed, it is published in the *Federal Register*. This step serves as the notification step in the permanent adoption process. At this point, the standard is subjected to all of the other adoption steps outlined in the preceding section.

How to Appeal a Standard?

After a standard has been passed, it becomes effective on the date prescribed. This is not necessarily the final step in the appeals process, however. A standard, either permanent or temporary, may be appealed by any person who is opposed to it.

An appeal must be filed with the U.S. Court of Appeals serving the geographic region in which the complainant lives or does business. Appeal paperwork must be initiated within 60 days of a standard's approval. However, the filing of one or more appeals does not delay the enforcement of a standard unless the court of appeals handling the matter mandates a delay. Typically, the new standard is enforced as passed until a ruling on the appeal is handed down.

Requesting a Variance

Occasionally, an employer may be unable to comply with a new standard by the effective date of enforcement. In such cases, the employer may petition OSHA at the state or federal level for a variance. The following different types of variances can be granted.

Temporary Variance. When an employer advises that it is unable to comply with a new standard immediately, but may be able to if given additional time, a **temporary variance** may be requested. OSHA may grant such a variance for up to a maximum of one year. To be granted a temporary variance, employers must demonstrate that they are making a concerted effort to comply and taking the steps necessary to protect employees while working toward compliance.

Application procedures are very specific. Prominent among the requirements are (1) identification of the parts of the standard with which the employer cannot comply, (2) explanation of the reasons why compliance is not possible, (3) detailed explanations of the steps that have been taken so far to comply with the standard, and (4) explanation of the steps that will be taken to comply fully.

According to the U.S. Department of Labor, employers are required to keep their employees informed; they must "certify that workers have been informed of the variance application, that a copy has been given to the employees' authorized representative, and that a summary of the application has been posted wherever notices are normally posted. Also, employees must be informed that they have the right to request a hearing on the application."[6]

Variances are not granted simply because an employer cannot afford to comply. For example, if a new standard requires employers to hire a particular type of specialist, but there is a shortage of people with the requisite qualifications, a temporary variance might be granted. However, if the employer simply cannot afford to hire such a specialist, the variance will probably be denied. Once a temporary variance is granted, it may be renewed twice. The maximum period of each extension is six months.

Permanent Variance. Employers who feel that they already provide a workplace that exceeds the requirements of a new standard may request a **permanent variance**. They present their evidence, which is inspected by OSHA. Employees must be informed of the application for a variance and notified of their right to request a hearing. Having reviewed the evidence and heard testimony (if a hearing has been held), OSHA can award or deny the variance. If a permanent variance is awarded, it comes with a detailed explanation of the employer's ongoing responsibilities regarding the variance. If, at any time, the company does not meet these responsibilities, the variance can be revoked.

Other Variances. In addition to temporary and permanent variances, an experimental variance may be awarded to companies that participate in OSHA-sponsored experiments to test the effectiveness of new health and safety procedures. Variances also may be awarded in cases in which the secretary of labor determines that a variance is in the best interest of national defense.

When applying for a variance, employers are required to comply with the standard until a decision has been made. If this is a problem, the employer may petition OSHA for an interim order. If granted, the employer is released from the obligation to comply until a decision is made. In such cases, employees must be informed of the order.

SAFETY FACTS & FINES

Companies should understand that poor safety practices will not go unnoticed. Employees who are concerned about the conditions in which they work can give OSHA an anonymous tip. That is what happened in Bellefonte, Pennsylvania. An employee complaint to OSHA resulted in an inspection that, in turn, led to 18 citations for serious violations and a fine of $154,700. The serious citations included (1) failure to provide adequate communication while crane loads were being attached, (2) failure to implement an emergency evacuation plan, and (3) failure to provide adequate eye and face protection for employees.

RECORD KEEPING AND REPORTING

One of the breakthroughs of the OSH Act was the centralization and systematization of **record keeping** and **reporting**. This has simplified the process of collecting health and safety statistics for the purpose of monitoring problems and taking the appropriate steps to solve them (Figure 6–1).

Over the years, OSHA has made substantial changes to its record-keeping and reporting requirements. Employers had complained for years about the mandated injury and illness record-keeping system. Their complaints can be summarized as follows:

- Original system was cumbersome and complicated.
- OSHA record-keeping rule had not kept up with new and emerging issues.
- There were too many interpretations in many of the record-keeping documents.
- Record-keeping forms were too complex.
- Guidelines for record keeping were too long and difficult to understand.

In response to these complaints, OSHA initiated a dialogue among stakeholders to improve the record-keeping and reporting process. Input was solicited and received from employers, unions, trade associations, record keepers, OSHA staff, state occupational safety and health personnel, and state consultation program personnel. OSHA's goals for the new record-keeping and reporting system were as follows:

- Simplify all aspects of the process.
- Improve the quality of records.
- Meet the needs of a broad base of stakeholders.
- Improve access for employees.
- Minimize the regulatory burden.
- Reduce vagueness—give clear guidance.
- Promote the use of data from the new system in local safety and health programs.

In recording and reporting occupational illnesses and injuries, it is important to have common definitions. The Department of Labor uses the following definitions for record-keeping and reporting processes:

> An occupational injury is any injury such as a cut, fracture, sprain, or amputation that results from a work-related accident or from exposure involving a single incident in the work environment. An occupational illness is any abnormal condition or disorder other than one resulting from an occupational injury caused by exposure to environmental factors associated with employment. Included are acute and chronic illnesses or diseases which may be caused by inhalation, absorption, ingestion, or direct contact with toxic substances or harmful agents.[7]

KEEPING EMPLOYEES INFORMED

One of the most important requirements of the OSH Act is *communication*. Employers are required to keep employees informed about safety and health issues that concern them (Figure 6–2). Most of OSHA's requirements in this area concern the posting of material. Employers are required to post the following material at locations where employee information is normally displayed:

- OSHA Poster 2203, which explains employee rights and responsibilities as prescribed in the OSH Act. The state version of this poster may be used as a substitute.
- Summaries of variance requests of all types.
- Copies of all OSHA citations received for failure to meet standards. Unlike other informational material, citations must be posted near the site of

FIGURE 6–1 Systematic record keeping is an important aspect of OSHA compliance.
Source: Tashatuvango/Fotolia

FIGURE 6–2 OSHA compliance requires that workers be kept informed.
Source: Lakeviewimages/Fotolia

the violation. They must remain until the violation is corrected or for a minimum of three days, whichever period is longer.

- The summary page of OSHA's Summary of Work-Related Injuries and Illnesses Report. Each year, the new summary page must be posted by February 1 and must remain posted until April 30.

In addition to the posting requirements, employers must also provide employees who request them with copies of the OSH Act and any OSHA rules that may concern them. Employees must be given access to records of exposure to hazardous materials and medical surveillance that has been conducted.

OSHA AUDITS AND WORKPLACE INSPECTIONS

One of the methods OSHA uses for enforcing its rules is the **workplace inspection**. OSHA personnel may conduct workplace inspections unannounced, and, except under special circumstances, giving an employer prior notice is a crime punishable by fine, imprisonment, or both (Figure 6–3).

When OSHA compliance officers arrive to conduct an inspection, they are required to present their credentials to the person in charge. Having done so, they are authorized to enter, at reasonable times, any site, location, or facility where work is taking place. They may inspect, at reasonable times, any condition, facility, machine, equipment, materials, and so on. Finally, they may question, in private, any employee or other person formally associated with the company.

FIGURE 6–3 Workplace inspections are used to enforce OSHA's rules.
Source: Lisa F. Young/Fotolia

Under special circumstances, employers may be given up to 24 hours' advance notice of an inspection. These circumstances are as follows:

- When imminent danger conditions exist.
- When special preparation on the part of the employer is required.
- When inspection must take place at times other than during regular business hours.
- When it is necessary to ensure that the employer, employee representative, and other pertinent personnel are present.
- When the local area director of OSHA advises that advance notice results in a more effective inspection.

Employers may require that OSHA have a judicially authorized warrant before conducting an inspection. The U.S. Supreme Court handed down this ruling in 1978 (*Marshall v. Barlow's Inc.*).[8] However, having obtained a legal warrant, OSHA personnel must be allowed to proceed without interference or impediment.

The OSH Act applies to approximately 6 million work sites in the United States. Sheer volume dictates that OSHA establish priorities for conducting inspections. These priorities, from most to least important, are as follows: imminent danger situations, catastrophic fatal accidents, worker complaints, planned high-hazard inspections, and follow-up inspections.

After being scheduled, the inspection proceeds in the following steps:

1. The OSHA compliance officer presents his or her credentials to a company official.
2. The compliance officer conducts an **opening conference** with pertinent company officials and employee representatives. The following information is explained during the conference: why the site was selected for inspection, the purpose of the inspection, its scope, and applicable standards.
3. After choosing the route and duration, the compliance officer makes the **inspection tour**. During the tour, the compliance officer may observe, interview pertinent personnel, examine records, take readings, and make photographs.
4. The compliance officer holds a **closing conference**, which involves open discussion between the officer and company and employee representatives. OSHA personnel advise the company representatives of problems noted, actions planned as a result, and assistance available from OSHA.

CITATIONS AND PENALTIES

Based on the findings of the compliance officer's workplace inspections, OSHA is empowered to issue citations and assess penalties. A citation informs the employer of

OSHA violations. Penalties are typically fines assessed as a result of citations. The types of citations are as follows:

- *Other-than-serious violation.* A violation that has a direct relationship to job safety and health, but probably would not cause death or serious physical harm. A **proposed penalty** for each violation is discretionary. A penalty for an **other-than-serious violation** may be adjusted downward by as much as 95 percent—depending on the employer's good faith (demonstrated efforts to comply with the act), history of previous violations, and size of business.
- *Willful violation.* A violation that the employer intentionally and knowingly commits. The employer either knows that what he or she is doing constitutes a violation or is aware that a hazardous condition exists and has made no reasonable effort to eliminate it. A proposed penalty for a **willful violation** may be adjusted downward—depending on the size of the business and its history of previous violations. Usually, no credit is given for good faith. If an employer is convicted of a willful violation of a standard that has resulted in the death of an employee, the offense is punishable by a court-imposed fine or by imprisonment of the responsible company representative for up to six months, or both.
- *Repeat violation.* A violation of any standard, regulation, rule, or order for which, upon reinspection, a substantially similar violation is found. To be the basis of a **repeat violation** citation, the original citation must be final; a citation under contest may not serve as the basis for a subsequent repeat citation.
- *Failure to correct previous violation.* Failure to correct a previous violation may bring a civil penalty that increases each day the violation continues beyond the prescribed abatement date.[9]

In addition to the citations and penalties described in the preceding paragraphs, employers may also be penalized by additional fines or prison, if convicted of any of the following offenses: (1) falsifying records or any other information given to OSHA personnel, (2) failing to comply with posting requirements, and (3) interfering in any way with OSHA compliance officers in the performance of their duties.

THE APPEALS PROCESS

Employee Appeals

Employees may not contest the fact that citations were or were not awarded or the amounts of the penalties assessed. However, they may initiate the **appeals process** on the basis of the following aspects of OSHA's decisions regarding their workplace: (1) the amount of time (**abatement period**) given to an employer to correct a hazardous condition that has been cited and (2) an employer's request for an extension of an abatement period. Such appeals must be filed within 10 working days of a posting. Although opportunities for formal appeals by employees are unlimited, employees may request an informal conference with OSHA officials to discuss any issues a relating to the findings and results of a workplace inspection.

Employer Appeals

Employers may appeal a citation, an abatement period, or the amount of a proposed penalty. Before actually filing an appeal, however, an employer may ask for an informal meeting with OSHA's area director. The area director is empowered to revise citations, abatement periods, and penalties to settle disputed claims. If the situation is not resolved through this step, an employer may formalize the appeal. Formal **employer appeals** are of two types: (1) a petition for modification of abatement or (2) a notice of contest. The specifics of both are explained in the following paragraphs.

Petition for Modification of Abatement (PMA). The PMA is available to employers who intend to correct the situation for which a citation was issued, but need more time. As a first step, the employer must make a good-faith effort to correct the problem within the prescribed time frame. Having done so, the employer may file a petition for modification of abatement. The petition must contain the following information:

- Descriptions of steps taken to comply so far
- How much additional time is needed for compliance and why
- Descriptions of the steps being taken to protect employees during the interim
- Verification that the PMA has been posted for employee information and that the employee representative has been given a copy

Notice of Contest. An employer who does not wish to comply may contest a citation, an abatement period, or a penalty. The first step is to notify OSHA's area director in writing. This is known as filing a **notice of contest**. It must be done within 15 working days of receipt of a citation or penalty notice. The notice of contest must clearly describe the basis for the employer's challenge and contain all of the information about what is being challenged (e.g., amount of proposed penalty or abatement period).

Once OSHA receives a notice of contest, the area director forwards it and all pertinent materials to the **Occupational Safety and Health Review Commission (OSHRC)**. OSHRC is an independent agency that is not associated with OSHA or the Department of Labor. The Department of Labor describes how OSHRC handles an employer's claim:

> The Commission assigns the case to an administrative law judge. The judge may disallow the contest if it is found to be legally invalid, or a hearing may be scheduled for a public place near the employer's workplace. The employer and the employees have the right to participate in the hearing; the OSHRC does not require that they be represented by attorneys. Once the administrative law judge has ruled, any party to the case may request further review by OSHRC. Any of the three OSHRC commissioners also may, at his or her own motion, bring a case before the Commission for review. Commission rulings may be appealed to the appropriate U.S. Court of Appeals.[10]

STATE-LEVEL OSHA PROGRAMS

States are allowed to develop their own safety and health programs.[11] In fact, the OSH Act encourages it. As an incentive, OSHA will fund up to 50 percent of the cost of operating a state program for states with approved plans. States may develop comprehensive plans covering public and private employers or limit their plans to coverage of public employers only. In such cases, OSHA covers employers not included in the state plan.

To develop an OSHA-approved safety and health plan, a state must have adequate legislative authority and must demonstrate the ability to develop procedures for (1) setting standards, enforcement, and appeals within three years, (2) public employee protection, (3) a sufficient number of qualified enforcement personnel, and (4) education, training, and technical assistance programs. When a state satisfies all of these requirements and accomplishes all developmental steps,

> OSHA then certifies that a state has the legal, administrative, and enforcement means necessary to operate effectively. This action renders no judgment on how well or poorly a state is actually operating its program, but merely attests to the structural completeness of its program. After this certification, there is a period of at least 1 year to determine if a state is effectively providing safety and health protection.[12]

Figure 6–4 lists the telephone numbers for state and federal OSHA offices.

Regional Offices for States under Federal Jurisdiction

Region	States	Telephone
Region 1	Connecticut Massachusetts Maine New Hampshire Rhode Island	617-565-9860
Region 2	New York New Jersey	212-337-2378
Region 3	District of Columbia Delaware Pennsylvania West Virginia	215-596-1201
Region 4	Alabama Florida Georgia Mississippi	404-562-2300
Region 5	Illinois Ohio Wisconsin	312-353-2220
Region 6	Arkansas Louisiana Oklahoma Texas	214-767-4731
Region 7	Kansas Missouri Nebraska	816-426-5861
Region 8	Colorado Montana North Dakota South Dakota	303-844-1600
Region 9		415-975-4310
Region 10	Idaho	201-533-5930

Offices for State Jurisdictions

State	Telephone	State	Telephone
Alaska	907-269-4957	New Mexico	505-827-4230
Arizona	602-542-5795	New York*	518-457-2574
California	415-703-5100	North Carolina	919-807-2875
Connecticut*	860-566-4380	Oregon	503-378-3272
Hawaii	808-586-9100	Puerto Rico	787-754-2171
Indiana	317-232-3325	South Carolina	803-734-9632
Iowa	515-281-3606	Tennessee	615-741-2793
Kentucky	502-564-3070	Utah	801-530-6901
Maryland	410-767-2371	Vermont	802-828-2765
Michigan	517-322-1851	Virginia	804-786-6613
Minnesota	651-296-2116	Virgin Islands	340-772-1315
Nevada	702-687-3250	Washington	360-902-5554
New Jersey*	609-292-2313		

FIGURE 6–4 Telephone numbers for state and federal OSHA offices.

*Public sector only.

Source: From Occupational Safety & Health Administration. Published by U.S. Department of Labor.

SERVICES AVAILABLE FROM OSHA

In addition to setting standards and inspecting for compliance, OSHA provides services to help employers meet the latest safety and health standards. The services are typically offered at no cost. Three categories of services are available from OSHA: consultation, voluntary protection programs, and training and education services.

Consultation services provided by OSHA include assistance in (1) identifying hazardous conditions, (2) correcting identified hazards, and (3) developing and implementing programs to prevent injuries and illnesses. To arrange consultation services, employers contact the consultation provider in their state (Figure 6–5).

State	Telephone	State	Telephone
Alabama	205-348-3033	Nebraska	401-471-4717
Alaska	907-264-2599	Nevada	701-789-0546
Arizona	602-255-5795	New Hampshire	603-271-3170
Arkansas	501-682-4522	New Jersey	609-984-3517
California	415-557-2870	New Mexico	505-827-2885
Colorado	303-491-6151	New York	518-457-5468
Connecticut	203-566-4550	North Carolina	919-733-3949
Delaware	302-571-3908	North Dakota	701-224-2348
District of Columbia	202-576-6339	Ohio	614-644-2631
Florida	850-488-3044	Oklahoma	405-528-1500
Georgia	404-894-8274	Oregon	503-378-3272
Guam	9-011671-646-9246	Pennsylvania	800-381-1241
Hawaii	808-548-7510		412-357-2561 (toll-free)
Idaho	208-385-3283	Puerto Rico	809-754-2134-2171
Illinois	312-917-2339	Rhode Island	401-277-2438
Indiana	317-232-2688	South Carolina	803-734-9579
Iowa	515-281-5352	South Dakota	605-688-4101
Kansas	913-296-4386	Tennessee	615-741-7036
Kentucky	502-564-6895	Texas	512-458-7254
Louisiana	504-342-9601	Utah	801-530-6868
Maine	207-289-6460	Vermont	801-828-2765
Maryland	310-333-4219	Virginia	804-367-1986
Massachusetts	616-727-3463	Virgin Islands	809-772-1315
Michigan	517-335-8250 (Health)	Washington	206-586-0961
	517-322-1814 (Safety)	West Virginia	304-348-7890
Minnesota	612-297-2393	Wisconsin	608-266-8579 (Health)
Mississippi	601-987-3961		414-512-5063 (Safety)
Missouri	314-751-3403	Wyoming	307-777-7786
Montana	406-444-6401		

FIGURE 6–5 State consultation project directory.

The actual services are provided by professional safety and health consultants, who are not OSHA employees. They typically work for state agencies or universities and provide consultation services on a contract basis; OSHA provides the funding. OSHA Publication 3047, entitled *Consultation Services for the Employer*, may be obtained from the nearest OSHA office.

Voluntary Protection Programs

OSHA's **Voluntary Protection Programs (VPPs)** were created to serve the following basic purposes:

1. To recognize companies that have incorporated safety and health programs into their overall management system
2. To motivate companies to incorporate health and safety programs into their overall management system
3. To promote positive, cooperative relationships among employers, employees, and OSHA

OSHA currently operates three programs under the VPP umbrella. Companies participating in any of the VPPs are exempt from regular programmed OSHA inspections. However, employee complaints, accidents that result in serious injury, or major chemical releases are "handled according to routine enforcement procedures."[13] These programs are discussed in the following paragraphs.

Star Program. The **Star Program** recognizes companies that have incorporated safety and health into their regular management system so successfully that their injury rates are below the national average for their industry. This is OSHA's most strenuous program. To be part of the Star Program, a company must demonstrate the following[14]:

- Management commitment
- Employee participation
- An excellent work-site analysis program
- A hazard prevention and control program
- A comprehensive safety and health training program

Merit Program. The **Merit Program** is less strenuous than the Star Program. It is seen as a stepping stone to recognize companies that have made a good start toward Star Program recognition. OSHA works with such companies to help them take the next step and achieve Star Program recognition.

Demonstration Program. The Department of Labor describes the **Demonstration Program** as follows: "For companies that provide Star-quality worker protection in industries where certain Star requirements can be changed to include these companies as Star participants."[15]

Training and Education Services

Training and education services available from OSHA take several forms. OSHA operates a training institute in Des Plaines, Illinois, that offers a wide variety of services. The institute has a full range of facilities, including classrooms and laboratories, in which it offers more than 60 courses.

To promote training and education in locations other than the institute, OSHA awards grants to nonprofit organizations. Colleges, universities, and other nonprofit organizations apply for funding to cover the costs of providing workshops, seminars, or short courses on safety and health topics currently high on OSHA's list of priorities. Grant funds must be used to plan, develop, and present instruction. Grants are awarded annually and require that employers match at least 20 percent of the total grant amount.

EMPLOYER RIGHTS AND RESPONSIBILITIES

OSHA is very specific in delineating the rights and responsibilities of employers regarding safety and health. These rights and responsibilities, as set forth in OSHA Publication 2056, are summarized in this section.

Employer Rights

The following list of **employer rights** under the OSH Act is adapted from OSHA 2056. Employers have the right to do the following:

- Seek advice and consultation as needed by contacting or visiting the nearest OSHA office.
- Request proper identification of the OSHA compliance officer before an inspection.
- Be advised by the compliance officer of the reason for the inspection.
- Have an opening and closing conference with the compliance officer in conjunction with an inspection.
- Accompany the compliance office on the inspection.
- File a notice of contest with the OSHA area director within 15 working days of receipt of a notice of citation and proposed penalty.
- Apply for a temporary variance from a standard if unable to comply because the materials, equipment, or personnel needed to make necessary changes within the required time are not available.
- Apply for a permanent variance from a standard if able to furnish proof that the facilities or methods of operation provide employee protection at least as effective as that required by the standard.
- Take an active role in developing safety and health standards through participation in OSHA Standards Advisory Committees, nationally recognized standards-setting organizations, and evidence and views presented in writing or at hearings.
- Be assured of the confidentiality of any trade secrets observed by an OSHA compliance officer during an inspection.
- Ask NIOSH for information concerning whether any substance in the workplace has potentially toxic effects.[16]

Employer Responsibilities

In addition to the rights set forth in the previous subsection, employers have prescribed responsibilities.[17] The following list of **employer responsibilities** under the OSH Act is adapted from OSHA 2056. Employers must do the following:

- Meet the general duty responsibility to provide a workplace free from hazards that are causing or are likely to cause death or serious physical harm to employees and to comply with standards, rules, and regulations issued under the OSH Act.
- Know mandatory standards and make copies available to employees for review upon request.
- Keep employees informed about OSHA.
- Continually examine workplace conditions to ensure that they conform to standards.
- Minimize or reduce hazards.
- Make sure employees have and use safe tools and equipment (including appropriate personal protective equipment [PPE]) that is properly maintained.
- Use color codes, posters, labels, or signs as appropriate to warn employees of potential hazards.
- Establish or update operating procedures and communicate them so that employees follow safety and health requirements.
- Provide medical examinations when required by OSHA standards.
- Provide the training required by OSHA standards.
- Report any fatal accident or one that results in the hospitalization of five or more employees to the nearest OSHA office within 48 hours.

- Keep OSHA-required records of injuries and illnesses, and post a copy of the totals as required.
- At a prominent location within the workplace, post OSHA Poster 2203, which informs employees of their rights and responsibilities.
- Provide employees, former employees, and their representatives access to the Log and Summary of Work-Related Injuries and Illnesses.
- Give employees access to medical and hazard-exposure records.
- Give the OSHA compliance officer the names of authorized employee representatives who may be asked to accompany the compliance officer during an inspection.
- Do not discriminate against employees who properly exercise their rights under the act.
- Post OSHA citations at or near the work site involved. Each citation or copy must remain posted until the violation has been abated or for three working days, whichever is longer.
- Abate cited violations within the prescribed period.

EMPLOYEE RIGHTS AND RESPONSIBILITIES

Employee Rights

Section 11(c) of the OSH Act delineates **employee rights**. These rights are actually protection against punishment for employees who exercise their right to pursue any of the following courses of action:

- Complaint to an employer, union, OSHA, or any other government agency about job safety and health hazards
- Filing safety or health grievances
- Participation in a workplace safety and health committee or in union activities concerning job safety and health
- Participation in OSHA inspections, conferences, hearings, or related activities[18]

Employees who feel they are being treated unfairly because of actions they have taken in the interest of safety and health have 30 days in which to contact the nearest OSHA office. Upon receipt of a complaint, OSHA conducts an investigation and makes recommendations based on its findings. If an employer refuses to comply, OSHA is empowered to pursue legal remedies at no cost to the employee who filed the original complaint.

In addition to those just set forth, employees have a number of other rights. Employees must be allowed to

- review copies of OSHA standards and requirements,
- receive information from employers about hazards that may be present in the workplace,
- ask employers for information on emergency procedures,
- receive safety and health training,
- receive updated information about safety and health issues,
- anonymously ask OSHA to conduct an investigation of hazardous conditions at the work site,
- know of actions taken by OSHA as a result of a complaint,
- observe during an OSHA inspection and respond to the questions asked by a compliance officer,
- see records of hazardous materials in the workplace,
- see their medical records,
- review the annual Log and Summary of Work-Related Injuries,
- have an exit briefing with the OSHA compliance officer after an OSHA inspection,
- anonymously ask NIOSH to provide information about toxicity levels of substances used in the workplace,
- challenge the abatement period given to employers to correct hazards discovered in an OSHA inspection,
- participate in hearings conducted by the OSHRC,
- know when an employer requests a variance to a citation or any OSHA standard,
- testify at variance hearings,
- appeal decisions handed down at OSHA variance hearings, and
- give OSHA input concerning the development, implementation, modification, or revocation of standards.[19]

Employee Responsibilities

Employees have a number of specific responsibilities. The following list of **employee responsibilities** is adapted from OSHA 2056. Employees must

- read the OSHA poster at the job site and be familiar with its contents,
- comply with all applicable OSHA standards,
- follow safety and health rules and regulations prescribed by the employer and use required PPE while engaged in work,
- report hazardous conditions to the supervisor,
- report any job-related injury or illness to the employer and seek treatment promptly,
- cooperate with the OSHA compliance officer who is conducting an inspection, and
- exercise their rights under the OSH Act in a responsible manner.[20]

KEEPING UP-TO-DATE ON OSHA STANDARDS

OSHA's standards, rules, and regulations are always subject to change. The development, modification, and revocation of standards is an ongoing process. It is important for prospective and practicing construction professionals to stay up to date with the latest actions and activities of OSHA. The following is an annotated list of strategies that can be used to keep current:

- Establish contact with the nearest regional or area OSHA office, and periodically request copies of new publications or contact the OSHA Publications Office at the following address:

  OSHA Publications Office
  200 Constitution Avenue, N.W.
  Room N-3101
  Washington, D.C. 20210
  http://www.osha.gov

- Review professional literature in the safety and health field. Numerous periodicals carry OSHA updates that are helpful.
- Establish and maintain relationships with safety and health professionals for the purpose of sharing information, and do so frequently.
- Join professional organizations, review their literature, and attend their conferences.

PROBLEMS WITH OSHA

Federal agencies are seldom without their detractors. Resentment of the federal bureaucracy is intrinsic in the American mind-set. Consequently, complaints about OSHA are common. Even supporters occasionally join the ranks of the critics. Often, the criticisms leveled against OSHA are valid.

Criticisms of OSHA take many different forms. Some characterize OSHA as an overbearing bureaucracy with little or no sensitivity to the needs of employers, who are struggling to survive in a competitive marketplace. At the same time, others label OSHA as timid and claim that it does too little. At different times and in different cases, both points of view have probably been at least partially accurate.

Most criticism of OSHA comes in the aftermath of major accidents or a workplace disaster. Such events typically attract a great deal of media attention, which, in turn, draws the attention of politicians. In such cases, the criticism tends to focus on the question, "Why didn't OSHA prevent this disaster?" At congressional hearings, detractors typically answer this question by claiming that OSHA spends too much time and too many resources dealing with matters of little consequence while ignoring real problems. Supporters of OSHA typically answer the question by claiming that a lack of resources prevents the agency from being everywhere at once. There is a measure of validity in both answers.

There is evidence that OSHA has made a positive difference since the inception of the OSH Act. In the first 20 years of OSHA's experience, occupational fatalities dropped by 25 percent—from 12,000 to 9,000 annually.

Statistics show that OSHA has made a difference in the condition of the workplace in this country. On the other hand, large, centralized bureaucratic agencies rarely achieve a high level of efficiency; this is compounded in OSHA's case by the fact that the agency is subject to the ebb and flow of congressional support, particularly in the area of funding. Consequently, OSHA is likely to continue to be an imperfect organization that is subject to ongoing criticism.

OTHER FEDERAL AGENCIES AND ORGANIZATIONS

Although OSHA is the most widely known safety and health organization in the federal government, it is not the only one. Figure 6–6 lists government agencies and organizations (including OSHA) with safety and health as part of their mission. Of those listed, the most important to modern safety and health professionals are NIOSH and OSHRC.

NIOSH

NIOSH is part of the Department of Health and Human Services (whereas OSHA is part of the Department of Labor). NIOSH has two broad functions: research and education. The main focus of the agency's research is on toxicity levels and human tolerance levels of hazardous substances. NIOSH prepares recommendations along these lines for OSHA standards dealing with hazardous substances. NIOSH studies are also published and made available to employers. Every year, NIOSH publishes updated lists of toxic materials and recommended tolerance levels. These publications represent the educational component of NIOSH's mission. The Department of Health and Human Services describes NIOSH as follows:

> In 1973, NIOSH became a part of the Centers for Disease Control (CDC), an arm of the Public Health Service in the Department of Health and Human Services. NIOSH is unique among federal research institutions because it has the authority to conduct research in the workplace, and to respond to requests for assistance from employers and employees.[21]

NIOSH also consults with the Department of Labor and other federal, state, and local government agencies to promote occupational safety and health and makes recommendations to the department about worker exposure limits.

American Public Health Association
800 I Street
Washington, D.C. 20005

Bureau of Labor Statistics
U.S. Department of Labor 2 Massachusetts Ave NE
Washington, D.C. 20212

Bureau of National Affairs, Inc.
Occupational Safety and Health Reporter
1231 25th Street, N.W.
Washington, D.C. 20037

Commerce Clearing House
Employee Safety and Health Guide
4205 W. Peterson Avenue
Chicago, IL 60646

Mine Safety and Health Administration
201 12th Street S
Arlington, VA 22202

Environmental Protection Agency
1200 Pennsylvania Ave NW
Washington, D.C. 20460

National Institute for Occupational Safety and Health
395 E Pearl Street
Patriots Plaza - Suite 9200
Washington, D.C. 20201

Occupational Safety and Health Administration
U.S. Department of Labor
200 Constitution Avenue
Washington, D.C. 20210

Occupational Safety and Health Review Committee
200 Constitution Avenue
Washington, D.C. 20210

U.S. Consumer Product Safety Commission
4330 East West Highway
Bethesda, MD 20814

FIGURE 6–6 Agencies that deal with safety and health.

In addition to workplace death and injuries, it is estimated that more than 10 million men and women are exposed to hazardous substances on their jobs that can eventually cause fatal or debilitating diseases. To help establish priorities in developing research and control of these hazards, NIOSH developed a list of the 10 leading work-related diseases and injuries:

1. Occupational lung disease
2. Musculoskeletal injuries
3. Occupational cancers
4. Occupational cardiovascular disease
5. Severe occupational traumatic injuries
6. Disorders of reproduction
7. Neurotoxic disorders
8. Noise-induced hearing loss
9. Dermatological problems
10. Psychological disorders[22]

OSHRC

OSHRC is not a government agency. Rather, it is an independent board whose members are appointed by the president and given quasi-judicial authority to handle contested OSHA citations. When a citation, proposed penalty, or abatement period issued by an OSHA area director is contested by an employer, OSHRC hears the case. OSHRC is empowered to review the evidence as well as approve or reject the recommendations of the OSHA area director or revise them by assigning substitute values. For example, if an employer contests the amount of a proposed penalty, OSHRC is empowered to accept the proposed amount, reject it completely, or change it.

Mining Enforcement and Safety Administration (MESA)

The mining industry is exempt from OSHA regulations. Instead, mining is regulated by the Metal and Non-Metallic Mine Safety Act. OSHA does regulate those aspects of the industry that are not directly involved in actual mining work. There is a formal memorandum of understanding between OSHA and MESA—the agency that enforces the Metal and Non-Metallic Mine Safety Act.

In 1977, Congress passed the Mine Safety and Health Amendment, which established the Mine Safety and Health Administration (MSHA) as a functional unit within the U.S. Department of Labor. MSHA works with MESA to ensure that the two agencies do not become embroiled in jurisdictional disputes.

Federal Railroad Administration

Railroads, for the most part, fall under the jurisdiction of OSHA. The Federal Railroad Administration (FRA) exercises limited jurisdiction over railroads in situations involving working conditions. Beyond this, railroads

SAFETY FACTS & FINES

A high injury and illness rate can lead to more careful scrutiny by OSHA. A company in Claremont, New Hampshire, was selected for inspection under OSHA's site-specific targeting plan because of its poor safety and health record. By the time OSHA inspectors completed their work, the company had been cited for six willful and eight serious violations. The fine assessed was $301,050. The willful violations included (1) failure to develop and use hazardous energy control and (2) failure to properly use guards on powered machines. The serious violations included (1) failure to develop and implement an emergency response plan and (2) failure to ensure that employees use personal protective clothing when appropriate.

must adhere to the standards for General Industry in CFR Part 1910. OSHA and FRA personnel coordinate to ensure that jurisdictional disputes do not arise.

STANDARDS AND CODES

A **standard** is an operational principle, criterion, or requirement—or a combination of these. A **code** is a set of standards, rules, or regulations related to a specific area. Standards and codes play an important role in modern safety and health management and engineering. These written procedures detail the safe and healthy way to perform job tasks and, consequently, to make the workplace safer and healthier.

Having written standards and codes that employees carefully follow can also decrease a company's exposure to costly litigation. Courts tend to hand down harsher rulings to companies that fail to develop or adapt, implement, and enforce appropriate standards and codes. Consequently, construction professionals should be familiar with the standards and codes of their company.

Numerous organizations develop standards for different industries. These organizations can be categorized broadly as government and professional organizations and technical and trade associations.

Organizations that fall within these broad categories develop standards and codes in a wide variety of areas, including—but not limited to—the following: dust hazards, electricity, emergency electricity systems, fire protection, first aid, hazardous chemicals, instrumentation, insulation, lighting, lubrication, materials, noise or vibration, paint, power, wiring, pressure relief, product storage and handling, piping materials, piping systems, radiation exposure, safety equipment, shutdown systems, and ventilation.

LAWS AND LIABILITIES

The body of law pertaining to workplace safety and health grows continually as a result of a steady stream of liability litigation. Often a company's designated "competent person" (the person assigned responsibility for safety and health) is a key player in litigation alleging negligence on the part of the company when an accident or health problem has occurred. Because health and safety litigation has become so prevalent today, construction professionals need to be familiar with certain fundamental legal principles related to such litigation. These principles are explained in the following section.

Applicable Legal Principles

The body of law that governs safety and health litigation evolves continually. However, even though cases that set new precedents and establish new principles continue to occur, a number of fundamental legal principles surface frequently in the courts. The most important of these, along with several related legal terms, are summarized in the following paragraphs.

Negligence. **Negligence** means failure to take reasonable care or failure to perform duties in a way that prevents harm to humans and damage to property. The concept of *gross negligence* means failure to exercise even slight care or intentional failure to perform duties properly, regardless of the potential consequences. *Contributory negligence* means that an injured party contributed in some way to his or her own injury. In the past, this concept was used to protect defendants against negligence charges because the courts awarded no damages to plaintiffs who had contributed in any way to their own injury. Modern court cases have rendered this approach outdated with the introduction of *comparative negligence*. This concept distributes the negligence assigned to each party involved in litigation according to the findings of the court.

Liability. **Liability** is a duty to compensate as a result of being held responsible for an act or omission. A newer, related concept is *strict liability*. This means that a company is liable for damages caused by a product that it produces, regardless of negligence or fault.

Care. Several related concepts fall under the heading of care. *Reasonable care* is the care that would be taken by a prudent person in exercising his or her legal obligations toward others. *Great care* means that amount of care that would be taken by an extraordinarily prudent person in exercising his or her legal obligations toward others. *Slight care* represents the other extreme: a measure of care that is less than what a prudent person would take. A final concept in this category is the *exercise of due care*. This means that all people have a legal obligation to exercise the amount of care necessary to avoid, to the extent possible, bringing harm to others and damage to their property.

Ability to Pay. The concept of **ability to pay** applies when there are a number of defendants in a case, but not all have the ability to pay financial damages. It allows the court to assess all damages against the defendant or defendants who have the ability to pay. For this reason, it is sometimes referred to as the "deep pockets" principle.

Damages. **Damages** are financial awards assigned to injured parties in a lawsuit. *Compensatory damages* are awarded to compensate for injuries suffered and for those that will be suffered. *Punitive damages* are awarded to ensure that a guilty party is disinclined to engage in negligent behavior in the future.

Proximate Cause. **Proximate cause** is the cause of an injury or damage to property. It is that action or lack of action that ties one person's injuries to another's lack of reasonable care.

Willful or Reckless Conduct. Behavior that is even worse than gross negligence is **willful or reckless conduct**. It involves intentionally neglecting one's responsibilities to exercise reasonable care.

Tort. A **tort** is an action involving a failure to exercise reasonable care that may, as a result, lead to civil litigation.

Foreseeability. The concept of **foreseeability** holds that a person can be held liable for actions that result in damages or injury only when risks could have been reasonably foreseen.

The types of questions around which safety and health litigation often revolve are these: Does the company keep employees informed of rules and regulations? Does the company enforce its rules and regulations? Does the company provide its employees with the necessary training? The concepts set forth in this section come into play as both sides in the litigation deal with these questions from their respective points of view.

Construction professionals can serve their companies best by: (1) making sure that a policy and corresponding rules and regulations are in place, (2) keeping employees informed about rules and regulations, (3) encouraging proper enforcement practices, and (4) ensuring that employees get the education and training they need to perform their jobs safely.

OSHA'S CONSTRUCTION STANDARDS

OSHA standards apply to employers involved in construction, alteration, and repair activities. To further identify the scope of the applicability of its construction standards, OSHA took the terms *construction*, *alteration*, and *repair* directly from the text of the Davis–Bacon Act. This act provides minimum wage protection for employees working on construction projects. The implication is that if the Davis–Bacon Act applies to an employer, OSHA construction standards also apply.

These standards are contained in CFR Part 1926 Subparts A–Z. OSHA does not base citations on material contained in Subparts A and B. Consequently, those subparts have no relevance here. The remaining subpart subjects are as follows:

Subpart C	General safety and health provisions
Subpart D	Occupational health and environmental controls
Subpart E	Personal protective and lifesaving equipment
Subpart F	Fire protection and prevention
Subpart G	Signs, signals, and barricades
Subpart H	Materials handling, storage, use, and disposal
Subpart I	Tools—hand and power
Subpart J	Welding and cutting
Subpart K	Electrical
Subpart L	Scaffolding
Subpart M	Fall protection
Subpart N	Hoists, elevators, and conveyors
Subpart O	Motor vehicles, mechanized equipment, and marine operations
Subpart P	Excavations
Subpart Q	Concrete and masonry construction
Subpart R	Steel erection
Subpart S	Tunnels and shafts, caissons, cofferdams, and compressed air
Subpart T	Demolition
Subpart U	Blasting and use of explosives
Subpart V	Power transmission and distribution
Subpart W	Rollover protective structures; overhead protection
Subpart X	Stairways and ladders
Subpart Y	Commercial diving operations
Subpart Z	Toxic and hazardous substances
Subpart AA	Confined spaces in construction
Subpart CC	Crane and Derricks in Construction

Summary

The impetus for passing the OSH Act was that workplace accidents were causing an average of 14,000 deaths every year in the United States. Every year, 2.5 million workers were disabled in workplace accidents, and approximately 300,000 new cases of **occupational diseases** were reported annually.

The mission of OSHA is to ensure, to the extent possible, that every working person in the United States has a safe and healthy working environment so that valuable human resources are preserved and protected. The Department of Labor breaks down this mission statement further into the following specific purposes: (1) encourage employers and employees to reduce workplace hazards, (2) implement new safety and health programs, (3) improve existing safety and health programs, (4) encourage research that leads to innovative ways of dealing with workplace safety and health problems, (5) establish the rights of employers and employees regarding the improvement of workplace safety and health, (6) monitor job-related illnesses and injuries through a system of reporting and record keeping, (7) establish training programs to increase the number of safety and health professionals and to continually improve their competence, (8) establish and enforce mandatory workplace safety and health standards, (9) provide for the development and approval of state-level workplace safety and health programs, and (10) monitor, analyze, and evaluate these state-level programs.

The OSH Act covers all employers and all 50 states, the District of Columbia, Puerto Rico, and all other territories that fall under the jurisdiction of the U.S. government, with the following exceptions: (1) persons who are self-employed, (2) family farms that employ only immediate members of the family, (3) federal agencies covered by other federal statutes, and (4) state and local governments.

OSHA develops standards on the basis of its perception of need and at the request of other federal agencies, state and local governments, other standards-setting agencies, labor organizations, or even individual private citizens. OSHA uses the committee approach for developing standards. OSHA's standing committees are the National Advisory Committee on Occupational Safety and Health (NACOSH) and the Advisory Committee on Construction Safety and Health.

OSHA can take three different types of action on standards: a standard may be adopted, amended, or revoked. Before any of these actions can be undertaken, OSHA must publish its intentions in the *Federal Register*. OSHA has two options for meeting this requirement: a notice of proposed rule making and an advance notice of proposed rule making.

Once the standard has been passed, it becomes effective on the date prescribed. However, any person who is opposed to a standard may file an appeal in the court of appeals that serves the geographic region in which the complainant lives or does business. Appeal paperwork must be initiated within 60 days of a standard's approval.

When an employer is unable to comply with a new standard immediately, but may be able to if given time, a temporary variance may be requested. OSHA grants such a variance for a maximum of one year. Employers must demonstrate that they are making a concerted effort to comply and must take the steps necessary to protect employees while working toward compliance.

Employers who feel that their workplace already exceeds the requirements of a new standard may request a permanent variance and must present their evidence to OSHA for inspection. Employees must be informed of the application for a variance and notified of their right to request a hearing.

OSHA provides for the centralization and systematization of record-keeping and reporting requirements of the OSH Act to employers of 11 or more workers. Both exempt and nonexempt employers must report the following types of accidents within 48 hours: (1) those that result in deaths and (2) those that result in the hospitalization of five or more employees.

All occupational illnesses and injuries must be reported if they result in one or more of the following: (1) death to one or more workers, (2) one or more days away from work for the employee, (3) restricted motion or restrictions to the work an employee can do, (4) loss of consciousness to one or more workers, (5) transfer of an employee to another job, and (6) medical treatment needed beyond in-house first aid or if they appear in Appendix B of the OSH Act.

All records required by OSHA can be maintained by using the following forms: **OSHA Form 300, OSHA Form 300A,** and **OSHA Form 301**.

Employers are required to post the following material at locations where employee information is normally displayed: OSHA Poster 2203, summaries of variance requests of all types, copies of all OSHA citations received for failure to meet standards, and the summary page of OSHA Form 300A.

OSHA compliance officers are authorized to take the following actions with regard to workplace inspections: (1) enter at reasonable times any site, location, or facility in which work is taking place; (2) inspect at reasonable times any condition, facility, machine, equipment, or materials; and (3) question in private any employee or other person formally associated with the company.

OSHA is empowered to issue citations and set penalties. Citations are for: (1) other-than-serious violations, (2) willful violations, (3) repeat violations, and (4) failure to correct previous violations.

Employees may appeal the following aspects of OSHA's decisions regarding their workplace: (1) the amount of time (abatement period) given an employer to correct a hazardous condition that has been cited and (2) an employer's request for an extension of an abatement period.

Employers may petition for modification of abatement or contest a citation, an abatement period, or a penalty.

States are allowed to develop their own safety and health programs. As an incentive, OSHA funds up to 50 percent of the cost of operating a state program for states with approved plans. States may develop comprehensive plans covering public and private sector employers or limit their plans to coverage of public employers only.

In addition to setting standards and inspecting for compliance, OSHA provides services to help employers meet the latest safety and health standards. Services are typically offered at no cost and are intended for smaller companies, particularly those with especially hazardous processes or materials. Services available from OSHA include consultation, voluntary inspection programs, and training and education.

OSHA is not without its detractors. Criticisms of OSHA take many forms, depending on the perspective of the critic. Some characterize OSHA as an overbearing bureaucracy with little or no sensitivity to the needs of employers who are struggling to survive in a competitive marketplace. Others label OSHA as timid and claim it does not do enough. At different times and different places, both points of view have probably been at least partially accurate. Other federal agencies and organizations that play important roles with regard to workplace safety and health are the National Institute for Occupational Safety and Health (NIOSH), which is part of the Department of Health and Human Services, and the Occupational Safety and Health Review Commission (OSHRC), which is an independent board consisting of members appointed by the president and given quasi-judicial authority to handle contested OSHA citations.

A standard is an operational principle, criterion, or requirement—or a combination of these. A code is a set of standards, rules, or regulations relating to a specific area. Standards and codes play an important role in modern safety and health. They provide practices for performing jobs safely, which, in turn, makes for a safer and healthier workplace.

Fundamental legal principles with which safety and health professionals should be familiar are negligence, liability, care, ability to pay, damages, proximate cause, willful or reckless conduct, tort, and foreseeability.

Key Terms and Concepts

Abatement period
Ability to pay
Appeals process
Closing conference
Code
Consultation services
Damages
Demonstration Program
Employee responsibilities
Employee rights
Employer appeals
Employer responsibilities
Employer rights
Foreseeability
Inspection tour
Liability
Merit Program
National Institute for Occupational Safety and Health (NIOSH)
Negligence
Notice of contest
Notice of proposed rule making
Occupational disease
Occupational Safety and Health Administration (OSHA)
Occupational Safety and Health Act (OSH Act)
Occupational Safety and Health Review Commission (OSHRC)
Opening conference
OSHA Form 300
OSHA Form 300A
OSHA Form 301
Other-than-serious violation
Permanent variance
Proposed penalty
Proximate cause
Record keeping
Repeat violation
Reporting
Standard
Star Program
State-level OSHA program
Temporary emergency standard
Temporary variance
Tort
Voluntary Protection Programs (VPPs)
Willful or reckless conduct
Willful violation
Workplace accident
Workplace inspection

Review Questions

1. Briefly explain the rationale for the OSH Act.
2. What is OSHA's mission or purpose?
3. List those persons/businesses that are exempted from coverage by OSHA.
4. Explain the difference between an OSHA standard and an OSHA regulation.
5. Explain how the following processes relating to OSHA standards are accomplished: (1) passage of a new standard, (2) request for a temporary variance, and (3) appeal regarding a standard.
6. Briefly describe OSHA's latest record-keeping requirements.
7. What are OSHA's reporting requirements?
8. Explain what employers are required to do to keep employees informed.
9. Describe how a hypothetical OSHA workplace inspection would proceed from the first step to the last.
10. List and explain the three different types of OSHA citations and the typical penalties that accompany them.
11. Describe the process for appealing an OSHA citation.
12. List and briefly explain OSHA's Voluntary Protection Programs.
13. List the five employer responsibilities.
14. List the five employee rights.
15. Describe the purpose and organization of NIOSH.
16. Define the following legal terms as they relate to workplace safety: *negligence, liability, ability to pay, and tort.*

Critical Thinking and Discussion Activities

1. Following an order from his supervisor, a carpenter working for Jones Construction Company removed all the guardrails from stairwell openings, floor openings, and open-sided stairs. Uncomfortable with this action, the carpenter telephoned his union representative, who, in turn, called OSHA.

Jones Construction Company was fined $15,000 for willful violations. Jones Construction Company, in turn, fired the carpenter. Was the firing lawful (under OSHA standards)? Was the action of the carpenter appropriate (under OSHA standards)?

2. "OSHA is a nightmare! It's nothing more than another federal department of government bureaucrats who bully private industry." This was the opening line in a debate on government safety regulations sponsored by the University of Eastern Florida. The OSHA advocate in the debate responded as follows: "OSHA is a model agency. Had private companies been responsive to the safety and health needs of employees, there would have been no need for government regulations in the first place." Join this debate. What is your opinion?

Application Activities

1. Find a local company that will work with you on this activity. Identify the person responsible for safety and health. Interview this person about OSHA's record-keeping and reporting requirements. Is the individual familiar with the requirements? What does this individual think about the requirements (good or bad)?
2. Contact the state consultation office for your state and two other states. Compare the services available to employers. Are they the same? Does one state office provide services that another does not offer? If so, what are those services?
3. Go to OSHA's website. Make a list of all the various types of information that are available at this site. Find OSHA's Construction Standard (29 CFR 1926), and print the contents page.

Endnotes

1. U.S. Department of Labor, *All About OSHA*, 2015 (revised), http://www.osha.gov.
2. Ibid.
3. Ibid., 2.
4. Ibid., 5.
5. Ibid., 7.
6. Ibid., 9.
7. Ibid., 12.
8. Ibid., 17.
9. Ibid., 24–25.
10. Ibid., 27.
11. Ibid., 28.
12. Ibid.
13. Ibid.
14. Ibid., 29.
15. Ibid., 32.
16. Ibid., 34–35.
17. Ibid., 35–36.
18. Ibid., 37.
19. Ibid., 39–40.
20. Ibid., 37.
21. U.S. Department of Health and Human Services, *Occupational Safety & Health under Public Law 91–596*, Public Health Service, Centers for Disease Control, National Institute for Occupational Safety and Health, 2001, Washington, D.C., 11.
22. Ibid., 12.

PART **TWO**

APPLICATION ON THE JOB

CHAPTER **SEVEN**

CONSTRUCTION SAFETY AND HEALTH: PROGRAMS, PLANS, AND POLICIES

LEARNING OBJECTIVES

- Explain the rationale for written safety and health plans.
- List the various components of a safety and health plan.
- Describe how to communicate the safety and health plan to employees.
- Summarize how to evaluate a safety and health program.

There is disagreement among construction professionals concerning the value of written safety plans. "Time spent writing safety plans could be put to better use at the job site." This is a frequently heard criticism in the profession. When the topic of discussion is written safety plans, such terms as *red tape*, *bureaucracy*, *paperwork*, and *counterproductive* often find their way into the conversation. However, there is hard factual evidence to support the opposite point of view. This evidence makes it clear that well-written safety plans can prevent accidents, injuries, illnesses, and associated expenses that can drain a company's productivity as well as its profits.

RATIONALE FOR WRITTEN SAFETY AND HEALTH PLANS

The Occupational Safety and Health Administration (OSHA) is very specific about requiring plans as well as what must be included in them. At a minimum, safety and health plans must include the following:

- Organizational structure
- Comprehensive plan
- Site-specific plan
- Training
- Medical surveillance
- Standard operating procedures
- Interfacing between the general plan and site-specific plan

There are several other reasons why construction companies should devote the resources necessary to develop and maintain comprehensive written safety and health plans. These reasons range from the philosophical to the practical, and all fall into one of the following broad categories.

Ethical Factors

Ethical factors are an important part of the rationale for maintaining a safe and healthy work environment. Philosophically speaking, the most important rationale for having a comprehensive written safety and health plan is because it is the right thing to do. Construction companies have an ethical obligation to provide the safest, healthiest work environment possible for their employees and subcontractors. One of the best ways to ensure a quality work environment is to have a comprehensive written plan. This can be seen in the fact that companies without such a plan experience 30 percent more accidents than those with plans.

Regulatory Factors

Regulatory factors are an important part of the rationale for maintaining a safe and healthy work environment. There are federal and state regulations that require construction companies to maintain a safe and healthy work environment. These regulations do not necessarily require a comprehensive written plan, although some federal regulations do require written plans that focus on specific issues, including the following:

- Fall protection plan (29 CFR 1926.500)
- Emergency action plan (29 CFR 1926.35)
- Blood-borne pathogens and exposure control plan (29 CFR 1910.1030)

Although a company may comply by developing only the written plans that are specifically called for regulations, a more comprehensive approach is better because comprehensive written plans do more than comply—they help to establish and maintain a high-quality work environment. The recommended approach is to develop a comprehensive master plan that includes plans related to specific terms as components.

Economic Factors

Economic factors are an important part of the rationale for maintaining a safe and healthy work environment. An unsafe, unhealthy work environment is a costly environment. Accidents and injuries are directly and indirectly expensive. The undeniable fact is that it costs less to maintain a high-quality work environment than to pay for accidents and injuries. A good safety and health plan that is effectively implemented can save money in a number of different ways, including the following:

- Holding down insurance costs
- Reducing costly litigation
- Reducing temporary and permanent disability claims
- Increasing the productivity of employees
- Leading to more contracts (good reputation)
- Reducing the number of compliance inspections and associated penalties

Practical Factors

The practical reason for having a comprehensive written plan is that effectively implementing such a plan is the best way to establish a high-quality work environment. The practical reasons for this include the following:

- The plan forces construction companies to put their commitment to safety and health in writing.
- The plan forces construction companies to establish policies and set goals for safety and health.
- The plan commits to writing the procedures that must be followed by all employees and subcontractors.
- The plan is an effective way to communicate policies, procedures, and goals related to the quality of the work environment.

COMPONENTS OF THE PLAN

To be effective, a construction safety and health plan must be written, and it must be comprehensive. There are no regulatory requirements governing the specific contents of a construction safety and health plan. However, if one accepts the full counsel of federal and state regulations, a plan should have at least the following components:

- Safety and health policy
- Safety and health goals
- Roles and responsibilities
- Discipline policy and procedures
- Job-site inspections
- Accident investigations
- Record keeping
- Training
- Medical response and first aid
- Emergency response
- Miscellaneous components (e.g., fall protection, blood-borne pathogens)

A written plan with these components is a comprehensive plan with all of the company's various safety and health plans contained in one document.

Safety and Health Policy

Management commitment to safe and healthy job sites is critical. Evidence of management commitment is a written safety and health plan, and the most important component of such a plan is the safety policy. In addition to the policy, a set of specific goals that translate the policy into measurable actions should be included.

A safety policy is a statement of management's commitment to a high-quality work environment. To emphasize the commitment, it should be signed by the company's chief executive officer. A well-written safety policy communicates at least the following messages (Figure 7–1):

- The company is committed to providing a safe and healthy work environment, and such an environment is a high priority.
- All employees are expected to work in a safe and healthy manner.
- Safety and health rules and regulations are enforced.

Safety and Health Goals

The safety and health policy describes management's commitment in general terms. **Safety and health goals** then take the next step and describe the commitment in more specific and measurable terms. Like any goals set by a company, the safety and health goals should be lofty enough to challenge the organization, but realistic enough to be credible. Goals set too low are viewed as "low-balled" goals that are not worthy of attention or effort. Goals set too high are viewed as "pie-in-the sky" goals that cannot possibly be attained and, therefore, have no credibility. Figure 7–2 contains examples of the types of goals that construction companies might include in their safety and health plans. Each of these goals is explained in the paragraphs that follow.

Safety and Health Policy:
Jones Construction Company

Jones Construction Company (JCC) is committed to providing a safe and healthy work environment for all employees and subcontractors. Consequently, no other factor is to take priority over safety and health on JCC job sites. All employees at all levels and all subcontractors are expected to work safely themselves and to encourage their fellow workers to do so. All are expected to comply with applicable safety rules and regulations and to speak out when others break them.

JCC provides employees with the knowledge (training) and equipment (such as hard hats, safety glasses) needed to work safely. Employees, in turn, are expected to apply their training and use their equipment to do their jobs safely. All JCC employees should "think safely" at all times, and when in doubt, err on the side of safety.

____
Chief Executive Officer

FIGURE 7–1 Construction companies should develop and adopt a **safety and health policy**.

Safety and Health Goals: Jones Construction Company
Zero fatal accidents during the year
Reduce job-related injuries by 25 percent
Reduce lost time due to accidents by 40 percent
Reduce workers' compensation claims by 30 percent
Reduce the cost of damage to JCC property and equipment by 20 percent
Reduce "near-miss" accidents by 25 percent
Increase participation of employees in safety training to 100 percent

FIGURE 7–2 Sample safety and health goals for a construction company.

It is necessary to have a goal related to fatalities because no company wants to have a fatal accident; yet such accidents are always a possibility. Since this is the case, the only acceptable goal related to fatalities is "zero." Few construction executives are going to publish a safety and health plan that implies acceptance of even one fatal accident. Reduction of job-related injuries should also be considered a mandatory goal. Few construction companies are going to maintain a perfect safety record over time. As long as there are people working together to meet deadlines, there will be accidents and injuries. The question then becomes "how many?" The answer has got to be "as few as possible." The percent of reduction that accompanies this goal depends on the company's current safety record and how much it can be realistically improved if the company is appropriately challenged.

This same rule of thumb applies to goals related to lost time for accidents, workers' compensation claims, and near-miss accidents. With all of these goals, it is important to avoid both the "low-ball" and the "pie-in-the-sky" syndromes. Striking a realistic balance between being sufficiently challenging on the one hand and fully credible on the other is the key. If, for example, the company had 100 workers' compensation claims last year, reducing this number by 75 percent might be unrealistic. On the other hand, reducing it by 40 percent would be challenging and might be sufficiently realistic.

Goals related to **training** are similar to those related to fatal accidents. If a company is going to set a goal in this area, it must be set high. Otherwise, the wrong message is sent. A 100 percent participation rate is required; to set the goal lower means that the company accepts the fact that employees might be working in a potentially hazardous situation without the proper training.

Responsibility and Roles

Everyone has a role to play in maintaining a safe work environment. This part of the comprehensive safety and health plan sets forth the responsibilities of managers, supervisors, safety and health professionals, and employees. It also explains how employees at all levels are held accountable for carrying out their responsibilities and what happens when they fail to do so.

Management's Responsibilities. Management's **responsibilities** include the following:

- Establishing and maintaining the company's commitment to safety and health
- Developing safety and health policies
- Setting goals
- Providing the necessary resources
- Setting a positive example
- Organizing, directing, controlling, evaluating, and revising the overall safety and health program

Supervisors' Responsibilities. Supervisors' **responsibilities** must be considered when planning for a safe and healthy work environment. Supervisors play a key role in maintaining a quality work environment. They represent the level of management closest to employees and subcontractors on a day-to-day basis. Consequently, supervisors carry the primary responsibility for ensuring that employees do the following:

- Comply with all applicable rules and regulations
- Put safety ahead of other factors (without using it as an excuse for not getting the job done)
- Attend periodic safety meetings and discussions
- Participate in safety training
- Properly use personal protective equipment (PPE)
- Promptly report accidents and near misses
- Do the necessary "housekeeping" chores every day
- Maintain a positive attitude toward safety
- Speak out when other employees are engaging in unsafe practices

Employees' Responsibilities. Employees may play the most important role of maintaining a quality work environment. This role involves working safely themselves and insisting that their fellow employees do the same. Specifically, **employees' responsibilities** include the following:

- Complying with all applicable safety rules and regulations
- Following instructions from managers, supervisors, and safety professionals
- Asking questions to clarify when in doubt about a course of action
- Calling hazardous conditions to the attention of the supervisor or safety professional
- Reporting all accidents and near misses to the supervisor or safety professional
- Reporting damage to company property and equipment to the supervisor
- Performing all necessary housekeeping chores promptly and regularly
- Having a positive attitude toward safety

SAFETY FACTS & FINES

Written procedures that explain the safe and healthy way to perform a dangerous task are important. Failure to provide written work procedures can have disastrous results. A company in Moss Point, Mississippi, was fined $118,500 when a worker was killed by the accidental release of toxic gas during the introduction of a new process. The company was cited for the following violations: (1) exposure of employees to toxic fumes and (2) lack of written procedures.

Safety Professionals' Responsibilities. Many medium-sized and larger construction companies employ one or more safety and health professionals and assign them responsibility for directing, coordinating, and facilitating the company's overall safety and health program. **Safety professionals' responsibilities** include the following:

- Safety audits
- Job-site inspections
- Hazard analysis
- Accident investigations
- Record keeping
- Reporting
- Training

Discipline and Accountability

Accountability is a critical component in a company's safety and health program. It is not enough to simply have rules and explain them. There must be consequences for failure to follow the rules and regulations. Consequences are set forth in the **discipline policy** section of the comprehensive safety and health plan. A plan without a discipline component is a plan without "teeth."

The discipline policy should explain that all employees at all levels are expected to adhere to applicable safety and health rules and regulations, that the "safe way" is the right way, and that, when in doubt, employees should err on the side of safety. The policy should then set forth the disciplinary consequences for failing to comply. Figure 7–3 is an example of a properly written discipline policy.

In Figure 7–3, Jones Construction Company (JCC) reserves the right to terminate an employee immediately for a flagrant violation. This is both well advised and legal. With nonflagrant violations, the company's policy follows the established legal standard for progressive discipline. This means that employees may commit minor violations without being immediately terminated. However, it also means that each violation results in a step being taken up the disciplinary ladder, and this ladder goes only one way. With progressive discipline, minor offenses are cumulative. Since they are recorded in the employee's file, they never go away—at least not in JCC. Some companies adopt a "second-chance" policy that allows an employee to erase a minor violation by working an established period of time without any kind of violation.

Such policies are left to the discretion of the individual company. The benefit of such a policy is that it adds a carrot to the stick when dealing with minor offenders. On the other hand, it has the potential to undercut the credibility of the discipline policy if employees see the second-chance option as being too forgiving. Consequently, when second-chance options are built into the discipline policy, they should be sufficiently difficult to challenge minor offenders. For example, consider the following second-chance option:

> Minor offenses may be removed from the employee's personnel file at a rate of one offense removed for every six months of uninterrupted work without a violation and the completion within that same period of at least eight hours of safety training on the employee's unpaid time.

Job-Site Inspections

Job-site inspections are an important component of a company's overall safety and health program. They represent a proactive approach to ensuring a high-quality workplace by identifying hazardous conditions and eliminating them before they cause accidents and injuries. The safety audit is an effective way to conduct an onsite inspection. It involves creating a checklist that is tailored to the specific job site and using it as a guide

Discipline Policy
Jones Construction Company

At JCC, the safe way is the right way. All employees and subcontractors are expected to follow applicable company rules as well as local, state, and federal regulations. Failure to do so will lead to disciplinary action as follows:

- **Termination**
  JCC reserves the right to terminate immediately any employee who fully or flagrantly endangers himself or herself, other workers, or company property through unsafe behavior.
- **Verbal Warning (First violation)**
  Nonflagrant first violations result in a verbal warning accompanied by an appropriate notation in the employee's personnel file.
- **Written Warning (Second violation)**
  Nonflagrant second violations result in a written warning—a copy of which is placed in the employee's personnel file.
- **Suspension (Third violation)**
  Nonflagrant third violations result in suspension without pay for a period of one week (five full workdays).
- **Dismissal (Final violation)**
  The next violation after a suspension results in immediate termination.

FIGURE 7–3 Sample discipline policy for a construction company.

Audit Checklist
Job-Site Inspection

Job Site ______________________ Date ______________

Check only those areas where problems exists. Then write an explanation of the hazardous condition that exists and what should be done to correct it. Attach explanation to this checklist.

__ Housekeeping
__ Personal protective equipment
__ First aid/medical services
__ Sanitation
__ Noise
__ Radiation (ionizing)
__ Radiation (nonionizing)
__ Gases, vapors, fumes, dust, mists
__ Illumination
__ Ventilation
__ Safety belts, lifelines, lanyards
__ Safety nets
__ Fire protection/prevention
__ Signs, signals, barricades
__ Material handling, storage, disposal
__ Hand tools/power tools
__ Welding and cutting
__ Electrical issues
__ Motorized vehicles, mechanical equipment, and machinery
__ Excavations
__ Concrete/masonry work
__ Steel erection
__ Demolition/blasting
__ Stairways and ladders
__ Other

FIGURE 7–4 Checklists like this one can improve job-site inspections.

when conducting the audit. Job-site inspections are typically conducted by the company's safety and health professional. However, supervisors can also conduct audits using checklists prepared by the safety professional. In smaller companies, supervisors prepare the checklist and conduct the audit.

Figure 7–4 is an example of an audit checklist for a job-site inspection. This is a comprehensive checklist. Not every item necessarily applies to every job site. In addition, this checklist assumes in-depth knowledge on the part of the person using it. Every item on this checklist could have a comprehensive sub-checklist developed for it.

Accident Investigations

How **accident investigations** are conducted and by whom should be explained in the company's comprehensive safety and health plan. A typical division of labor is to have supervisors conduct the investigations because they are closest to the work and to employees, and to provide supervisors with a standard form developed by a safety and health professional. A simple statement in the plan such as the following example is all that is necessary:

> Supervisors are required to conduct a detailed investigation whenever an accident occurs in their area of responsibility. Supervisors may and should contact the company's safety and health manager for advice and assistance. The standard accident investigation form is to be used.

Part of a supervisor's safety training should involve learning how to conduct comprehensive accident investigations and complete the necessary forms. Accident investigation is covered in greater depth in Chapter 9.

Record Keeping

The responsibility for **record keeping** is typically assigned to the company's safety and health professional in larger companies and to a designated employee in smaller companies. This component in the comprehensive safety and health plan should explain who is responsible for record keeping and what forms are to be used. Record keeping is explained in greater depth in Chapter 9. A statement such as the following is all that is necessary in the plan:

> Jones Construction Company follows the latest OSHA regulations related to record keeping. Accurate, comprehensive records of recordable injuries and illnesses are maintained and evaluated to identify trends and other causal data that can be used in their prevention. A summary of all recordable accidents and injuries is developed annually. This form is submitted to OSHA as required and posted for the information of all employees (as required by OSHA).

Training

No aspect of a comprehensive safety and health plan is more important than the training component. Safety training should be mandatory for all new hires as part of their orientation and for all employees any time they change positions or job sites. Orientation training should be generic and broadly applicable. In addition, job-site-specific training

SAFETY FACTS & FINES

When companies add new processes that are potentially harmful to the environment, they need to do more than just update their safety and health goals. Securing the proper permits from appropriate regulatory agencies and installing the necessary pollution control equipment are also important. A company in Portland, Oregon, was fined $11.2 million by the Environmental Protection Agency when it failed to secure the necessary permits and failed to install the necessary pollution prevention equipment before beginning new, potentially harmful processes. The fine applied to 13 different plants in four states. This is the largest fine ever assessed for violations of the Clean Air Act.

should be provided before a new employee is allowed to begin work. Updated training should be provided periodically so that new information is communicated to employees and old information is refreshed. Training requirements appear in numerous OSHA regulations. A statement such as the following is all that is necessary in the comprehensive safety and health plan.

> All employees are provided with the training needed to work safely. New employees receive generic training as part of their orientation and site-specific training provided by their supervisor, before beginning work. Periodic updated training is provided throughout the year and whenever an employee changes positions or job sites. Jones Construction Company complies with all applicable OSHA regulations related to training. From time to time, supervisors provide brief "chalk talk" training sessions at the beginning of a shift when circumstances indicate a need.

Medical Assistance and First Aid

Medical assistance and first aid must be available or easily accessible. All employees should know how to quickly call for medical assistance and first aid. Companies may provide medical assistance onsite or rely on third-party providers (e.g., fire departments, emergency medical agencies, ambulance services). Relying on third-party providers is the most widely used approach, but even when this is the preferred method, first aid supplies should be readily available at the job site. In addition, designated employees and supervisors should be trained in first aid fundamentals. The following statement is an example of the type that should be included in the safety and health plan:

> Jones Construction Company is committed to providing prompt, competent medical services when and where needed.

First Aid First aid supplies, as recommended by the National Safety Council, are readily available at all JCC job sites. In addition, designated personnel are trained to provide first aid and document all aid given in the company's first aid log.

Medical Assistance Medical assistance beyond first aid is provided by the designated first responder agency for each JCC job site. First responders are contacted by calling 911. In addition, direct-dial emergency numbers are posted conspicuously at all job sites.

Records Employees have full access to their personal medical and exposure records. Procedures for reviewing personal records are posted conspicuously at all job sites.

Figure 7–5 is an example of a first aid log of the type that should be available at all job sites. A blank copy should be appended to the comprehensive safety and health plan to ensure uniformity throughout the company.

Record of First Aid Treatment

Name of injured employee ____________________
Date and time of injury ____________________
Location (job site) ____________________
Body part injured ____________________
Treatment given ____________________
Cause of injury ____________________
Current status of employee ____________________
Comments:
____
____
____
____
____
____

FIGURE 7–5 Treatment records should be kept on file for at least five years.

Emergency Response

All employees must know what to do in the event of an emergency. The most common emergency at construction sites is fire. However, a structural collapse, tornado, earthquake, flood, or other natural or human-error disaster can also create an emergency in which the quality of the response is critical. The following example contains the elements that should be included in the company's plan:

> **Fire Response:** All employees at all job sites are trained in fire protection procedures and response. The firefighting equipment necessary for each specific job site is readily available and properly maintained. Emergency telephone numbers are conspicuously posted.
>
> **Job-Site Evacuation Response:** When it becomes necessary to evacuate a job site, a loud continuous siren is used to alert employees. The location of the "safe area" where personnel should collect for a head count is conspicuously posted and made known to all employees. Supervisors are responsible for conducting a head count of employees directly reporting to them at the safe area.

COMMUNICATING THE PLAN TO EMPLOYEES

The safety and health plan does not become a program until it is implemented. To implement the plan, employees have to know and understand its contents. **Communicating the plan** to employees is critical, and management cannot communicate too much. A key difference between having just a plan and having an effective program is communication.

Communication is always an imperfect process no matter how hard the company tries. Some employees do not listen well. Some do not pay attention. Others do not like to read. Consequently, when trying to communicate about the safety plan, it is important to build in repetition and use a variety of media. The following strategies can be used to communicate with employees about the safety and health plan:

- Face-to-face meetings
- Bulletin board notices
- "Tailgate talks" convened by supervisors at the beginning of a shift
- New-employee orientations
- Audiotapes that can be played while driving to and from work
- Videotapes
- Newsletters, memoranda, and bulletins
- Signs and posters
- Setting a positive example (the best way to communicate the importance of safety)

EVALUATING THE PROGRAM

The Deming cycle is shown in Figure 7–6. It was developed by W. Edwards Deming, the quality expert who helped Japanese manufacturers become dominant players in the global marketplace.[1] The cycle shows the relationship between planning, implementing, and evaluating. Step 1 (Plan) involves development of the comprehensive safety and health plan, as described in this chapter. Step 2 (Do) involves implementing the plan so that it becomes an actual functioning program. Step 3 (Check) involves **evaluating the program** periodically to determine its effectiveness. Step 4 (Act) involves acting on what is learned from the evaluation in such a way as to strengthen the program. After completing Step 4, the cycle repeats continually.

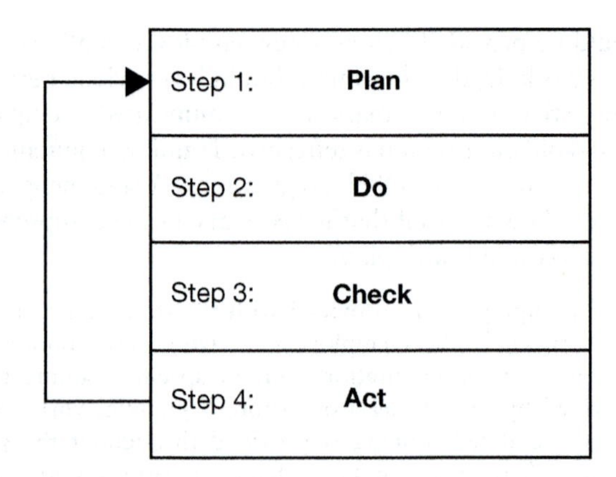

FIGURE 7–6 The Deming cycle.

Evaluation Checklist

A checklist, such as the one shown in Figure 7–7, may simplify the evaluation process. Note that this checklist mirrors the contents of the comprehensive safety and health plan: It makes no sense to plan and implement one thing and evaluate another. Checklists such as this one are used to evaluate the content of the safety and health program and its associated processes, rather than its performance. Evaluating the performance of the program is covered in later chapters.

Evaluation Checklist:
Safety and Health Program

The only correct response to all of the following questions (that apply) is "Yes."
Therefore, check only those questions that must be answered "No."

Safety and Health Policy

___ 1. Is there a written safety and health policy signed by the chief executive officer?
___ 2. Does the policy state clearly that the company is committed to providing a safe and healthy work environment?
___ 3. Does the policy state clearly that all employees are expected to work in a safe and healthy manner?
___ 4. Does the policy state clearly that safety and health rules are enforced?

Safety and Health Goals

___ 5. Is there a goal related to fatalities?
___ 6. Is there a goal related to lost time due to accidents?
___ 7. Is there a goal related to workers' compensation claims?
___ 8. Is there a goal related to property damage?
___ 9. Is there a goal related to "near-miss" accidents?
___ 10. Is there a goal related to participation of employees in safety training?
___ 11. Is there a goal related to the "safety image" of the company?
___ 12. Are goals tied to a specific year?
___ 13. Are goals updated every year?

Responsibilities and Rules

___ 14. Are the responsibilities of management described in writing?
___ 15. Are the responsibilities of supervisors described in writing?
___ 16. Are the responsibilities of employees described in writing?
___ 17. Are the responsibilities of safety professionals described in writing?

Discipline and Accounting

___ 18. Is there a written discipline policy?
___ 19. Does the policy describe the company's right of termination?
___ 20. Does the policy describe the progressive discipline process for minor violations?

Job-site Inspection

___ 21. Has a comprehensive audit checklist been developed for job-site inspection, and does it include the requirement material safety data sheets (MSDA) be available at the job site or electronically to employees?
___ 22. Are job-site inspections conducted on a regular basis?
___ 23. Are written reports made listing all discrepancies identified and the necessary follow-up actions?
___ 24. Are written reports used to correct discrepancies identified?
___ 25. Are written reports available to any employee who wants to see them?

FIGURE 7–7 Safety and health programs should be evaluated periodically.

Evaluation Checklist:
Safety and Health Program

Accident Investigations

___ 26. Has a standard accident investigation form been developed?
___ 27. Are all accidents and illness investigated promptly?
___ 28. Are all "near-miss" accidents investigated?
___ 29. Are investigations conducted whenever equipment is damaged in an incident or accident?
___ 30. Are accident reports filled out for all incidents?
___ 31. Are actions taken to prevent future incidents?
___ 32. Are accident reports monitored to identify trends?
___ 33. Are the trend data used to prevent future incidents?
___ 34. Are accident reports made available to employees who wish to view them?

Record Keeping

___ 35. Has an individual been assigned responsibility for record keeping?
___ 36. Are records maintained in accordance with the latest OSHA regulations (including OSHA Form 300/301)?

Training

___ 37. Do all employees receive safety and health training as part of their in-processing orientation?
___ 38. Do all employees receive job-site-specific training before being allowed to work?
___ 39. Do all employees receive task-specific training before being allowed to perform the tasks in question?
___ 40. Do employees receive periodic updated training?
___ 41. Do employees receive additional training when they change jobs or job sites? When new equipment or new methods are put to use?
___ 42. Do managers and supervisors receive the safety and health training they need?
___ 43. Are comprehensive, up-to-date training records maintained?

Medical Assistance and First Aid

___ 44. Are adequate first aid supplies available at all job sites?
___ 45. Are emergency telephone numbers for medical assistance posted conspicuously?
___ 46. Are up-to-date employee health and medical records maintained and made available to employees who want to receive them?

Emergency Response

___ 47. Are all employees trained in fire response procedures?
___ 48. Are all employees trained in job-site evacuation procedures?

Communication

___ 49. Is the program sufficiently communicated to all employees?
___ 50. Are employees allowed to give feedback about the program?

FIGURE 7–7 *(Continued)*

Summary

There is disagreement among construction professionals concerning the need for safety and health plans and programs. However, the evidence shows that well-written safety and health plans can prevent accidents, illnesses, injuries, and the expenses associated with them. There are ethical, regulatory, economic, and practical reasons for having comprehensive written safety and health plans. Such a plan should have at least the following components: (1) safety and health policy, (2) safety and health goals, (3) **roles and responsibilities,** (4) discipline policy and procedures, (5) job-site inspections, (6) accident investigations, (7) record keeping, (8) training, (9) medical response and first aid, (10) **emergency response,** and (11) miscellaneous components (e.g., fall protection, blood-borne pathogens). Communicating the safety and health plan to all employees is important and should include a variety of strategies. Effective strategies include the following: (1) face-to-face meetings, (2) bulletin board notices, (3)"tailgate talks" by supervisors, (4) new-employee orientations, (5) audiotapes, (6) videotapes, (7) newsletters, memoranda, and bulletins, (8) signs and posters, and (9) setting a positive example. It is important to evaluate the safety and health plan for content and processes. Checklists that follow the patterns of the plan can be developed to help simplify the evaluation process.

Key Terms and Concepts

Accident investigations
Accountability
Communicating the plan
Discipline policy
Economic factors
Emergency response
Employees' responsibilities
Ethical factors
Evaluating the program
Job-site inspections
Management's responsibilities
Medical assistance and first aid
Record keeping
Regulatory factors
Roles and responsibilities
Safety and health goals
Safety and health policy
Safety professionals' responsibilities
Supervisors' responsibilities
Training

Review Questions

1. Explain the rationale for a comprehensive written safety and health plan.
2. The safety and health policy should convey certain messages. What are they?
3. Write examples of three safety and health goals that might be used in the plan for any construction company.
4. Explain management's responsibilities in the safety and health program.

5. Explain the supervisor's responsibilities in the safety and health program.
6. Explain the employee's responsibilities in the safety and health program.
7. Explain the safety professional's responsibilities in the safety and health program.
8. What is "progressive discipline" as it relates to the safety and health program?
9. What is the purpose of the job-site inspection?
10. Who typically conducts accident investigations at the job site? Why this person?
11. How do you think the safety and health training that an employee receives during in-processing orientation and job-site-specific training would differ?
12. Describe the two different ways that medical assistance might be provided by a construction company.
13. The emergency response component of the safety and health plan has at least two critical elements. Explain both of them.
14. List the six different strategies that might be used to communicate the safety and health plan to employees.
15. Explain how construction professionals can use the Deming cycle in planning, implementing, and evaluating the safety and health program.

Critical Thinking and Discussion Activities

1. John Rodgers is vice president for business operations at Northwest Construction Company (NCC); Mark Danson is chief engineer. They are debating the issue of safety and health plans. Rodgers says, "Written plans for safety are a waste of time. You don't have to write things down in order to do them. Our supervisors should show employees the safe way to do their jobs and make sure they comply. It takes too much time and costs too much money to write plans." Mark Danson disagrees. He says, "You're wrong, John. If we don't write it down, we won't do it. You have all of NCC's business plans and procedures in writing. Why is safety any different?" Join this debate. Who do you think is right and why? Defend your opinion.
2. Two supervisors for Casey Construction, Inc. (CCI) are discussing the medical assistance and first aid component of CCI's safety and health plan. "I don't like this approach we have to use," said the first supervisor. "I don't want any of our employees playing doctor. We need to leave medical problems up to the medical professionals. I think we should revise the plan so that outside help is called in when we need medical assistance, and drop the first aid part." The second supervisor disagreed. He said, "What if somebody needs CPR? By the time the EMTs get here, he'll be dead. I think we should have at least one person on every crew who is first aid certified." What do you think? Take the side of one of these supervisors, and make his case for him.

Application Activities

1. Identify a construction company that has a comprehensive written safety and health plan and is willing to work with you in completing this activity. Using the information learned in this chapter, develop a program evaluation checklist and evaluate the company's plan. Make note of any shortcomings in the plan. Tactfully share any shortcomings identified with your contact in the company.
2. Develop a comprehensive safety and health plan that could be used as a model by a construction company in developing its own plan. Add an appendix to the plan that describes how it should be communicated to employees.

Endnote

1. Retrieved December 2, 2015, from http://deming.org/index.cfm?Content=66

CHAPTER EIGHT

JOB SAFETY AND HAZARD ANALYSIS

LEARNING OBJECTIVES

- Provide a comprehensive overview of hazard analysis.
- Compare and contrast preliminary and detailed hazard analysis.
- Explain detailed hazard analysis.
- Explain the concept of hazard prevention and reduction.
- Demonstrate how to conduct a risk assessment.

There is a saying, "An ounce of prevention is worth a pound of cure." This is certainly the case with workplace safety and health. Every accident that can be prevented should be. Every **hazard** that can be identified should be corrected or at least minimized through the introduction of appropriate safeguards. Careful analysis of potential hazards in the workplace has led to many of today's widely used safety measures and practices.

The key to preventing accidents is identifying and eliminating hazards. A hazard may be defined as follows:

> A hazard is a condition or combination of conditions that, if left uncorrected, may lead to an accident, illness, or property damage.

This chapter provides construction professionals with the information they need to analyze the workplace, identify hazards that exist there, and take the preventive measures necessary to neutralize the hazards.

OVERVIEW OF HAZARD ANALYSIS

If a hazard is a condition that could lead to an injury or illness, **hazard analysis** is a systematic process for identifying hazards and recommending corrective action. There are two approaches to hazard analysis: preliminary and detailed (Figure 8–1). A **preliminary hazard analysis (PHA)** is conducted to identify potential hazards and prioritize them according to: (1) the likelihood of an accident or injury from the hazard and (2) the **severity** of injury, illness, or property damage that may result if the hazard caused an accident.[1]

Key concepts are these terms that describe the likelihood of an accident occurring (very likely, likely, and not likely) and those that rate the probable level of injuries that could occur (catastrophic, critical, marginal, and nuisance). A key step in this process is rating the cost of correcting hazards. These concepts are covered later in this chapter.

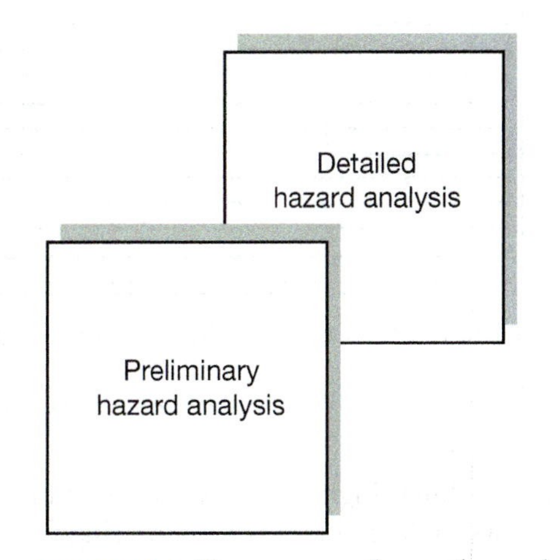

FIGURE 8–1 Two approaches to hazard analysis.

Whereas a preliminary analysis may involve just observation or pilot testing of new equipment and systems, the **detailed hazard analysis** involves the application of analytical, inductive, and deductive methods. Figure 8–2 lists some of the more widely used methods for conducting a detailed hazard analysis. Each of these methods is covered at length later in this chapter.

PRELIMINARY HAZARD ANALYSIS

It is not always feasible to wait until all the data are compiled from a detailed analysis before taking steps to identify and eliminate hazards. For example, when a new system or piece of equipment is installed, management probably wants to bring it on line as soon

Detailed hazard analysis methods
Failure mode and effects analysis (FMEA)
Fault tree analysis (FTA)
Hazard and operability review (HAZOP)
Human error analysis (HEA)
Risk analysis
Technic of operations review (TOR)

FIGURE 8–2 Detailed hazard analysis methods.

Operation:										Date:						
Number of Employees	Job Title	Exposure Substance	Form (Type of Hazard)						Route of Entry		Control Methods					
			Dust	Liquid	Vapor	Gas	Fume	Mist	Skin	Lungs	Local Ventilation	General Ventilation	Respirator	Gloves	Face Protection	Other Protection

FIGURE 8–3 Sample job hazard analysis survey.

as possible. In such cases, a PHA is in order. The PHA can serve two purposes:

- It can expedite bringing the new system on line, but at a substantially reduced risk of injuring workers.
- It can serve as a guide for a future detailed analysis.

PHA amounts to forming an ad hoc team of experienced personnel who are familiar with the equipment, material, substance, and process being analyzed. **Experience** and **related expertise** are important factors in conducting a preliminary review. For example, if a new piece of equipment is installed, the safety and health professional or supervisor may form a team that includes a variety of experienced personnel.

All members of the team are asked to examine the new equipment hazards. Then, they work as a group to play devil's advocate. Each team member asks the others a series of "what if" questions: What if a cutting bit breaks? What if the wrong command is entered? What if the material stock is too long? Depending on the nature of the process being analyzed, personnel from related processes should be added to the team.

Figure 8–3 is an example of a job hazard analysis survey adapted from one developed by the National Institute for Occupational Safety and Health (NIOSH). A preliminary analysis team would use this form to identify potential hazards associated with a painting process. Key elements include the substances to which workers will be exposed, the form that those substances will take, the probable route of entry, and recommended hazard-control strategies. A similar form can be developed for any process or operation that may be the focus of a PHA.

Cost and Benefit in Hazard Analysis

Typically, every hazard has several different remedies. Every remedy has a corresponding *cost* and corresponding *benefit*. Management is not likely to want to apply $10 solutions to $1 problems. Therefore, it is important to factor in cost when recommending corrective action regarding hazards. This amounts to listing all of the potential remedies along with their respective costs and then estimating the extent to which each will reduce the hazard (its benefit).

Going back to the earlier example of the new equipment, assume that the analysis team identified the following potential hazards:

- Lubricants spraying on the machine operator.
- Flying wood chips hitting the operator or other workers.
- Jammed wood stock kicking back into the operator.

Figure 8–4 is a matrix that may be developed by the analysis team to illustrate the cost of each hazard versus the benefit of each remedy. After examining this matrix,

Cost–Benefit Analysis Matrix

		Hazard and Impact of Remedy		
Possible Remedy	**Estimated Cost**	**Spraying Lubricants**	**Metal Chips**	**Jammed Stock**
Plexiglass door	$250	E	E	R
Flexible curtain	$75	R	R	N
Acme chip/ jam guard	$260	N	E	E
Acme spray guard	$260	N	E	E

Code for Impact of Remedy:
R = Reduces the hazard
E = Eliminates the hazard
N = No effect on the hazard
I = Increases the hazard
C = Creates new hazard

FIGURE 8–4 Sample cost–benefit analysis matrix for possible remedies to hazards associated with a piece of manufacturing equipment.

the remedy that makes the most sense from the perspectives of cost and **impact** on the hazards is the Plexiglass door. It eliminates two of the hazards and reduces the third. The flexible curtain costs less, but does not have a sufficient impact on the hazards. The third and fourth options cost more and have less impact on the hazards.

DETAILED HAZARD ANALYSIS

Typically, a PHA is sufficient. However, in cases where the potential exists for serious injury, multiple injuries, or catastrophic illness, a detailed hazard analysis is conducted. A number of different methods can be used for conducting detailed analyses. The most widely used of these are as follows:

- Failure mode and effects analysis
- Hazard and operability review
- Technic of operations review
- Human error analysis
- Fault tree analysis
- Risk analysis

Failure Mode and Effects Analysis

Failure mode and effects analysis (FMEA) is a formal step-by-step analytical method that is a spin-off of reliability analysis—a method used to analyze complex engineering systems. FMEA steps proceed as follows:

1. Critically examine the system in question.
2. Divide the system into its various components.
3. Examine each individual component and record all of the various ways in which the components may fail. Rate each potential failure according to the degree of hazard posed on a scale from 0 to 4: 0 = no hazard, 1 = slight, 2 = moderate, 3 = extreme, and 4 = severe.
4. Examine all potential failures for each individual component of the system, and decide what effect the failures could have.

Figure 8–5 is an example of an FMEA conducted on a concrete extrusion process. The process or system is broken down into seven components: die backer, die, billet, dummy block, pressing stem, container liner, and container fillet. The types of failures that may occur are identified as corrosion, cracking, shattering, bending, and surface wear. Of the various components, only the dummy block poses an extreme hazard and a corresponding hazard to workers.

An FMEA produces an extensive analysis of a specific process or system, as illustrated in Figure 8–5. However, FMEAs have their limitations. First, the element of **human error** is missing. This is a major weakness because human

Jones Prestressed Concrete, Inc.
1605 Highway 39
Fort Walton Beach, Florida 32548

Department Production **Process/System** Concrete Extrusion of Cored Floor Members **Date** January 15, 2002

		Potential Effect on														
Component	**Type of Potential Failure**	**Component**	**Related Components**	**Process/ System**	**Workers**	**0**	**1**	**2**	**3**	**4**	**H**	**M**	**L**	**U**	**Examination Method**	**Recommendation**
Diebacker	Corrosion	Shutdown to replace	None	Shutdown to replace	None	✓							✓			
Die	Cracking	Shutdown to replace	Damage to die backer	Shutdown to replace				✓				✓				
Billet	—	—	—	—	—	—	—	—	—	—	—	—	—	—	—	—
Dummy block	Shattering	Shutdown to replace	Could damage other	Shutdown all	Injuries from flying metal				✓			✓				
Pressing stem	Bending	Shutdown to replace	None	Shutdown to replace	None	✓							✓			
Container liner	Surface wear	Shutdown to replace	None	None	None	✓								✓		
Container fillet	—	—	—	—	—	—	—	—	—	—	—	—	—	—	—	—

Analysis conducted by: ______

FIGURE 8–5 Sample failure mode and effects analysis (FMEA).

error is more frequently at the heart of a workplace accident than is system or process failure. This weakness can be overcome by coupling human error analysis with an FMEA. Second, FMEAs focus on the components of a given system as if the components operate in a vacuum. They do not take into account the interface mechanisms between components or between systems. It is at these interface points that problems often occur.

Hazard and Operability Review

Hazard and operability review (HAZOP) is an analysis method that allows problems to be identified even before a body of experience has been developed for a given process or system. It is especially useful for operations involving chemicals or toxic materials. Although originally intended for use with new processes, it need not be limited to new operations. HAZOP works equally well with old processes and systems.

HAZOP involves forming a team of experienced, knowledgeable people from a variety of backgrounds related to the process or system and having team members brainstorm about potential hazards. The person responsible for safety and health should chair the team and serve as a facilitator. The chair's role is to elicit and record the ideas of team members, make sure that one member does not dominate or intimidate other members, encourage maximum participation from all members, and assist members in combining ideas, when appropriate, to form better ones.

A variety of approaches can be used with HAZOP. The most widely used approach is based on the following guide words: *no, less, more, part of, as well as, reverse,* and *other than.*[2]

These guide words relate to the operation of a specific component in the system or a specific part of an overall operation. They describe ways in which the component may deviate from its design or its intended mode of operation. For example, if a component that should rotate 38 degrees in a cycle fails to rotate at all, the *no* guide word applies. If it rotates fewer than 38 degrees, the *less* guide word applies. *More* would apply if the component's rotation exceeded 38 degrees. *Reverse* would be used if the component rotated 38 degrees in a direction opposite of the one intended. *As well as* is similar to *more* in that it indicates an increase in an intended amount. *Other than* is used when what actually occurs is something completely different from what was intended. For example, if the component fell off rather than rotating 38 degrees, *other than* would be used.

HAZOP proceeds in a step-by-step manner, which is summarized as follows:

1. Select the process or system to be analyzed.
2. Form the team of experts.
3. Explain the HAZOP process to all team members.
4. Establish goals and time frames.
5. Conduct brainstorming sessions.
6. Summarize all input.

Figure 8–6 is an example of a form that can be used to help organize and focus brainstorming sessions. It can also be used for summarizing the results of the brainstorming sessions. This particular example involves a plastic mixing process. Only one component in the process (flow-gate number 1) has been analyzed. If the flow-gate does not work as intended, there will be no flow, too little flow, or too much flow. Each condition results in a specific problem. Action necessary to correct each situation has been recommended. Every critical point, sometimes referred to as a *node*, in the process is analyzed in a similar manner.

HAZOPs have the same weaknesses as FMEAs: they do not factor human error into the equation. HAZOPs predict problems associated with system or process failures; however, these are technological failures. Since human error is so often a factor in accidents, this weakness must be addressed. The next section provides guidelines for analyzing human error.

Martin Contracting
1512 Airport Road
Crestview, Florida 32536
HAZOP SUMMARY FORM

Department Department B **System/Process** Termite Protection **Date** January 7, 2002

System/Process Component	Factor Analyzed	Guideword	Resulting Difference	Potential Problem	Recommended Remedy
Flow-gate number one	Amount of flow	No	No flow	No mix	Make sure flow-gate is open
		Less	Insufficient flow	Weak mix	Troubleshoot and repair the flow-gate
		More	Excess flow	Too strong mix	Troubleshoot and repair the flow-gate

FIGURE 8–6 Sample hazard and operability review (HAZOP).

Human Error Analysis

Human error analysis (HEA) is used to predict human error, not to review what has occurred. Although the records of past accidents can be studied to identify trends that can, in turn, be used to predict accidents, this should be done as part of an accident investigation. HEA should be used to identify hazards before they cause accidents.

Two approaches to HEA can be effective: (1) observing employees at work and noting hazards (the task analysis approach) and (2) actually performing job tasks to get a firsthand feel for hazards. Regardless of how HEA is conducted, it is a good idea to perform it in conjunction with FMEA and HAZOP. This will enhance the effectiveness of all three processes.

Technic of Operations Review

Technic of operations review (TOR) is an analysis method that allows supervisors and employees to work together to analyze workplace accidents, failures, and incidents. It answers the question, "Why did the system allow this incident to occur?" Like FMEA and HAZOP, this approach seeks to identify systemic causes. It does not seek to assign blame.

TOR is not new. It was originally developed in the early 1970s by D. A. Weaver of the American Society of Safety Engineers. However, for 20 years, user documentation on TOR was not readily available. Consequently, wide-scale use did not occur until the early 1990s, when documentation began to be circulated.

> The purpose of TOR is to allow construction professionals and safety professionals to identify the root causes of an operational failure. TOR is initiated by a specific incident that occurs at a specific time and place involving specific people.[3]

A weakness of TOR is that it is designed as an after-the-fact process. It is triggered by an accident or incident. A strength of TOR is its involvement of line personnel in the analysis. The process proceeds as follows:

1. *Establish the TOR team.* The team should consist of workers who were present when the accident or incident occurred, the supervisor, and the safety and health director. The safety and health professional should chair the team and serve as a facilitator.
2. *Conduct a roundtable discussion to establish a common knowledge base among team members.* At the beginning of the discussion, five team members may have five different versions of the accident or incident. At the end, there should be a consensus.
3. *Identify one major systematic factor that led to or played a significant role in causing the accident or incident.* This one TOR statement, about which there must be a consensus, serves as the starting point for further analysis.
4. *Use the group consensus to respond to a sequence of yes/no options.* Through this process, the team identifies a number of factors that contributed to the accident or incident.
5. *Evaluate identified factors carefully to make sure that there is a team consensus about each.* Then prioritize the contributing factors beginning with the most serious.
6. *Develop corrective or preventive strategies for each factor.* Include the **corrective or preventive strategies** in a final report that is forwarded through normal channels for appropriate action.

Fault Tree Analysis

Fault tree analysis (FTA) can be used to predict and prevent accidents or as an investigative tool after the fact. FTA is an analytical method that uses a graphic model to display the analysis process visually. A fault tree is built using special symbols, some derived from Boolean algebra. Consequently, the resultant model resembles a logic diagram or a flow chart. Figure 8–7 shows and describes the symbols used in constructing fault trees. Figure 8–8 shows how these symbols may be used to construct a fault tree. The top box in a fault tree represents the accident or incident that either could occur or has occurred.

All symbols below the top box represent events that contribute in some way to the ultimate accident or incident. The sample fault tree shown in Figure 8–8 is qualitative in nature. Fault trees can be made quantitative by assigning **probability** figures to the various events below the top box. However, this is rarely done because reliable probability figures are seldom available. A fault tree is developed using the following steps:

1. Decide on the accident or incident to be placed at the top of the tree.
2. Identify the broadest level of failure or fault event that could contribute to the top event. Assign the appropriate symbols.

SAFETY FACTS & FINES

Asbestos removal is a delicate, difficult process, and it can be tempting to do it the easy way. But the easy way can be more trouble than it is worth. A company in Milan, Illinois, was fined $1 million by the Environmental Protection Agency for improperly removing asbestos. It knowingly stripped material containing the dangerous substance without wetting it down first. Three individual employees were fined and were sentenced to either prison or home confinement.

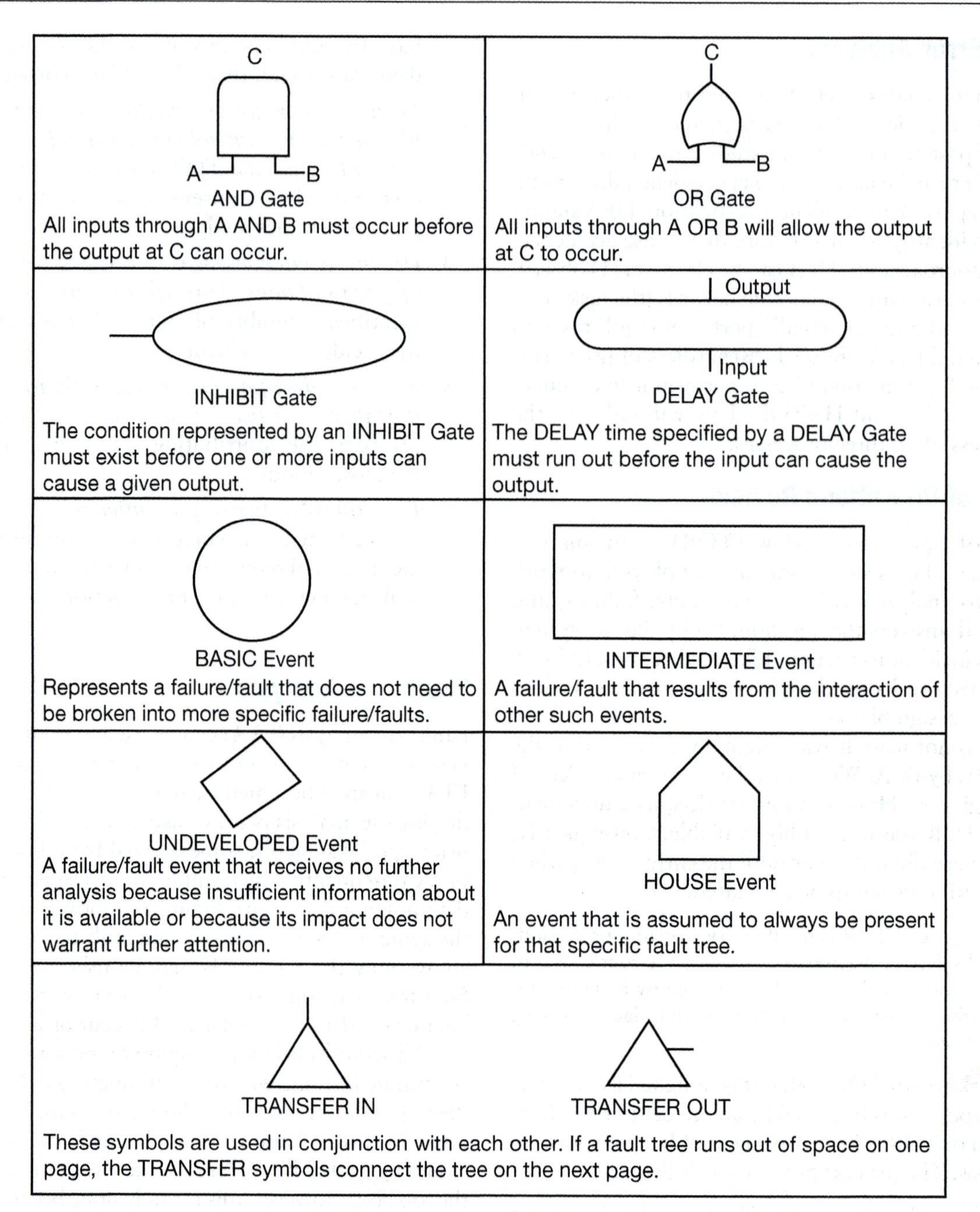

FIGURE 8–7 Symbols used in fault tree analysis (FTA).

3. Move downward through successively more specific levels until basic events are identified.

Experience, deliberate care, and systematic analysis are important in constructing fault trees. Once a fault tree has been constructed, it is examined to determine the various combinations of failure or fault events that could lead to the top event. With simple fault trees, this can be accomplished manually (without computer assistance); with more complex trees, this step is more difficult. However, computer programs are available to assist in accomplishing this step. The final step involves making recommendations for preventive measures.

Risk Analysis

Where are we at risk? Where are we at greatest risk? These are important questions for safety and health professionals involved in analyzing the workplace for the purpose of identifying and overcoming hazards. **Risk analysis** is an analytical method that is normally associated with insurance and investments. However, risk analysis can be used to analyze the workplace, identify hazards, and develop strategies for overcoming these hazards. The risk analysis process focuses on two key questions:

1. How frequently does a given event occur?
2. How severe are the consequences of a given event?

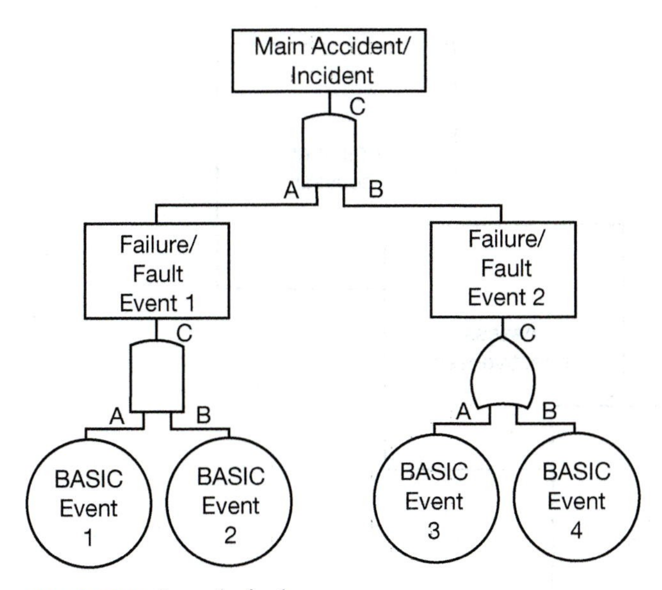

FIGURE 8–8 Sample fault tree.

The fundamental rule of thumb of risk analysis is that risk is lessened by decreasing the **frequency** and severity of hazard-related events. Construction professionals should understand the relationship that exists between the *frequency* and *severity* factors related to accidents. Historical data on accidents, injuries, and illness show that the less severe an injury or illness is, the more frequently it is likely to occur. Correspondingly, the more severe an injury or illness is, the less frequently it is likely to occur. For example, there are many more minor scrapes, bumps, and abrasions experienced in the workplace than major debilitating injuries, such as amputations or broken bones.

A number of different approaches can be used in conducting a risk analysis. One of the most effective approaches considers probability and frequency.[4]

Probability levels and corresponding frequency of occurrence ratings are as follows:

1 = Impossible (frequency of occurrence: 10^{-8}/day)

2 = Extremely unlikely (frequency of occurrence: 10^{-6}/day)

3 = Remote (frequency of occurrence: 10^{-5}/day)

4 = Occasional (frequency of occurrence: 10^{-4}/day)

5 = Reasonably probable (frequency of occurrence: 10^{-3}/day)

6 = Frequent (frequency of occurrence: 10^{-2}/day)[5]

The lowest rating (1) means it is impossible that a given error will be committed or a given failure will occur. The highest rating (6) means it is very likely that a given error will be committed frequently or a given failure will occur frequently. Note the quantification of frequency levels for each level of probability. For example, the expected frequency of occurrence for a probability level of 3 is 10^{-5}/day.

Severity levels can also be linked with the corresponding likely consequence of an accident or failure event. The least severe incidents: (1) are not likely to cause an injury or damage to property. The most severe incidents (4) are almost certain to cause death or serious property damage. Critical accidents (3) may cause severe injury or major loss. Marginal accidents (2) may cause minor injury, minor occupational illness, or minor damage.[6]

HAZARD PREVENTION AND REDUCTION

All of the methods and procedures discussed in this chapter have been concerned with identifying potential hazards. This section deals with using the information learned during analysis to prevent accidents and illnesses. The following hazard-control methods are recommended:

Eliminate, substitute, reduce, remove, isolate, and dilute. Other effective **hazard prevention** and reduction methods include using personal protective equipment (PPE), training, and good housekeeping.[7]

For every hazard identified during the analysis process, one or more of these hazard-control methods will apply. Figure 8–9 shows the steps involved in implementing hazard-control methods. The first step involves selecting the method or methods that are most likely to produce the desired results. Once selected, the method is applied and monitored to determine if the expected results are being achieved.

Monitoring and observing are informal procedures. They should be followed by a more formal, more structured assessment of the effectiveness of the method. If the method selected is not producing the desired results, adjustments should be made. This may mean changing the way in which the method is applied or dropping it and trying another.

The example of Mathew Construction Company's (MCC) problems with toxic paint illustrates how the process works. MCC constructs aluminum buildings on military bases. The buildings are painted as the last step in the process. Although the specified paint was supposed to be only slightly toxic—a problem that should have been resolved by using PPE—painters complained frequently of various negative side effects.

MCC's safety and health director, working with management, solved the problem by applying the following steps:

1. *Select a method.* Of the various methods available, the one selected involved eliminating the source of the hazard (the toxic paint). MCC personnel were asked to test various nontoxic paints until one was found that could match the problem paint in all categories (e.g., ease of application, drying time, quality of surface finish). After 40 different paints were tested, a nontoxic substitute was found.
2. *Apply the method.* The new paint was ordered and used on a portion of a building.

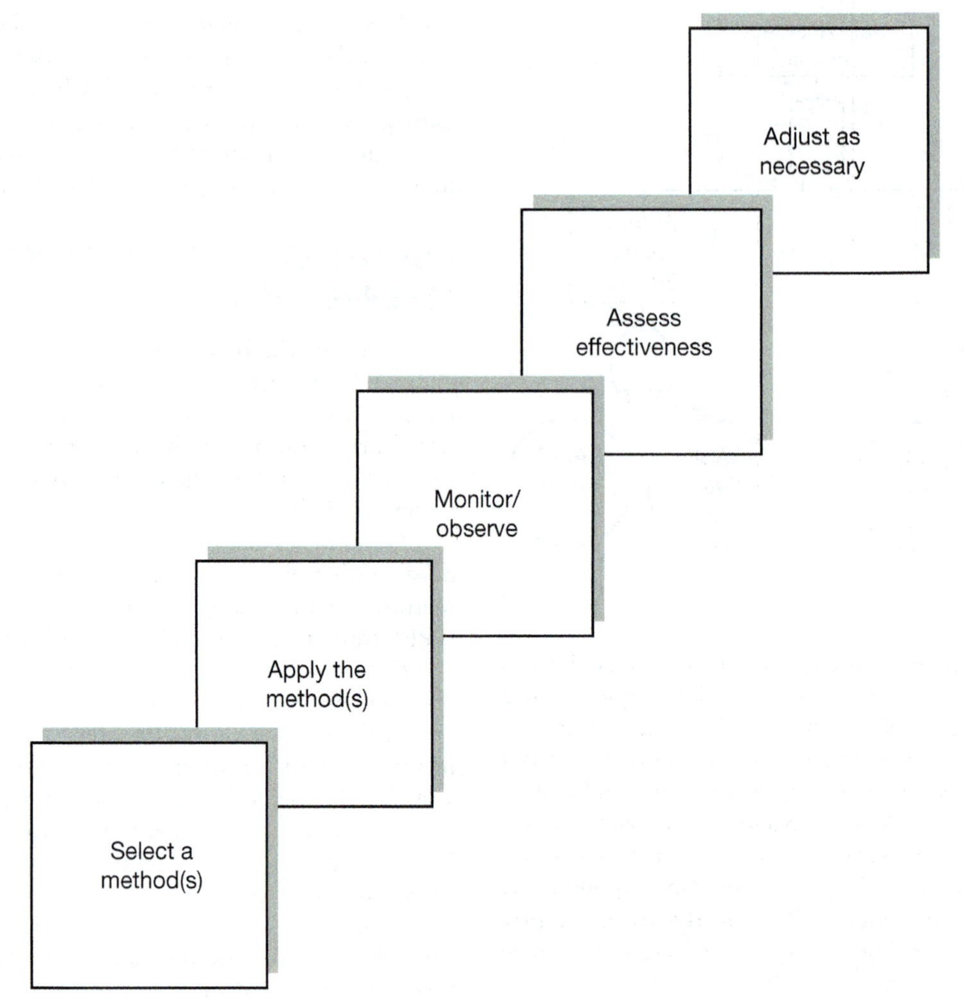

FIGURE 8–9 Steps for implementing hazard-control measures.

3. *Monitor and observe.* The safety and health director, along with MCC's painting supervisors, monitored both employee performance and employee complaints concerning the paint.
4. *Assess effectiveness.* To assess effectiveness, employee complaints were tabulated. The number of complaints was down to a negligible amount, and the complaints were not serious in nature. Productivity was also assessed. It was found that the new paint had no noticeable negative or positive effect on productivity.
5. *Adjust as necessary.* MCC found that no further adjustments were necessary.

RISK ASSESSMENT

Risk assessment in this context involves quantifying the level of risk associated with a given process. It should be a structured and systematic process that answers the following four specific questions:

1. How *severe* are potential injuries?
2. How *frequently* are employees exposed to the potential hazards?
3. What is the *possibility* of avoiding the hazard if it does occur?
4. What is the *likelihood* of an injury if a safety-control system fails?

The most widely used risk assessment technique is the decision tree, coupled with codes representing these four questions and defined levels of risk. Figure 8–10 is an example of a risk assessment decision tree. In this example, the codes and their associated levels of risk are as follows:

S = Severity

Question 1: Severity of potential injuries

S1 Slight injury (bruise, abrasion)

S2 Severe injury (amputation or death)

F = Frequency

Question 2: Frequency of exposure to potential hazards

F1 Infrequent exposure

F2 Frequent to continuous exposure

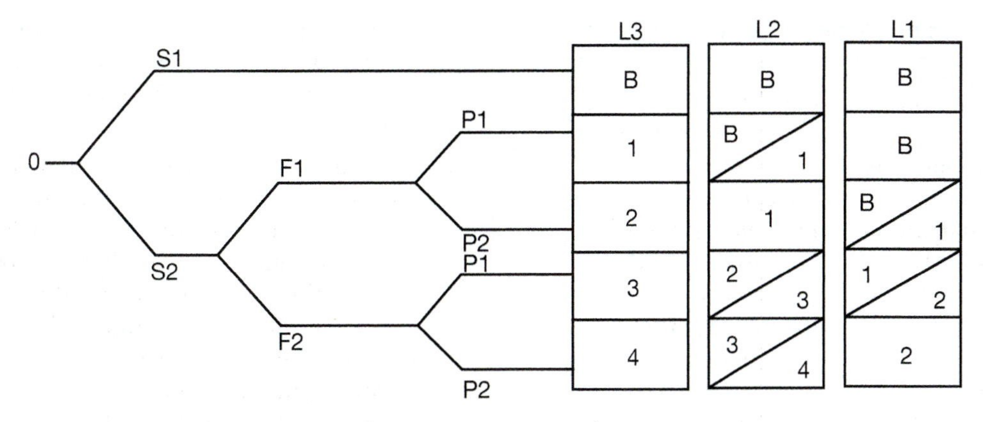

FIGURE 8–10 Risk assessment decision tree (see explanation in text).

P = Possibility

Question 3: Possibility of avoiding the hazard if it does occur

P1 Possible

P2 Less possible to not possible

L = Likelihood

Question 4: Likelihood that the hazard will occur

L1 Highly unlikely

L2 Unlikely

L3 Highly likely

Risk Levels = B, 1, 2, 3, or 4

Associated risk factors ranging from lowest (B) to highest (4)

By applying the decision tree in Figure 8–10 or a similar device, the risk associated with the operation of a given process can be quantified. This assists safety personnel in assigning logical priorities for hazard prevention.

Summary

A hazard is a condition or combination of conditions that, if left uncorrected, may lead to an accident, illness, or property damage. Hazards can be ranked as potentially catastrophic, critical, marginal, or nuisance.

Hazard analysis is a systematic process for identifying hazards and recommending corrective action. There are two approaches to hazard analysis: preliminary and detailed. A preliminary hazard analysis (PHA) involves forming an ad hoc team of experienced personnel who are familiar with the equipment, material substance, and process being analyzed. Experience and related expertise are critical in conducting a PHA.

Failure mode and effects analysis (FMEA) is a detailed hazard analysis method that involves dividing a system into its various components, examining each component to determine how it may fail, rating the probability of failure, and deciding what effect these failures would have.

Hazard and operability review (HAZOP) is a detailed hazard analysis method that was developed for use in the chemical industry. It involves forming a team of experts and brainstorming.

Human error analysis (HEA) is used to predict human error and its potential effects. It can be used in conjunction with FMEA and HAZOP to strengthen those approaches.

Technic of operations review (TOR) is a hazard analysis method that allows workers and supervisors to conduct the analysis. It uses a simple worksheet that allows team members to respond to a sequence of yes-or-no options.

Fault tree analysis (FTA) is a hazard analysis method that uses a graphic model to display the analytical process visually. The model resembles a logic diagram.

Risk analysis, although more commonly associated with the insurance industry, can be used for hazard or safety analysis. The process revolves around answering two questions: (1) How frequently does a given event occur? (2) How severe are the consequences of a given event? The fundamental rule of thumb of risk analysis is that risk is lessened by decreasing the frequency and severity of hazard-related events.

The fundamentals of hazard prevention and deterrence include the following strategies: eliminate the source of the hazard, substitute a less hazardous substance, reduce the hazard at the source, remove the employee from the hazard, isolate the hazard, dilute the hazard, apply appropriate management strategies, use personal protective equipment (PPE), provide employee training, and practice good housekeeping.

Risk assessment should answer four questions: (1) How severe are potential injuries? (2) How frequently are employees exposed to the potential hazards? (3) What is the possibility of avoiding the hazard if it does occur? (4)What is the likelihood of an injury if a safety-control system fails?

Key Terms and Concepts

Corrective or preventive strategies
Detailed hazard analysis
Experience
Failure mode and effects analysis (FMEA)
Fault tree analysis (FTA)
Frequency
Hazard
Hazard analysis
Hazard and operability review (HAZOP)
Hazard prevention or reduction
Human error
Human error analysis (HEA)
Impact
Preliminary hazard analysis (PHA)
Probability
Related expertise
Risk analysis
Risk assessment
Severity
Technic of operations review (TOR)

Review Questions

1. Define the term *hazard*.
2. What is the purpose of preliminary hazard analysis?
3. Explain why experience and related expertise are so important when conducting a preliminary hazard analysis.
4. Why is cost–benefit analysis such a critical part of hazard analysis and prevention?
5. Briefly describe the following detailed hazard analysis methods: FMEA, HAZOP, HEA, FTA, and TOR.
6. What is the most fundamental weakness of both FMEA and HAZOP? How can it be overcome?
7. Name and briefly explain the two approaches to HEA.
8. Why did it take so long for TOR to be adopted?
9. What is the most important strength of TOR?
10. Name the five widely applicable hazard-prevention strategies.
11. What is risk assessment? How is it used?

Critical Thinking and Discussion Activities

1. "This stuff is way too complicated. I don't see anybody using hazard analysis in construction," said one college student. "I disagree. I like this stuff. If I get a job with a large enough company, I'm going to use it," answered his classmate. Join this debate. Is it realistic to expect construction professionals to apply the various hazard analysis techniques described in this chapter? Why or why not? Do you think that the size of the construction company is a factor?
2. "The only hazard analysis technique I'm going to use is *human error analysis*," said one student. "Why do you say that?" asked her classmate. "Because human error is what's really behind every construction accident," replied the first student. Do you agree with the first student that human error is ultimately the cause in every construction accident? What might some other causes be? Could these causes be attributed at any level to human error?

Application Activities

1. Locate a construction company in your community that will allow you to visit a job site and observe the work there. Select a given operation and observe it in detail. Develop a *fault tree analysis* for the project.
2. Locate a construction company that will work with you and your classmates on this project. With the permission of the company, organize a team of students to develop a partial *hazards inventory* containing at least four different processes.

Endnotes

1. The Associated General Contractors of America, retrieved December 1, 2015, from http://www.agc.org/cs/safety_management_training_course
2. American Institute of Chemical Engineers, *Guidelines for Hazard Evaluation Procedures* (Chicago: American Institute of Chemical Engineers, 1985), 13.
3. Federal Aviation Administration. "Technic of Operations Review." Retrieved from http://www.hf.faa.gov/Workbenchtools/default.aspx?rPage=Tooldetails&SubCatId=43&toolID=268 on January 7, 2016.
4. United States Nuclear Regulatory Commission. "Tutorial on Probabilistic Risk Assessment (PRA)." Retrieved from www.nrc.gov/about-nrc/regulatory/risk-informed/rpp/nra-tutorial.pdf on January 15, 2016.
5. Ibid., 37.
6. Ibid., 38.
7. The Associated General Contractors of America, retrieved December 1, 2015, from http://www.agc.org/cs/safety_management_training_course

CHAPTER NINE

ACCIDENT INVESTIGATION, RECORD KEEPING, AND REPORTING

LEARNING OBJECTIVES

- Explain when an accident should be investigated.
- Explain what should be investigated after an accident.
- Explain who should investigate accidents.
- Describe how to conduct an accident investigation.
- Summarize the process of interviewing witnesses.
- Describe the requirements/process for reporting accidents.
- Describe the requirements/process for record keeping.

When an accident occurs, it must be investigated thoroughly. A comprehensive **accident report** can help construction professionals pinpoint the cause of the accident. This information can then be used to prevent future accidents, which is the primary purpose of **accident investigation**.

According to the Associated General Contractors of America, the importance of thoroughly investigating accidents is attested to by the following:

> *The purpose of accident investigation is to determine the cause for the purpose of preventing future accidents, not for assessing blame.*[1]

This chapter gives construction students and professionals the information they need to conduct thorough, effective accident investigations and prepare comprehensive accident reports.

WHEN TO INVESTIGATE?

Of course, the first thing to do when an accident takes place is to implement **emergency procedures**. This involves bringing the situation under control and caring for the injured worker. As soon as all emergency procedures have been accomplished, the accident investigation should begin. Waiting too long to complete an investigation can harm the results. This is one important rule of thumb to remember. Another is that *all* accidents, no matter how small, should be investigated. **Evidence** suggests that the same factors that cause minor accidents also cause major accidents.[2] Furthermore, a near miss should be treated like an accident and investigated thoroughly.

There are several reasons for conducting investigations immediately. First, immediate investigations are more likely to produce accurate information. The longer the time span between an accident and an investigation, the greater the likelihood of important facts becoming blurred as memories fade. Second, it is important to collect information before the **accident scene** is changed and before witnesses begin comparing notes. Human nature encourages people to change their stories to agree with those of other witnesses.[3] Finally, an immediate investigation is evidence of management's commitment to preventing future accidents. An immediate response shows that management cares.[4]

WHAT TO INVESTIGATE?

The purpose of an *accident investigation* is to collect facts, not to find fault. It is important that construction professionals make this distinction known to all involved. **Faultfinding** can cause reticence among witnesses who have valuable information to share. **Causes** of the accident should be the primary focus. The investigation should be guided by the following words: **who, what, when, where, why**, and **how**.[5]

This does not mean that mistakes and breaches of precautionary procedures by workers are not noted. Rather, when these things are noted, they are recorded as facts instead of faults. If fault must be assigned, that should come later—after all the facts are in. The distinction is a matter of emphasis.

In attempting to find the facts and identify causes, certain questions should be asked, regardless of the nature of the accident.

The Canadian Center for Occupational Health and Safety recommends that accident investigators focus on five areas of potential causation: task, material, environment, personnel, management, and task.[6] There are certain questions that should be asked in each of these areas. The most important of these are listed in the following paragraphs.

Task-Related Questions

- Were safe work procedures used?
- Had conditions changed in ways that made normal procedures unsafe?

- Were the appropriate tools and materials available? Were the tools properly maintained?
- Were the proper tools used and were they used properly?
- Were safety devices working properly?
- Was either lockout or tagout used where necessary?

Material-Related Questions

- Was there an equipment failure?
- What caused the failure?
- Was the machinery poorly designed?
- Were hazardous substances involved? If so, were they clearly identified? Was a less hazardous substance possible for use and available?
- Was the raw material substandard in some way?
- Should PPE have been used? Was it used? Was it the proper PPE? Was it worn/used properly? Were workers trained in the proper use of the PPE?

Environment-Related Questions

- What were the weather conditions? Was it too hot or too cold? Was it wet or dry?
- What was the state of housekeeping?
- Was noise a problem?
- Was there sufficient light for the task being performed?
- Were toxic or hazardous dusts, gases, or fumes present?
- Were there obstacles or impediments to workers in the environment?

Personnel-Related Questions

- Were workers trained to perform the assigned task(s) safely?
- Were workers experienced in the assigned task(s)?
- Were workers physically capable of performing the assigned task(s)?
- Were workers fatigued?
- Were workers in a healthy physical condition?
- Were workers overly stressed from work, personal, or other factors?

Management-Related Questions

- Is there a safety policy and was it followed?
- Were there written procedures and were they followed?
- Were safety procedures clearly communicated to all workers?
- Were safety procedures enforced?
- Were workers properly supervised?
- Were workers properly trained?
- Had hazards been identified previously? If so, had action been taken to eliminate or mitigate the hazards?
- Were unsafe conditions corrected?
- Were tools, equipment, and machinery properly maintained?
- Were safety inspections carried on a systematic and regular basis?

The answers to these questions should be carefully and copiously recorded. You may find it helpful to dictate your findings into a microcassette recorder. This approach allows you to focus more time and energy on investigating and less on taking written notes. Regardless of how the findings are recorded, it is important to be thorough. What may seem like a minor, unrelated fact at the moment could turn out to be a valuable fact later—when all of the evidence has been collected and is being analyzed.

WHO SHOULD INVESTIGATE?

Who should conduct the accident investigation? Should it be the supervisor responsible? A higher-level manager? An outside specialist? There is no simple answer to this question, and there is disagreement among professional people of goodwill.

In some companies, the supervisor of the injured worker conducts the investigation. In others, a safety and health professional performs the job. Some companies form an investigative team; others bring in outside specialists. There are several reasons for the various approaches used. Factors considered in deciding how to approach accident investigations include the following:

- Size of the company.
- Structure of the company's safety and health program.
- Type of accident.
- Seriousness of the accident.
- Technical complexity.
- Number of times similar accidents have occurred.
- Company's management philosophy.
- Company's commitment to safety and health.

After considering all of the variables just listed, it is difficult to envision a scenario in which a safety and health professional would not be involved in conducting an accident investigation. If the accident in question is very minor, the injured employee's supervisor may conduct the investigation, but a safety and health professional should at least study the accident report and be consulted regarding recommendations for corrective action.

If the accident is so serious that it has widespread negative implications in the community and beyond, responsibility for the investigation may be given to a high-level manager or corporate executive. In such cases, a safety and health professional should assist in conducting the investigation. If a company prefers the team approach, a safety and health professional should be a member of the team and, in most cases, chair it. Regardless of the approach preferred by a given company, a safety and health professional should play a leadership role in collecting and analyzing the facts and developing recommendations.

CONDUCTING THE INVESTIGATION

The questions in the previous section summarize what to look for when conducting accident investigations. Figure 9–1 lists five steps to follow when conducting an accident investigation.[7] These steps are explained in the following paragraphs.

Isolate the Accident Scene

You may have seen a crime scene that was sealed off by the police. The entire area surrounding such a scene is typically blocked off by barriers or heavy yellow tape. This is done to keep curious onlookers from removing, disturbing, or unknowingly destroying vital evidence. This same approach should be used when conducting an accident investigation. As soon as emergency procedures have been completed and the injured worker has been removed, qualified personnel should **isolate the accident scene** until all pertinent evidence has been collected or observed and recorded. Furthermore, nothing but the injured worker should be removed from the scene. If necessary, a security guard should be posted to maintain the integrity of the accident scene. The purpose of isolating the scene is to maintain as closely as possible the conditions that existed at the time of the accident.

Record All Evidence

Make a permanent record of all *pertinent evidence* as quickly as possible. There are three reasons for this:

1. Certain types of evidence may be perishable.
2. The longer an accident scene must be isolated, the more likely it is that evidence will be disturbed, whether knowingly or unknowingly.
3. If the isolated scene contains a critical piece of equipment or a critical component in a larger process, pressure will quickly mount to get it back into operation.

Evidence can be recorded in a variety of ways, including written notes, sketches, photography, videotape, dictated

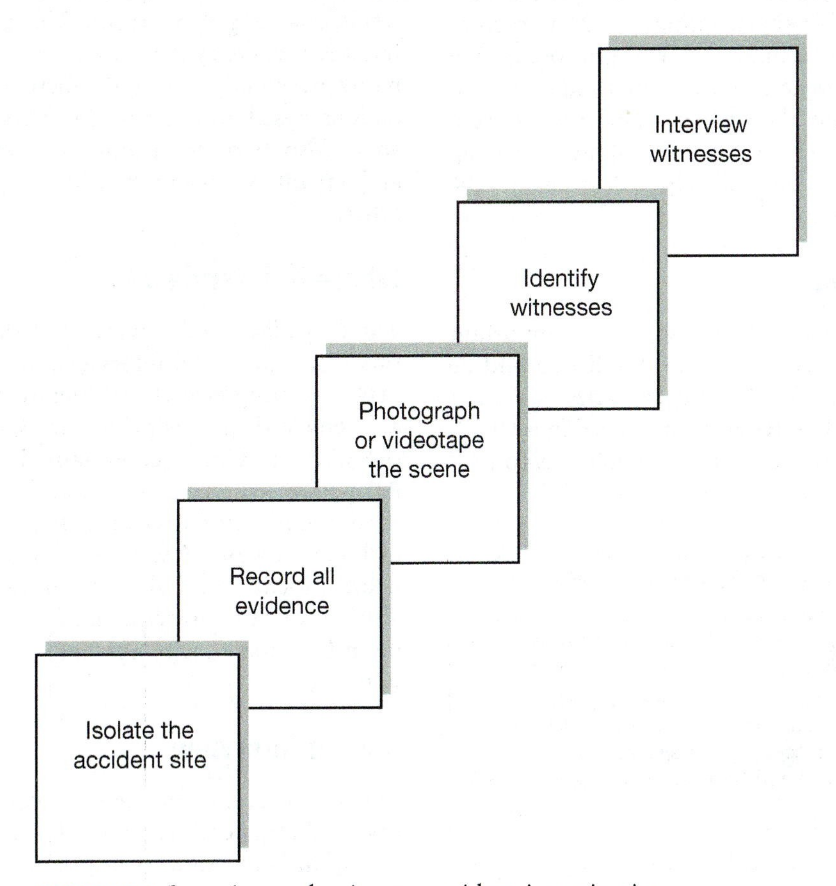

FIGURE 9–1 Steps in conducting an accident investigation.

observations, and diagrams. In deciding what to record, a good rule of thumb is "If in doubt, record it." It is better to record too much than to skip evidence that may be needed later—after the accident scene has been disturbed.

Photograph or Videotape the Scene

This step is actually an extension of the previous step. Photographic and videotaping technology has simplified the task of observing and recording evidence. Construction professionals who are responsible for safety should be proficient in the operation of a still camera, even if it is just an instant camera, and a videotaping camera.

The advent of the digital camera has introduced a new meaning to the term *instant photographs*. Using a digital camera in conjunction with a computer, photographs of accident scenes can be viewed immediately and transmitted instantly to numerous locations. Digital camera equipment is especially useful when photographs of accident scenes in remote locations are needed.

Both still and video cameras should be on hand, loaded, and ready to use immediately should an accident occur. As with the previous step, a good rule of thumb in photographing and videotaping is "If in doubt, shoot it." When recording evidence, it is better to have more shots than necessary than to risk missing a vital piece of evidence.

A problem with photographs is that, by themselves, they don't always reveal objects in their proper perspective. To overcome this shortcoming, the NSC recommends the following technique. When photographing objects involved in the accident, be sure to identify and measure them to show the proper perspective. Place a ruler or coin next to the object when making a close-up photograph. This technique will help to demonstrate the object's size or perspective.[8]

Identify Witnesses

During the process to **identify witnesses,** it is important to compile a witness list. Names on the list should be recorded in three categories: (1) **primary witnesses,** (2) **secondary witnesses,** and (3) **tertiary witnesses** (Figure 9–2). When compiling the witness list, ask employees to provide names of all three types of witnesses.

- Primary witnesses are eyewitnesses to the accident.
- Secondary witnesses are witnesses who did not actually see the accident happen, but were in the vicinity and arrived on the scene immediately or very shortly after the accident.
- Tertiary witnesses are witnesses who were not present at the time of the accident nor afterward, but may still have relevant evidence to present (e.g., an employee who had complained earlier about a problem with equipment involved in the accident).

FIGURE 9–2 Categories of accident witnesses.

Interview Witnesses

Every witness on the list should be interviewed, preferably in the following order: primary witnesses first, secondary next, and tertiary last. After all witnesses have been interviewed, it may be necessary to re-interview witnesses for clarification or corroboration. Interviewing witnesses is such a specialized process that the next major section is devoted to it.

INTERVIEWING WITNESSES

The techniques for interviewing accident witnesses are designed to ensure that the information is objective and accurate and can be corroborated in a manner as untainted by the personal opinions and feelings of witnesses as possible. For this reason, it is important to understand the *when*, *where*, and *how* of interviewing the accident witnesses.

When to Interview?

Immediacy is important. Interviews should begin as soon as the witness list has been compiled and, once begun, should proceed expeditiously. There are two main reasons for this. First, a witness's recollections will be best right after the accident. The more time that elapses between the accident and the interview, the more blurred the witness's memory will become. Second, immediacy avoids the possibility of witnesses comparing notes and, as a result, changing their stories. This is just human nature, but it is a tendency that can undermine the value of testimony given and, in turn, the facts collected. Recommendations based on questionable facts are not likely to be valid. Also, witnesses should be interviewed individually and separately, preferably before they have talked to each other.

Where to Interview?

The best place to interview is at the accident scene. If this is not possible, interviews should take place in a private setting elsewhere. Ensure that all distractions are removed, interruptions are guarded against, and the witness is not accompanied by other witnesses. All persons interviewed should be allowed to relate their recollections without fear of contradiction or influence by other witnesses or employees. Select a neutral location in which witnesses will feel comfortable. Avoid the **principal's office syndrome** by selecting a location that is not likely to be intimidating to witnesses.

How to Interview?

The key to getting at the facts is to put the witness at ease and to listen. Listen to what is said, how it is said, and what is not said. Ask questions that will get at the

SAFETY FACTS & FINES

Where to Conduct Accident Interviews

To ensure that employees are willing to give accurate information, safety and health professionals should conduct accident interviews in the privacy of their office, right? Not necessarily. Experience has shown that the best way to promote accuracy is to interview witnesses at the site of the accident. This puts the accident interview in context in a setting that will help stimulate the memory. To ensure privacy and confidentiality, interview witnesses one at a time.

information listed earlier in this chapter, but phrase them in an **open-ended question** format. For example, instead of asking "Did you see the victim pull the red lever?", phrase your question as follows: "Tell me what you saw." Don't lead witnesses with your questions or influence them with gestures, facial expressions, tone of voice, or any other form of nonverbal communication. Interrupt only if absolutely necessary to seek clarification on a critical point. Remain nonjudgmental and objective.

The information being sought in an accident investigation can be summarized as *who, what, when, where, why*, and *how* (Figure 9–3).

As information is given, it may be necessary to take notes. If you can keep your note taking to a minimum during the interview, your chances of getting uninhibited information are increased. Note taking can distract and even frighten a witness.

An effective technique is to listen during the interview and make mental notes of critical information. At the end of the interview, summarize what you have heard and have the witness verify your summary. After the witness leaves, develop your notes immediately.

A question that sometimes arises is "Why not tape the interview?" Safety and health professionals disagree on the effectiveness and advisability of taping. Those who favor taping claim that it allows the interviewer to concentrate on listening without having to worry about forgetting a key point or having to interrupt the witnesses to jot down critical information. It also preserves everything that is said for the record and the tone of voice in which it is said. A complete transcript of the interview also ensures that information is not taken out of context.

Those opposed to taping say that taping devices tend to inhibit witnesses so that they are not as forthcoming as they would be without taping. Taping also slows down the investigation while the taped interview is transcribed, and the interviewer must wade through voluminous testimony trying to separate critical information from irrelevant information.

In any case, if the interview is to be taped, the following rules should be applied:

- Use the smallest, most unobtrusive taping device available, such as a microcassette recorder.
- Inform the witness that the interview will be taped.
- Make sure the taping device is working properly and that the tape it contains can run long enough so that you do not have to interrupt the witness to change it.
- Take time at the beginning of the interview to discuss unrelated matters long enough to put the witness at ease and overcome the presence of the taping device.
- Make sure that the personnel are available to transcribe the tapes immediately.
- Read the transcripts as soon as they are available, and highlight critical information.

An effective technique to use with **eyewitnesses** is to ask them to reenact the accident for you. Of course, the effectiveness of this technique is enhanced if the **reenactment** can take place at the accident site. However, even when this is not possible, an eyewitness reenactment can yield valuable information.

In using the *reenactment technique*, a word of caution is in order. If an eyewitness does exactly what the

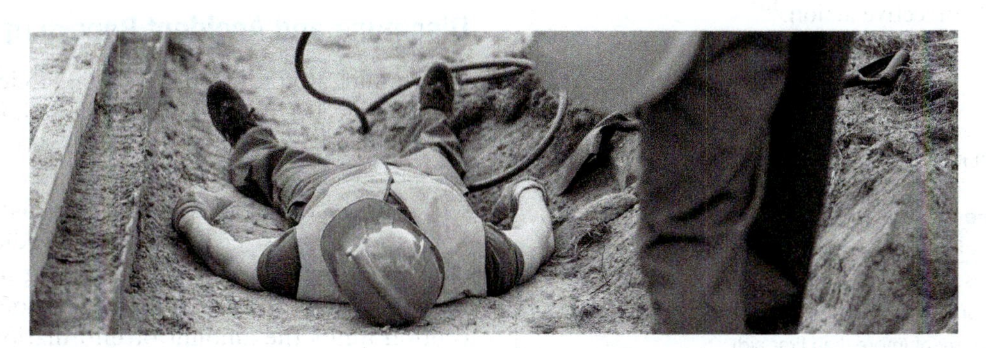

FIGURE 9–3 It is important to ask witnesses who, what, when, where, why, and how.

Source: Photographee.eu/Fotolia

SAFETY FACTS & FINES

Insufficient guarding devices on power tools and machines can introduce amputation hazards. A company in Trenton, Georgia, was fined $68,900 for failing to use machine guards in a way that would protect workers from amputation hazards. The company was cited for the following violations: (1) use of unguarded machines with exposed chains, (2) no lockout or tagout procedures, (3) improper use of compressed gas for cleaning, and (4) failure to provide guardrails for stairs.

victim did, there may be another accident. Have the eyewitnesses explain what they are going to do, before letting them do it. Then, have them *simulate,* rather than actually perform, the steps that led up to the accident.

REPORTING ACCIDENTS

Before presenting the details of accident reporting, it is important to present a pertinent Occupational Safety and Health Administration (OSHA) requirement: any case of a fatality or of three or more employees being hospitalized requires notification of the appropriate regional office of OSHA within eight hours of the incident (reference 29 CFR 1903 and 1904).

An accident investigation should culminate in a comprehensive accident report. The purpose of the report is to record the findings of the accident investigation, the cause or causes of the accident, and the recommendations for corrective action.

All injuries and illnesses should be recorded, regardless of severity, if they result in any of the outcomes shown in Figure 9–4. If an accident results in the death of an employee or hospitalization of three or more employees, a report must be submitted to the nearest OSHA office within eight hours. This rule applies regardless of the size of the company.[9]

Accident report forms vary from company to company. However, the information contained in them is fairly standard.

There are numerous examples of excellent accident report forms available for downloading on the Internet. All of them contain requirements for such information concerning the victim, injuries, witnesses, location, whether medical treatment was required (if so where it was provided by and whom), supervisor, cause of the accident, corrective action taken, and recommendations for additional corrective action.[10]

Injuries and illnesses should be recorded if they result in any of the following:

- Death
- One or more lost workdays
- Restriction of motion or work
- Loss of consciousness
- Transfer to another job
- Medical treatment (more than first aid)

FIGURE 9–4 Record-keeping requirements.

In addition to these items, you may want to record such additional information as the list of witnesses; dates, times, and places of interviews; historical data related to similar accidents; information about related corrective actions that had been taken in the past, but had not yet been followed up; and any other information that might be relevant. Figure 9–5 is OSHA Form 301, "Injury and Illness Incident Report." It satisfies the requirements for information, but is not a mandatory form. Companies may develop their own forms, as long as they contain all of the necessary information.

Why Some Accidents Are Not Reported?

Many accidents go unreported.[11]

There are several reasons why accidents go unreported. Be familiar with these reasons so that you can do your part to overcome them. The main reasons are red tape, ignorance, embarrassment, record spoiling, fear of repercussions, and lack of feedback.[12]

1. *Red tape.* Some people see the paperwork involved in accident reporting as red tape and, therefore, do not report accidents just to avoid paperwork.

Clearly, these reasons for not reporting accidents present construction professionals with a challenge. To overcome these inhibitors, it is necessary to develop a simple reporting system that will not be viewed as too much bureaucratic paperwork. Construction professionals must educate personnel at all levels concerning the purpose of accident reporting and why it is important. An important step is to communicate the fact that faultfinding is not the purpose. Another important step is to follow up to ensure that recommendations are enacted or that employees are told why they are not being enacted. This helps to ensure the integrity of the process.

Discipline and Accident Reporting

Faultfinding is not the purpose of an accident investigation.[13] However, an investigation sometimes reveals that an employee has violated or simply overlooked safety regulations. Should such violations be condoned?

There is a built-in dilemma here that construction professionals must be prepared to handle. On the one hand, it is important that faultfinding not be seen as the purpose of an accident investigation. Such a perception limits the amount of information that can be collected. On the other hand, if those workers whose behavior leads to accidents are not disciplined, the

OSHA's Form 301

Injury and Illness Incident Report

Attention: This form contains information relating to employee health and must be used in a manner that protects the confidentiality of employees to the extent possible while the information is being used for occupational safety and health purposes.

U.S. Department of Labor
Occupational Safety and Health Administration

Form approved OMB no. 1218-0176

This *Injury and Illness Incident Report* is one of the first forms you must fill out when a recordable work-related injury or illness has occurred. Together with the *Log of Work-Related Injuries and Illnesses* and the accompanying *Summary*, these forms help the employer and OSHA develop a picture of the extent and severity of work-related incidents.

Within 7 calendar days after you receive information that a recordable work-related injury or illness has occurred, you must fill out this form or an equivalent. Some state workers' compensation, insurance, or other reports may be acceptable substitutes. To be considered an equivalent form, any substitute must contain all the information asked for on this form.

According to Public Law 91-596 and 29 CFR 1904, OSHA's recordkeeping rule, you must keep this form on file for 5 years following the year to which it pertains.

If you need additional copies of this form, you may photocopy and use as many as you need.

Completed by ____________

Title ____________

Phone (____)____-____ Date ___/___/___

Information about the employee

1) Full name ____________

2) Street ____________

City ____________ State ______ ZIP ______

3) Date of birth ___/___/___

4) Date hired ___/___/___

5) ☐ Male
☐ Female

Information about the physician or other health care professional

6) Name of physician or other health care professional ____________

7) If treatment was given away from the worksite, where was it given?

Facility ____________

Street ____________

City ____________ State ______ ZIP ______

8) Was employee treated in an emergency room?
☐ Yes
☐ No

9) Was employee hospitalized overnight as an in-patient?
☐ Yes
☐ No

Information about the case

10) Case number from the *Log* ____________ *(Transfer the case number from the Log after you record the case.)*

11) Date of injury or illness ___/___/___

12) Time employee began work ________ AM / PM

13) Time of event ________ AM / PM ☐ Check if time cannot be determined

14) ***What was the employee doing just before the incident occurred?*** Describe the activity, as well as the tools, equipment, or material the employee was using. Be specific. *Examples:* "climbing a ladder while carrying roofing materials"; "spraying chlorine from hand sprayer"; "daily computer key-entry."

15) ***What happened?*** Tell us how the injury occurred. *Examples:* "When ladder slipped on wet floor, worker fell 20 feet"; "Worker was sprayed with chlorine when gasket broke during replacement"; "Worker developed soreness in wrist over time."

16) ***What was the injury or illness?*** Tell us the part of the body that was affected and how it was affected; be more specific than "hurt," "pain," or sore." *Examples:* "strained back"; "chemical burn, hand"; "carpal tunnel syndrome."

17) ***What object or substance directly harmed the employee?*** *Examples:* "concrete floor"; "chlorine"; "radial arm saw." *If this question does not apply to the incident, leave it blank.*

18) ***If the employee died, when did death occur?*** Date of death ___/___/___

Public reporting burden for this collection of information is estimated to average 22 minutes per response, including time for reviewing instructions, searching existing data sources, gathering and maintaining the data needed, and completing and reviewing the collection of information. Persons are not required to respond to the collection of information unless it displays a current valid OMB control number. If you have any comments about this estimate or any other aspects of this data collection, including suggestions for reducing this burden, contact: US Department of Labor, OSHA Office of Statistics, Room N-3644, 200 Constitution Avenue, NW, Washington, DC 20210. Do not send the completed forms to this office.

FIGURE 9–5 OSHA Form 301: Injury and Illness Incident Report.

Source: From Occupational Safety & Health Administration. Published by U.S Department of Labor.

credibility of the safety program is undetermined. Kane and Cunningham recommend the following procedures for handling this dilemma: *Never* discipline an employee because he or she had an accident. *Always* discipline employees for noncompliance with safety regulations.[14]

Such an approach applied with consistency will help maintain the integrity of both the accident investigation process and the overall safety program.

Electronic Reporting Requirements

OSHA's rule requiring certain employers to file reports on workplace injuries and illnesses electronically went into effect on January 1, 2017. Information from electronically filed reports will, in turn, be posted on OSHA's website (although personal information about injured employees will be redacted). However, the employer will be identified on the website. The electronic reporting requirement applies to construction firms with 20 or more employees. In addition to the electronic reporting requirement, the rule contains the following elements:

- ***Anti-retaliation measures.*** OSHA's anti-retaliation measures are strengthened to protect employees who report unsafe conditions. Employees who report unsafe conditions are protected from retaliation or discrimination by termination, reduced pay, assignment to less favorable jobs, or other adverse actions.
- ***Employee reporting measures.*** Employers must establish a reasonable procedure for employees to report injuries and illnesses (although OSHA does not define what is considered "reasonable").
- ***Safety incentives.*** OSHA began discouraged safety incentives in 2012 with a guidance memorandum believing they may inadvertently lead to underreporting of injuries and illnesses. The electronic reporting rule does not prohibit the use of incentives, but the language in the rule makes it clear that OSHA does not encourage them and will scrutinize their use closely.

RECORD KEEPING

OSHA has specific requirements for record keeping related to injuries and illnesses. OSHA Form 300, "Log of Work-Related Injuries and Illnesses," is used to classify work-related injuries and illnesses and to describe the extent or severity of each case (Figure 9–6).[15] OSHA Form 300A, "Summary of Work-Related Injuries and Illnesses," is used to show the total number of injuries and illnesses in each category in a given year (Figure 9–7). Employers are required to maintain a separate "log" for each location or site. They must also maintain a separate "summary" for each location that is expected to be in operation for one year or more. Entering a case on the "log" does not imply any guilt or negligence on the part of the employer, nor does it mean that an employee has qualified for workers' compensation.

"Work-Related" Injury or Illness Defined

OSHA requires employers to keep records of work-related injuries and illnesses, but how does an employer know if a case is work related? An injury or illness is work related if an event or exposure in the work environment caused or contributed to the injury or illness or significantly aggravated a pre-existing condition. Injuries or illnesses that result from events or exposures in the workplace can be assumed to be work related, unless a specific exception applies. The work environment includes establishments, job sites, or other locations where one or more employees are working or are present as a condition of employment.

Injuries and Illnesses That Must Be Recorded

Companies are required to record work-related injuries and illnesses that result in any of the following:

- Death
- Loss of consciousness
- Days away from work
- Restricted work activity or transfer
- Medical treatment beyond first aid
- Diagnosis by a physician or other licensed health-care professional
- Cancer
- Fractured or cracked bone
- Chronic irreversible disease
- Punctured eardrum

In addition to these conditions, the following must be recorded: (1) any puncture or cut from a sharp object that is contaminated with another person's blood or any other infectious substance, (2) any condition requiring an employee to be removed under the requirements of an OSHA health standard, (3) any standard threshold shift in hearing, and (4) tuberculosis infection as diagnosed by a physician or a licensed health-care professional or an approved skin test.

Medical Treatment versus First Aid

In deciding whether or not an injury or illness should be recorded or in deciding how it should be recorded, it is important to understand the difference between medical treatment and first aid. Medical treatment involves managing and caring for a patient for the purpose of combating a disease or disorder. The following are *not* considered medical treatment: (1) visits to a physician or health-care professional for the purpose of observation

OSHA's Form 300

Log of Work-Related Injuries and Illnesses

Attention: This form contains information relating to employee health and must be used in a manner that protects the confidentiality of employees to the extent possible while the information is being used for occupational safety and health purposes.

Year 20__ __

U.S. Department of Labor
Occupational Safety and Health Administration

Form approved OMB no. 1218-0176

You must record information about every work-related death and about every work-related injury or illness that involves loss of consciousness, restricted work activity or job transfer, days away from work, or medical treatment beyond first aid. You must also record significant work-related injuries and illnesses that are diagnosed by a physician or licensed health care professional. You must also record work-related injuries and illnesses that meet any of the specific recording criteria listed in 29 CFR Part 1904.8 through 1904.12. Feel free to use two lines for a single case if you need to. You must complete an Injury and Illness Incident Report (OSHA Form 301) or equivalent form for each injury or illness recorded on this form. If you're not sure whether a case is recordable, call your local OSHA office for help.

Establishment name ________________

City ____________ State ______

Identify the person			Describe the case			Classify the case												
						Using these four categories, check ONLY the most serious result for each case:				**Enter the number of days the injured or ill worker was:**		**Check the "Injury" column or choose one type of illness:** (M)						
(A) Case no.	(B) Employee's name	(C) Job title (e.g., *Welder*)	(D) Date of injury or onset of illness	(E) Where the event occurred (*e.g., Loading dock north end*)	(F) Describe injury or illness, parts of body affected, and object/substance that directly injured or made person ill (*e.g., Second degree burns on right forearm from acetylene torch*)	Death (G)	Days away from work (H)	Remained at work: Job transfer or restriction (I)	Remained at work: Other record-able cases (J)	On job transfer or restriction (K)	Away from work (L)	Injury (1)	Musculoskeletal disorder (2)	Skin disorder (3)	Respiratory condition (4)	Poisoning (5)	Hearing loss (6)	All other illnesses (7)
			/ month/day			☐	☐	☐	☐	___ days	___ days	☐	☐	☐	☐	☐	☐	☐
			/ month/day			☐	☐	☐	☐	___ days	___ days	☐	☐	☐	☐	☐	☐	☐
			/ month/day			☐	☐	☐	☐	___ days	___ days	☐	☐	☐	☐	☐	☐	☐
			/ month/day			☐	☐	☐	☐	___ days	___ days	☐	☐	☐	☐	☐	☐	☐
			/ month/day			☐	☐	☐	☐	___ days	___ days	☐	☐	☐	☐	☐	☐	☐
			/ month/day			☐	☐	☐	☐	___ days	___ days	☐	☐	☐	☐	☐	☐	☐
			/ month/day			☐	☐	☐	☐	___ days	___ days	☐	☐	☐	☐	☐	☐	☐
			/ month/day			☐	☐	☐	☐	___ days	___ days	☐	☐	☐	☐	☐	☐	☐
			/ month/day			☐	☐	☐	☐	___ days	___ days	☐	☐	☐	☐	☐	☐	☐
			/ month/day			☐	☐	☐	☐	___ days	___ days	☐	☐	☐	☐	☐	☐	☐
			/ month/day			☐	☐	☐	☐	___ days	___ days	☐	☐	☐	☐	☐	☐	☐
			/ month/day			☐	☐	☐	☐	___ days	___ days	☐	☐	☐	☐	☐	☐	☐
			/ month/day			☐	☐	☐	☐	___ days	___ days	☐	☐	☐	☐	☐	☐	☐
					Page totals ➤	___	___	___	___	___	___	Injury (1)	Musculoskeletal disorder (2)	Skin disorder (3)	Respiratory condition (4)	Poisoning (5)	Hearing loss (6)	All other illnesses (7)

Be sure to transfer these totals to the Summary page (Form 300A) before you post it.

Page ___ of ___

Public reporting burden for this collection of information is estimated to average 14 minutes per response, including time to review the instructions, search and gather the data needed, and complete and review the collection of information. Persons are not required to respond to the collection of information unless it displays a currently valid OMB control number. If you have any comments about these estimates or any other aspects of this data collection, contact: US Department of Labor, OSHA Office of Statistics, Room N-3644, 200 Constitution Avenue, NW, Washington, DC 20210. Do not send the completed forms to this office.

FIGURE 9–6 OSHA Form 300: Log of Work-Related Injuries and Illnesses.

Source: From Occupational Safety & Health Administration. Published by U.S Department of Labor.

OSHA's Form 300A

Summary of Work-Related Injuries and Illnesses

Year 20__ __

U.S. Department of Labor
Occupational Safety and Health Administration

Form approved OMB no. 1218-0176

All establishments covered by Part 1904 must complete this Summary page, even if no work-related injuries or illnesses occurred during the year. Remember to review the Log to verify that the entries are complete and accurate before completing this summary.

Using the Log, count the individual entries you made for each category. Then write the totals below, making sure you've added the entries from every page of the Log. If you had no cases, write "0."

Employees, former employees, and their representatives have the right to review the OSHA Form 300 in its entirety. They also have limited access to the OSHA Form 301 or its equivalent. See 29 CFR Part 1904.35, in OSHA's recordkeeping rule, for further details on the access provisions for these forms.

Number of Cases

Total number of deaths	Total number of cases with days away from work	Total number of cases with job transfer or restriction	Total number of other recordable cases
____	____	____	____
(G)	(H)	(I)	(J)

Number of Days

Total number of days of job transfer or restriction	Total number of days away from work
____	____
(K)	(L)

Injury and Illness Types

Total number of . . .
(M)

(1) Injuries	____	(4) Respiratory conditions	____
		(5) Poisonings	____
(2) Musculoskeletal disorders	____	(6) Hearing loss cases	____
(3) Skin disorders	____	(7) All other illnesses	____

Establishment information

Your establishment name ____________

Street ____________

City ____________ State ____ ZIP ____

Industry description *(e.g., Manufacture of motor truck trailers)* ____________

Standard Industrial Classification (SIC), if known *(e.g., SIC 3715)* __ __ __ __

Employment information *(If you don't have these figures, see the Worksheet on the back of this page to estimate.)*

Annual average number of employees ____

Total hours worked by all employees last year ____

Sign here

Knowingly falsifying this document may result in a fine.

I certify that I have examined this document and that to the best of my knowledge the entries are true, accurate, and complete.

______________ Company executive ______________ Title

( ) ______________ Phone ______________ Date

Post this Summary page from February 1 to April 30 of the year following the year covered by the form.

Public reporting burden for this collection of information is estimated to average 58 minutes per response, including time to review the instructions, search and gather the data needed, and complete and review the collection of information. Persons are not required to respond to the collection of information unless it displays a currently valid OMB control number. If you have any comments about these estimates or any other aspects of this data collection, contact: US Department of Labor, OSHA Office of Statistics, Room N-3644, 200 Constitution Avenue, NW, Washington, DC 20210. Do not send the completed forms to this office.

FIGURE 9–7 OSHA Form 300A: Summary of Work-Related Injuries and Illnesses.

Source: From Bureau of Labor Statistics. Published by U.S. Department of Labor.

or counseling, (2) diagnostic procedures, including the administration of prescription medicines used solely for diagnostic purposes, and (3) any procedures that qualify as first aid.

First aid procedures do not qualify as medical treatment. Consequently, injuries and illnesses that require only first aid do not have to be recorded. First aid includes the following:

- Administering nonprescription medications at nonprescription strengths.
- Administering tetanus immunizations.
- Cleaning, flushing, or soaking wounds on the skin's surface.
- Using wound coverings, such as bandages, gauze pads, adhesive strips, or butterfly bandage.
- Administering hot or cold therapy.
- Using any totally nonrigid means of support, such as elastic bandages and wraps.
- Using temporary immobilization devices while transporting an accident victim (splints, slings, neck collars, or back boards).
- Drilling a fingernail or toenail to relieve pressure or draining fluid from blisters.
- Using eye patches.
- Using simple irrigation or a cotton swab to remove foreign bodies from the eye that are not embedded in or adhered to the eye.
- Using irrigation, tweezers, cotton swabs, or other simple means to remove splinters or foreign material from areas other than the eye.
- Using finger guards.
- Using massages.
- Drinking fluids to relieve heat stress.

Privacy Cases

In some cases involving injuries and illnesses, privacy becomes an issue. The following types of cases should be recorded as "privacy cases," rather than using the employee's name:

- An injury or illness involving the reproductive system or an intimate part of the body.
- An injury or illness resulting from sexual assault.
- A mental illness.
- A case involving HIV, hepatitis, or tuberculosis infection or transmission.
- A needlestick injury from a cut or a sharp object that is contaminated with blood or some other potentially infectious substance.
- An injury or illness in which the employee in question specifically requests that his or her name not be entered into the log.

Posting of the Summary

It is not necessary to post the log (OSHA Form 300), but the summary (OSHA Form 300A) must be posted according to specific criteria. The summary must be posted—in a conspicuous location accessible to employees—no later than February 1, covering the previous year that ended in December. The summary is prepared and posted according to calendar years rather than fiscal years. Once posted, the summary must remain posted until April 30. Both the log and the summary are to be retained on file for five years after the year they cover.

Summary

Accidents are investigated for the purpose of identifying causal factors that could lead to other accidents if not corrected. The purpose is *not* to assign blame.

It is important to begin an accident investigation as soon as possible after an accident occurs so that evidence and the memories of witnesses are still fresh.

Facts to be uncovered in an accident investigation can be summarized as *who, what, when, where, why*, and *how*.

Who conducts the accident investigation can vary according to circumstances. However, regardless of how it is done, a safety and health professional should play an active role in the process.

Steps for conducting an accident investigation are as follows: (1) isolate the accident scene, (2) record all evidence, (3) photograph or videotape the accident scene, (4) identify witnesses, and (5) interview witnesses.

Witnesses to accidents fall into one of three categories: primary (eyewitnesses); secondary (were present at the scene, but did not see the accident); and tertiary (were not present, but have information that may be relevant).

Interviews should take place at the accident site whenever possible. When this is not practical, interviews should take place at a neutral location that is private and where the witness is comfortable.

The keys to getting at the facts in an interview are as follows: (1) put the witness at ease, (2) ask open-ended questions, and (3) listen. Interrupt only if absolutely necessary.

When possible, let eyewitnesses reenact the accident through simulation at the job site. Do not let them actually perform the tasks that led up to the accident.

The purpose of an accident report is to record the findings of the accident investigation, the cause or causes of the accident, and the recommendations for corrective action. Report forms should meet the record-keeping specifications of OSHA.

Key Terms and Concepts

Accident investigation
Accident report
Accident scene
Causes
Emergency procedures
Evidence
Eyewitnesses
Faultfinding
How
Identify witnesses
Immediacy
Isolate the accident scene
Open-ended question
Primary witnesses
Principal's office syndrome
Reenactment
Secondary witnesses
Tertiary witnesses
What
When
Where
Who
Why

Review Questions

1. Explain the rationale for investigating accidents.
2. When should an investigation be reported? Why?
3. What are the terms that should guide the conduct of an accident investigation?
4. What role should the safety and health professional play in conducting an accident investigation?
5. List and explain the steps for conducting an accident investigation.
6. Why is it important to record all pertinent evidence relating to an accident immediately after the accident has occurred?
7. What can you do when taking close-up photographs to put them in the proper perspective?
8. List and differentiate among the three categories of witnesses to an accident.
9. Briefly explain the *when* and *where* of interviewing witnesses.
10. Briefly explain the *how* of interviewing witnesses.
11. What is the purpose of an accident report?

Critical Thinking and Discussion Activities

1. "I'm no investigator," said Mack Jones, project superintendent for Indies Construction Company (ICC). He had just put a seriously injured employee, Wanda Burke, in an ambulance and sent it speeding away to the hospital. "I'll get someone to help me investigate what happened tomorrow or the next day," said Jones. "Right now, we need to get everyone back to work. Then I'm going to the hospital to check on Wanda." Critique the approach Jones is taking to accident investigation. How effective do you think it will be? What, if anything, would you do differently?
2. "We've got to get to the bottom of this accident. I want to know who is responsible and I want him fired today!" You are the safety director for Bell Contracting Company, and these are the orders your boss gave you just 10 minutes ago. Should you do what he ordered, or should you give him time to cool down and then try to convince him to take a different approach? If you think a different approach is called for, what do you recommend?

Application Activities

1. Use the material in this chapter to develop a detailed list of questions that a medium-sized construction company can use to interview witnesses after an accident.
2. Use the material in this chapter to develop a comprehensive "Standing Operating Procedures Manual" for accident investigation that could be used by any construction company. Include the list of questions from Activity 1 in the manual.

Endnotes

1. The Associated General Contractors of America, retrieved January 5, 2016, from http://www.agc.org/cs/safety_managment_training_course
2. Ibid.
3. Ibid.
4. Ibid.
5. National Safety Council, *Supervisor's Safety Manual 9th edition* (Chicago: NSC, 1997), 69–70.
6. Canadian Center for Occupational Health and Safety. "What should be looked at as the cause of an accident?" Retrieved from http://ccohs.ca/oshanswers/hsprograms/investig.html on January 2, 2016.
7. National Safety Council, *Supervisor's Safety Manual,* 71.
8. Ibid.
9. OSHA 2056, (Revised), U.S. Department of Labor, 11.
10. National Safety Council, *Supervisor's Safety Manual,* 76–77.
11. Cunningham, J., and A. Kane, "Accident Reporting—Part I: Key to Prevention," *Safety & Health* 139, no. 4: 70.
12. Ibid., 70–71.
13. Kane, A., and J. Cunningham, "Accident Reporting—Part II: Consistent Discipline Is Vital," *Safety & Health* 139, no. 5: 78.
14. Ibid.
15. Occupational Safety and Health Administration, "OSHA Forms for Recording Work-Related Injuries and Illnesses," U.S. Department of Labor, January 6, 2016. Available at http://www.osha.gov

CHAPTER TEN

EMERGENCY RESPONSE PLAN

LEARNING OBJECTIVES

- Explain the rationale for emergency preparation.
- Summarize the key elements of the Emergency Planning and Community Right-to-Know Act.
- Describe how emergency responses should be organized and coordinated.
- Explain the role OSHA standards play in emergency planning and response.
- Describe how to customize emergency plans for local use.
- Summarize the uses of first aid in emergencies.
- Define the concept of the Emergency Response Team.
- Explain the role computers can play in emergency response.
- Summarize the key elements for dealing with the psychological trauma of emergencies.

Despite the best efforts of all involved, emergencies do sometimes occur. Fire, hurricanes, tornadoes, earthquakes, random acts of violence, hazardous waste spills, or accidental releases of toxic materials can create crisis conditions. It is important to respond to such emergencies in a way that minimizes harm to people and damage to property. To do so requires plans that can be implemented without delay. This chapter provides construction professionals and students with the information they need to prepare for emergencies in the workplace.

RATIONALE FOR EMERGENCY PREPARATION

An *emergency* is a potentially life-threatening situation, usually occurring suddenly and unexpectedly. Emergencies may be the result of natural or human causes. Have you ever witnessed the timely, organized, and precise response of a professional emergency medical crew to an automobile accident? While passersby and spectators may wring their hands and wonder what to do, the emergency response professionals quickly organize, stabilize, and administer. Their ability to respond in this manner is the result of preparation. As shown in Figure 10–1, preparation involves a combination of **planning, practicing, evaluating,** and **adjusting** to specific circumstances.

Emergency Response Preparation	
✓	Plan
✓	Practice
✓	Evaluate
✓	Adjust

FIGURE 10–1 Components of emergency response preparation.

When an emergency occurs, immediate response is essential. Speed in responding can mean the difference between life and death or between minimal damage and major damage. Ideally, all those involved should be able to respond properly with a minimum of hesitation. This can happen only if all exigencies have been planned for and planned procedures have been practiced, evaluated, and revised as necessary.

A quick and proper response—which results because of proper preparation—can prevent panic, decrease the likelihood of injury and damage, and bring the situation under control in a timely manner. Since no workplace is immune to emergencies, preparing for them is critical. An important component of preparation is planning. The **emergency response plan** should become a section in the company's comprehensive safety and health plan.

EMERGENCY PLANNING AND COMMUNITY RIGHT-TO-KNOW ACT

Title III of the Superfund Amendments and Reauthorization Act (SARA) is also known as the **Emergency Planning and Community Right-to-Know Act.** This law is designed to make information about hazardous chemicals available to a community where they are being used so that residents can protect themselves in the case of an emergency. It applies to all companies that use, make, transport, or store chemicals.

Safety and health professionals involved in developing emergency response plans for their companies should be familiar with the act's requirements related to emergency planning. The Emergency Planning and Community

Emergency Response Preparation	
✓	Emergency planning
✓	Emergency notification
✓	Toxic chemical release reporting
✓	Information requirements

FIGURE 10–2 Four categories of requirements in the Emergency Planning and Right-to-Know Act.

Right-to-Know Act has four major components, and those are discussed in the following paragraphs (Figure 10–2).

Emergency Planning

In developing emergency plans, refer to 29 CFR 1926 subpart C. The emergency planning component requires that communities form **local emergency planning committees (LEPCs)** and that states form **state emergency response commissions (SERCs)**. LEPCs are required to develop emergency response plans for the local communities, host public forums, select a planning **coordinator** for the community, and work with the coordinator in developing local plans. SERCs are required to oversee LEPCs and review their emergency response plans. Plans for individual companies in a given community should be part of that community's larger plan. Local emergency response professionals should use their community's plan as the basis for simulating emergencies and practicing their response.

Emergency Notification

The **emergency notification** component requires that chemical spills or releases of toxic substances that exceed established allowable limits be reported to appropriate LEPCs and SEPCs. Immediate notification may be verbal as long as a written notification is filed promptly. Such a report must contain at least the following information: (1) the names of the substances released, (2) where the release occurred, (3) when the release occurred, (4) the estimated amount of the release, (5) known hazards to people and property, (6) recommended precautions, and (7) the name of a contact person in the company.

Information Requirements

Information requirements dictate that local companies must keep their LEPCs, SERCs, and, through them, the public informed about the hazardous substances that the companies store, handle, transport, and use. This includes keeping comprehensive records of the substances on file, up-to-date, and readily available; providing copies of material safety data sheets for all hazardous substances; giving general storage locations for all hazardous substances; providing estimates of the amount of each hazardous substance on hand on a given day; and estimating the average annual amount of hazardous substances kept on hand.

Toxic Chemical Release Reporting

The **toxic chemical release reporting** component requires that local companies report the total amount of toxic substances released into the environment as either emissions or hazardous waste. Reports go to the Environmental Protection Agency (EPA) and the state-level environmental agency.

ORGANIZATION AND COORDINATION

Responses to emergencies are typically from several people or groups of people, including medical, firefighting, security, and safety personnel, as well as specialists from a variety of fields. People in each of these areas have a different—but interrelated and often interdependent—role to play in responding to the emergency. Because of the disparate backgrounds and roles of these people, organization and coordination are critical.

A company's emergency response plan should clearly identify the different personnel or groups that respond to various types of emergencies and, in each case, who is in charge. One person should be clearly identified and accepted by all emergency responders as the coordinator. Such a person should be knowledgeable—at least in a general sense—of the responsibilities of each individual emergency responder and how they relate to those of all other responders. This knowledge must include the **order of response** for each type of emergency set forth in the plan.

A company's designated safety and health professional is the obvious person to organize and coordinate emergency responses. However, regardless of who is designated, it is important that: (1) one person is in charge, (2) everyone involved knows who is in charge, and (3) everyone who has a role in responding to an emergency is given ample opportunities to practice in simulated conditions that come as close as possible to real conditions.

OSHA STANDARD AND EMPLOYER EMERGENCY ACTION PLANS

All **Occupational Safety and Health Administration (OSHA) standards** are written for the purpose of promoting a safe, healthy, accident-free, and, hence, emergency-free workplace. Therefore, OSHA standards do play a role in emergency prevention and should be considered when developing emergency plans. For example, 29 CFR 1926.35 contains requirements for employee **emergency action plans (EAPs)**. Such plans are required to be in writing (except in the case of companies with 10 or fewer employees) and to cover the actions employers and employees must take to ensure employee safety during emergencies.

Elements

At the very least, the following six elements must be included in an EAP:

1. Emergency escape procedures and emergency escape route assignments.
2. Procedures for employees who stay behind to run critical plant operations before evacuating.
3. Procedures to account for all employees after the emergency evacuation have been completed.
4. Rescue and medical duties for those employees who are to perform them.
5. Preferred procedures for reporting fires and other emergencies.
6. Contact personnel and information for anyone who needs more information or clarification about any aspect of the plan.

Alarm System

Employers are required to provide an alarm system that complies with 29 CFR 1026.159. If the employee alarm system is used for alerting members of the fire brigade or for any other purpose, a distinctive signal for each purpose must be decided on, used, and made known to all employees.

Evacuation

Evacuation procedures are critical. They must be developed and made known to all employees. Although the standard does not specifically require it, construction companies are well advised to practice emergency evacuations with all employees.

Training

Before implementing the EAP, employers must designate and train a sufficient number of workers to assist in the safe and orderly evacuation of employees. In addition, employers must review the EAP with all employees covered by the plan at the following times: (1) initially, when the plan is developed, (2) whenever the employee's responsibilities or designated actions under the plan change, and (3) whenever the plan itself is changed. Upon initial assignment, employers must review with every employee affected those parts of the plan that the employee must know in the event of an emergency. Figure 10–3 is a checklist that can be used for emergency planning.

Type of Emergency

_____ Fire	_____ Explosion
_____ Chemical spill	_____ Toxic emission
_____ Train derailment	_____ Hurricane
_____ Tornado	_____ Lightning
_____ Flood	_____ Earthquake
_____ Volcanic eruption	_____ Other __________

Procedures for Emergency Response

1. Controlling and isolating
2. Communication
3. Emergency assistance
4. First aid
5. Shutdown, evacuation, and protection of workers
6. Protection equipment and property
7. Egress, ingress, exits
8. Emergency equipment (i.e., fire extinguishers)
9. Alarms
10. Restoration of normal operations

Coordination

1. Medical care providers
2. Fire service providers
3. LEPC personnel
4. Environmental protection personnel
5. Civil defense personnel (in the case of public evacuations)
6. Police protection providers
7. Communication personnel

Assignments and Responsibilities

1. Who cares for the injured?
2. Who calls for emergency assistance?
3. Who shuts down power and operations?
4. Who coordinates communication?
5. Who conducts the evacuation?
6. Who meets emergency responders and guides them to the emergency site?
7. Who contacts coordinating agencies and organizations?
8. Who is responsible for ensuring the availability and upkeep?
9. Who is responsible for ensuring that alarms are in proper working order?
10. Who is responsible for organizing cleanup activities?

Accident-Prevention Strategies

1. Periodic safety inspections
2. Industrial hygiene strategies
3. Personal protective equipment
4. Ergonomic strategies
5. Machine safeguarding
6. Hand or portable power tool safeguarding
7. Material handling and storage strategies
8. Electrical safety strategies
9. Fire safety strategies
10. Chemical safety strategies

Schedules

1. Dates of practice drills
2. Times of practice drills
3. Duration of practice drills

FIGURE 10–3 Emergency planning checklist.

CUSTOMIZING PLANS TO MEET LOCAL NEEDS

EAPs must be **location specific**. General plans developed centrally and used at all job-site locations have limited effectiveness. The following rules can be used to ensure that EAPs are location specific:

SAFETY FACTS & FINES

Failure to provide training has serious consequences. A company in Niles, Illinois, that had not trained its employees in safe operating procedures experienced a devastating explosion. One employee was killed and five were seriously injured. The company was fined $99,000 by OSHA and cited for the following violations: failing to train workers sufficiently and failing to keep hazardous material away from open flames and sparks. In addition to paying the fine, the company had to agree to permanently stop the process that had led to the explosion.

1. *A map in the plan.* A map of the specific job site or sites helps to localize an EAP. The map should include the locations of exits, access points, evacuation routes, alarms, emergency equipment, a central control or command center, first aid kits, emergency shutdown buttons, and any other important element of the EAP.
2. *Chain of command.* An organizational chart illustrating the **chain of command**—who is responsible for what and who reports to whom—also helps to localize an EAP. The chart should contain the names and telephone numbers (internal and external) of everyone involved in responding to an emergency. It is critical to keep the organizational chart up-to-date as personnel changes occur and to have a designated backup person shown for every position on the chart.
3. *Coordination information.* All telephone numbers and contact names of people in agencies with which the company coordinates emergency activities should be listed. Periodic contact should be maintained with all such people so that the EAP can be updated as personnel changes occur.
4. *Local training.* All training should be geared toward the types of emergencies that might occur on the job site. In addition, practice drills should take place onsite and in the specific locations where emergencies are most likely to happen.

FIRST AID IN EMERGENCIES

Workplace emergencies often require a medical response. The immediate response is usually first aid. First aid consists of life-saving measures taken to assist an injured person until medical help arrives. According to Kelly,[1]

> This gap in time between when the injury happens and when medical care can be administered is when first aid proves itself. A trained co-worker or supervisor can begin cardiopulmonary resuscitation (CPR), stop bleeding at a pressure point, or use the Heimlich maneuver on a choking employee during lunch. In each of these cases, if first aid were not given within 4 to 6 minutes, the victim could die.

Since there is no way to predict when first aid might be needed, providing **first aid training** to employees should be part of preparing for emergencies. Figure 10–4 contains a list of the topics that might be covered in a first aid class for construction workers.

Basic First Aid

- Cardiopulmonary resuscitation (CPR)
- Severe bleeding
- Broken bones and fractures
- Burns
- Choking on an obstruction
- Head injuries and concussions
- Cuts and abrasions
- Electric shock
- Heart attack
- Stroke recognition
- Moving an injured person
- Drug overdose
- Unconscious victim
- Eye injuries
- Chemical burns
- Rescue
- Site-specific topics

FIGURE 10–4 Course outline for first aid class.

First Aid Training Programs

First aid programs are usually available in most communities. The continuing education departments of community colleges and universities typically offer first aid training. Classes can often be provided onsite and customized to meet the specific needs of individual companies.

The American Red Cross provides training programs in first aid specifically geared toward the workplace. For more information about these programs, construction professionals may contact the national office of the American Red Cross at (202) 639-3200.

The National Safety Council (NSC) is also a provider of first aid training materials. Its First Aid and Emergency Care Teaching Package contains a presentation, overhead transparencies, a test bank, and an instructor's guide. In addition, the NSC produces a book entitled *First Aid Essentials*. For more information about these materials, construction professionals may contact the NSC at (800) 832-0034 or http://www.nsc.org.

Beyond Training

Training employees in first aid techniques is an important part of preparing for emergencies. However, there is more to being prepared to administer first aid than just training. In addition, it is important to do the following:

• Sterile gauze dressings (individually wrapped)	• Aromatic spirits of ammonia
• Triangular bandages	• Scissors
• Roll of gauze bandages	• Tweezers
• Assorted adhesive bandages	• Needles
• Adhesive bandages	• Sharp knife or stiff-backed razor blades
• Adhesive tape	• Medicine dropper (eyedropper)
• Absorbent cotton	• Measuring cup
• Sterile saline solution	• Oral thermometer
• Mild antiseptic for minor wounds	• Rectal thermometer
• Ipecac syrup to induce vomiting	• Hot-water bag
• Powdered, activated charcoal to absorb swallowed poisons	• Wooden safety matches
• Petroleum jelly	• Flashlight
• Baking soda (bicarbonate of soda)	• Rubber surgical gloves
	• Face masks or mouthpieces for CPR

FIGURE 10–5 Minimum recommended contents of workplace first aid kits.

1. *Have well-stocked first aid kits available.* First aid kits should be placed throughout the workplace in clearly visible, easily accessible locations. They should be properly and fully stocked and periodically checked to ensure that they stay fully stocked. Figure 10–5 lists the minimum recommended contents for a workplace first aid kit.
2. *Have appropriate personal protective devices available.* Because of the concern about AIDS and hepatitis, administering first aid has become more complicated than in the past. The main concerns have to do with exposure to blood and other body fluids. Consequently, a properly stocked first aid kit for the modern workplace should contain rubber surgical gloves, face masks, and mouthpieces for CPR.
3. *Postemergency telephone numbers.* The advent of 911 service has simplified the process of calling for medical care, police, or firefighting assistance. If 911 services are not available, emergency numbers for ambulance, hospital, police, fire department, LEPC, and appropriate internal personnel should be posted at clearly visible locations near all telephones in the workplace.
4. *Keep all employees informed.* Some companies require all employees to undergo first aid training, although others choose to train one or more employees for each job site. Regardless of the approach used, employees must be informed and kept up-to-date concerning basic first aid information. Figures 10–6 and 10–7 are first aid fact sheets of the type used to keep employees informed.

Millwood Construction, Inc.

1200 Anchors Street, Industrial Park, Fort Walton Beach, Florida 32548

First Aid Fact Sheet No. 16
Moving an Injured Person

If a victim has a neck or back injury, do not move him or her unless it must be done to prevent additional injuries. If it is absolutely essential to move the victim, remember the following rules:

1. Call for professional medical help.
2. Always pull the body lengthwise, never sideways.
3. If there is time, slip a blanket under the victim and use it to pull him or her to safety.
4. If the victim must be lifted, support all parts of the body so that it does not bend or jack-knife.

FIGURE 10–6 First aid fact sheet.

Millwood Construction, Inc.

1200 Anchors Street, Industrial Park, Fort Walton Beach, Florida 32548

First Aid Fact Sheet No. 12
ABCs of First Aid

If a fellow employee is injured and you are the first person to respond, remember the ABCs of first aid.

A = Airway
Is the airway blocked? If so, clear it quickly.

B = Breathing
Is the victim breathing? If not, begin artificial respiration.

C = Circulation
Is the victim bleeding severely? If so, stop the bleeding. Is there a pulse? If not, begin administering CPR.

FIGURE 10–7 First aid fact sheet.

EMERGENCY RESPONSE TEAM

An **emergency response team** (ERT) is a special team that responds "to general and localized emergencies to facilitate personnel evacuation and safety; shut down building services, and utilities, as needed; work with responding civil authorities; protect and salvage company property; and evaluate areas for safety before reentry of personnel."[2] The actual composition of the ERT depends on the size of the company. The ERT should be contained in the assignments and responsibilities section of the EAP.

Not all ERTs are company based. Communities also have ERTs to respond to emergencies that occur outside of a company environment. Such teams should be included in a company's EAP in the coordinating organizations section. This is especially important for companies that use hazardous materials.

COMPUTERS AND EMERGENCY RESPONSE

Advances in chemical technology have made responding to certain types of emergencies particularly complicated. The following sample scenario makes the point:

> An employee accidentally crashes a forklift into several barrels of various chemicals. While you have been generally trained to handle certain spills, you cannot find specific information on this combined spill, what protective equipment should be used, and who should be contacted. Your company has an emergency plan, but it is 1½ inches thick and you need to take action now.

Fortunately, the complications brought by technology can also be simplified by technology. Expert computer systems especially programmed for use in emergency situations can simplify the challenge of responding to a mixed-chemical emergency or any other type of emergency involving multiple hazards interacting.

An **expert system** is a computer programmed to solve problems. Such systems rely on a database of knowledge about a particular subject area, an understanding of the problems addressed in that area, and skill at solving problems. Talking to an expert system is like sitting at a terminal and keying in questions to an expert who is sitting at another terminal responding to the questions. The expert, in this case, is a computer program that pulls information from a database and uses it to make decisions based on **heuristic** or suppositional rules stated in an "if-then" format. Laptops can be used at job sites to ensure access to expert systems.

An expert system used for responding to chemical emergencies provides such information as the following:

- Personal protective equipment needed for controlling and cleaning up.
- Methods to be used in cleaning up the spill or toxic release.
- Procedures to be followed for decontamination.
- Estimation of the likelihood that employees or the community will be exposed to hazard.
- Reactions that might result from interaction of chemicals.
- Combustibility of chemicals and other materials on hand.

SAFETY FACTS & FINES

Without proper approval from federal agencies, expanding can cost a company incredible amounts of money. In Portland, Oregon, a firm was fined $11.2 million for failing to obtain federal permits when it expanded. The EPA and U.S. Justice Department also cited a lack of air-pollution control equipment as a reason for the fine. Because the company has facilities in four states, it will spend a total of $93.2 million in environmental upgrades and projects. This is the largest fine ever imposed for a violation of the Clean Air Act.

- Evacuation information.
- Impact of different weather conditions on the situation.
- Recommended first aid procedures.

DEALING WITH THE PSYCHOLOGICAL TRAUMA OF EMERGENCIES

In addition to the physical injuries and property damage that can occur in emergencies, construction professionals must also be prepared to deal with potential psychological damage. Psychological **trauma** among employees involved in workplace disasters is as common as it is among combat veterans. Traumatic incidents affect not only the immediate survivors and witnesses. Most incidents result in victims beyond those who were actually injured.

Trauma is psychological stress. It occurs as the result of an event—typically a disaster or an emergency of some sort—that is so shocking that it impairs a person's sense of security, well-being, or reality. **Traumatic events** are typically unexpected and horrifying; they involve the reality or the threat of death.

Dealing with Emergency-Related Trauma

The typical approach to an emergency can be described as follows: control it, take care of the injured, clean up the mess, and get back to work. Often the psychological aspect is ignored. This leaves witnesses and other co-workers to deal on their own with the trauma they have experienced.

It is important to respond to trauma quickly—within 24 hours if possible and within 72 hours in all cases. The purpose of the response is to help employees get themselves back to normal by helping them come to grips with what they have experienced. This is best accomplished by a team of people who have had special training. Such a team is typically called the **trauma response team** (TRT).

Trauma Response Team

A company's TRT might consist of health and safety personnel who have undergone special training or fully credentialed counseling personnel, depending on the size of the company. In any case, the TRT should be included in the assignments and responsibilities section of the EAP.

The job of the TRT is to intervene as early as possible, help employees acknowledge what they have experienced, and give them opportunities to express how they feel about it to people who are qualified to help. The *qualified to help* aspect is very important. TRT members who are not counselors or mental health professionals should never attempt to provide care they are not qualified to give. Part of the trauma training that construction professionals receive involves recognizing the symptoms of employees who need professional care and referring them to qualified care providers.

In working with employees who need to deal with what they have experienced, but are not so traumatized as to require referral for outside professional care, a group approach is best. According to Johnson, the group approach offers several advantages, including the following:[3]

- It facilitates public acknowledgment of what the employees have experienced.
- It keeps employees informed, thereby cutting down on the number of unfounded rumors and horror stories that inevitably make the rounds.
- It encourages employees to express their feelings about the incident. This alone is often enough to get people back to normal and functioning properly.
- It allows employees to see that they are not alone in experiencing traumatic reactions (i.e., nightmares, flashbacks, disturbing memories) and that these reactions are normal.

Convincing Companies to Respond

Construction professionals responsible for safety may find themselves in a position of having to convince higher management of the need to have a TRT. Some corporate officials may not believe that trauma even exists. Others may acknowledge it, but view trauma as a personal problem that employees should handle on their own.

In reality, psychological trauma that is left untreated can manifest itself as **posttraumatic stress disorder**—the same syndrome that is common in combat veterans and many other people who have experienced trauma. This disorder is characterized by "intrusive thoughts and flashbacks of the stressful event, the tendency to avoid stimulation, paranoia, concentration difficulties, and physiological symptoms, such as rapid heartbeat and irritability."[4]

The American Psychiatric Association has included posttraumatic stress disorder in its *Diagnostic and Statistical Manual* since 1980.[5] Jeffrey T. Mitchell, president of the American Critical Incident Stress Foundation, likens the prevention of posttraumatic stress disorder to working with cement. Wet cement can be molded, shaped, manipulated, and even washed away. But once it hardens, there is not much one can do with it.[6] This is a rationale for early intervention.

In today's competitive marketplace, companies need all of their employees operating at peak performance levels. Employees experiencing trauma-related disorders are not at their best. This is one rationale that health and safety professionals should use when it is necessary to convince higher management of the need to provide a company-sponsored TRT.

Summary

An emergency is a potentially life-threatening situation, usually occurring suddenly and unexpectedly. Emergencies may be the result of natural or human causes. Preparing for emergencies involves planning, practicing, evaluating, and adjusting. An immediacy response is critical in emergencies.

The Emergency Planning and Community Right-to-Know Act has the following four main components: emergency planning, emergency notification, information requirements, and toxic chemical release reporting. For proper coordination of the internal emergency response, there must be one person in charge, and everyone involved must know who that person is.

Because there is no way to predict when first aid might be needed, part of preparing for emergencies should be preparing employees to administer first aid. In certain cases, the Occupational Safety and Health Administration (OSHA) requires that companies have at least one employee onsite who has been trained in first aid. In addition to providing first aid training, companies must ensure that well-stocked first aid kits are handy, personal protective devices and postemergency telephone numbers are available, and all employees are kept informed.

A company's emergency action plan (EAP) should be a collection of small plans for each anticipated emergency. These plans should have the following components: procedures, coordination, assignments and responsibilities, accident-prevention strategies, and schedules. EAPs should be customized so that they are location specific. This can be done by including a map, an organizational chart, local coordination information, and local training schedules.

An emergency response team (ERT) is a special team to handle general and localized emergencies, to facilitate evacuation and shutdown, to protect and salvage company property, and to work with civil authorities. An emergency response network is a network of ERTs that covers a designated geographical area.

Trauma is psychological stress that typically results from exposure to a disaster or an emergency that is so shocking that it impairs a person's sense of security, reality, or well-being. Trauma, left untreated, can manifest itself as posttraumatic stress disorder. This disorder is characterized by intrusive thoughts, flashbacks, paranoia, concentration difficulties, rapid heartbeat, and irritability.

Key Terms and Concepts

Adjusting
Chain of command
Coordinator
Emergency action plan (EAP)
Information requirements
Emergency Planning and Community Right-to-Know Act
Emergency response plan
Emergency response team (ERT)
Planning
Evaluating
Expert system
First aid training
Heuristics
State emergency response commission (SERC)
Local emergency planning committee (LEPC)
Location specific
Order of response
Occupational Safety and Health Administration (OSHA) standards
Posttraumatic stress disorder
Practicing
Emergency notification
Toxic chemical release reporting
Trauma
Trauma response team (TRT)
Traumatic event

Review Questions

1. Define the term *emergency*.
2. Explain the rationale for emergency preparation.
3. List and explain the four main components of the Emergency Planning and Community Right-to-Know Act.
4. Describe how a company's emergency response effort should be coordinated.
5. How do OSHA standards relate to emergency preparation?
6. Explain how you would provide first aid training if you were responsible for setting up a program at your company.
7. What are the steps related to first aid that a company should take beyond providing training?
8. Describe the essential components of an EAP.
9. How can a company localize its EAP?

10. Define the following emergency response concepts: ERT and TRT.
11. What is trauma?
12. Why should a company include trauma response in its EAP?
13. Describe how a company might respond to the trauma resulting from a workplace emergency.

Critical Thinking and Discussion Activities

1. Robert Martin and Andrew Morgan are construction superintendents for Baker Construction Company (BCC). The home office for BCC is in Orlando, Florida, but the company has projects in 12 states. Each project has its own superintendent. Martin and Morgan are attending a quarterly meeting at the home office with all of the company's construction superintendents. During a coffee break earlier in the day, they had gotten into a debate over the need for emergency plans. Martin summarized his point of view as follows: "Emergency planning is a waste of time and scarce resources. Chances are less than one in a thousand that an emergency will even occur. But if an emergency situation does occur, all we have to do is call 911." Morgan disagreed. He summarized his point of view as follows: "We need to have a comprehensive plan for emergency response that is standardized company-wide and that can be easily customized for each job site. Furthermore, we need to drill our employees in the proper execution of the plan until the proper emergency response becomes automatic." Join this debate. Who is right? Why?
2. Robert Martin and Andrew Morgan continued their debate over lunch. This time the topic was the psychological trauma of emergencies. Martin stated his views as follows: "Posttraumatic stress disorder is nonsense. Employees who claim to suffer from this so-called disorder are faking. They just want to claim workers' compensation or have an excuse to stay home from work." Morgan differed strongly. He rebutted Martin's argument as follows: "Posttraumatic stress disorder is real. I've read about how some combat veterans suffer the disorder for years after being subjected to situations that are psychologically traumatic. I'm sure construction employees can be just as traumatized by a disaster as soldiers are by combat." Join this debate. Who is right? Why?

Application Activities

1. Identify a construction company that has an emergency response plan. Review and critique the plan. Make a list of material you would add to the plan and other changes you would make.
2. Develop a "model" emergency response plan that could be easily adapted for use by any construction company.

Endnotes

1. Kelly, S., "First Aid Is Emergency Care," *Safety & Health* 139, no. 5: 47.
2. National Safety Council, *Introduction to Occupational Health and Safety* (Chicago: NSC, 1998), 341.
3. Johnson, E., "Where Disaster Strikes," *Safety & Health* 145, no. 2: 32.
4. National Safety Council, "Trained for Trauma," *Safety & Health* 145, no. 2: 32.
5. Ibid.
6. Ibid.

PART THREE

OTHER SAFETY AND HEALTH ISSUES AND PRACTICES

CHAPTER **ELEVEN**

PREVENTING WORKPLACE VIOLENCE AND RESPONDING TO TRAUMATIC EVENTS IN CONSTRUCTION

LEARNING OBJECTIVES

- Explain the relationship between occupational safety and workplace violence.
- Define the key concepts associated with workplace violence.
- Summarize the size of the workplace violence problem.
- Explain the legal considerations relating to workplace violence.
- Describe the social and cultural factors relating to workplace violence.
- Explain how conflict resolution can prevent workplace violence.
- Summarize the supervisor's role in preventing workplace violence.
- List the various components of an emergency preparedness plan for workplace violence.

Workplace violence has emerged as a critical safety and health issue. Homicide is the second leading cause of death among American workers. Although more than 80 percent of workplace homicide victims are men, workplace violence is not just a male problem. In fact, workplace homicide is the leading cause of death on the job for women in the United States.[1]

Almost one million people are injured or killed in workplace violence incidents every year in the United States, and the number of incidents is on the rise. In fact, the workplace is the most dangerous place in the United States.[2] Clearly, workplace violence is an issue of concern to construction professionals.

OCCUPATIONAL SAFETY AND WORKPLACE VIOLENCE: THE RELATIONSHIP

The prevention of workplace violence is a natural extension of the responsibilities of construction professionals. Hazard analysis, record analysis and tracking, trend monitoring, incident analysis, and prevention strategies based on administrative and **engineering controls** are all fundamental to both occupational safety and workplace violence prevention. In addition, emergency response and employee training are key elements of both. Consequently, construction professionals must be able to add the prevention of workplace violence to their normal safety and health promotion duties.

WORKPLACE VIOLENCE: DEFINITIONS

Construction professionals should be familiar with the language that has developed around the issue of workplace violence. This section contains the definitions of several concepts as they relate specifically to workplace violence.

- *Workplace violence.* Violent acts, behavior, or threats that occur in the workplace or are related to it. Such acts are harmful or potentially harmful to people, property, or organizational capabilities.
- *Occupational violent crime (OVC).* Intentional battery, rape, or homicide during the course of employment.[3]
- *Employee.* An individual with an employment-related relationship (present or past) with the victim of a workplace violence incident.

SAFETY FACTS & FINES

Fines are not the only types of costs associated with inadequate precautions on the job. A company in Los Angeles suffered untold costs to morale, employee performance, and its public image when an electrician shot and killed four supervisors. The electrician was angry because he faced possible dismissal for poor performance. After shooting the supervisors, the electrician quietly waited for police to arrive and arrest him. When asked why he had murdered four people, he responded that he felt he "was being picked on and singled out."

- *Outsider.* An individual with no relationship of any kind with the victim of a workplace violence incident or with the victim's employer.
- *Employee-related outsider.* An individual with some type of personal relationship (past or present) with an employee, but who has no work-related relationship with the employee.
- *Customer.* An individual who receives products or services from the victim of a workplace violence incident or from the victim's employer.

Each of these terms also has other definitions; those presented here reflect how the terms are used in the language that has evolved specifically around workplace violence.

SIZE OF THE PROBLEM

Violence in the workplace no longer amounts to just isolated incidents that are simply aberrations. In fact, workplace violence should be considered a common hazard worthy of the attention of construction professionals. Consider the following information about violent crimes in a typical year[4]:

- About 1 million individuals are the direct victims of some form of violent crime in the workplace every year. This represents approximately 15 percent of all violent crimes committed annually in America. Approximately 60 percent of these violent crimes committed in the workplace are categorized as simple assaults by the Department of Justice.
- Of all workplace violent crimes reported, over 80 percent are committed by males, 40 percent by complete strangers to the victims, 35 percent by casual acquaintances, 19 percent by individuals well known to the victims, and 1 percent by relatives of the victims.
- More than half of the incidents (56 percent) are not reported to police, although 26 percent are reported to at least one official in the workplace.
- In 62 percent of violent crimes, the perpetrator is not armed; in 30 percent of the incidents, the perpetrator is armed with a handgun.
- In 84 percent of the incidents, there are no reported injuries; 10 percent require medical intervention.
- More than 60 percent of violent incidents occur in private companies; 30 percent in government agencies; and 8 percent to self-employed individuals.
- It is estimated that violent crime in the workplace causes 500,000 employees to miss 1,751,000 days of work annually, or an average of 3.5 days per incident. This missed work equates to approximately $55 million in lost wages.

LEGAL CONSIDERATIONS

Most issues related to construction safety and health have legal ramifications, and workplace violence is no exception. The legal aspects of the issue revolve around the conflicting rights of violent employees and their co-workers. These conflicting rights create potential liabilities for employers.

Rights of Violent Employees

It may seem odd to be concerned about the rights of employees who commit violent acts on the job. After all, logic suggests that in such situations, the only concern would be the protection of other employees. However, even violent employees have rights. Remember, the first thing that law enforcement officers must do after taking criminals into custody is read them their rights.[5]

This does not mean that an employer cannot take the immediate action necessary to prevent a violent act or the recurrence of such an act. In fact, failure to act prudently in this regard can subject an employer to charges of negligence. However, before taking long-term action that will adversely affect the violent individual's employment, employers should follow applicable laws, contracts, policies, and procedures. Failure to do so can serve to exacerbate an already difficult situation.

In addition to complying with all applicable laws, policies, and procedures, it is important to apply them consistently. Dealing with one person one way and another person in a different way puts the employer at a disadvantage legally. Consequently, employers must be prepared to deal promptly with violent employees.[6]

Employer Liability for Workplace Violence

Having to contend with the rights of both violent employees and their co-workers, employers often feel as if they are caught between a rock and a hard place. Fortunately, the situation is less bleak than it may first appear, due primarily to the **exclusivity provision** of workers' compensation laws. This provision makes workers' compensation the employee's exclusive remedy for injuries that are work related. This means that even in cases of workplace violence—as long as the violence occurs within the scope of the victim's employment—the employer is protected from civil lawsuits and the excessive jury verdicts that have become so common.

The key to enjoying the protection of the exclusivity provision of workers' compensation laws lies in determining that violence-related injuries are within the scope of the victim's employment—a more difficult undertaking than one might expect. For example, if the violent act occurred at work but resulted from a non-work-related dispute, does the exclusivity provision apply? What if the dispute was work-related, but the violent act occurred away from the workplace?

Making Work-Related Determinations

The Society for Human Resource Management (SHRM) developed the following guidelines for categorizing an injury as being work related.[7]

If the violent act occurred on the employer's premises, it is considered an on-the-job event if one of the following criteria applies:

- The victim was engaged in work activity, apprenticeship, or training.
- The victim was on break, in a hallway, restroom, cafeteria, or storage area.
- The victim was in the employer's parking lot while working, arriving at, or leaving work (including construction-site parking lots).

If the violent act occurred off the employer's premises, it is still considered an on-the-job event if one of the following criteria applies:

- The victim was working for pay or compensation at the time, including working at home.
- The victim was working as a volunteer, emergency services worker, law enforcement officer, or firefighter.
- The victim was working in a profit-oriented family business, including farming.
- The victim was traveling on business, including to and from customer business contacts.
- The victim was in a vehicle engaged in work activity in which the vehicle is part of the work environment (e.g., truck, forklift, bulldozer).

Risk-Reduction Strategies

There are ethical, legal, practical, and economic reasons for doing everything feasible to reduce the possibility of violence in the workplace. The most fundamental concepts of prevention are as follows:

- Respond immediately to imminent threats or dangerous situations.
- Take threats seriously and investigate. Even apparently frivolous allegations might turn out to be real.
- Take disciplinary action when it is called for.
- Provide support for victims (both primary and secondary victims).
- Do everything necessary to return the work environment to normal following an incident.
- Deal with aggressive employees through counseling, aggression management training, or even termination.

Figure 11–1 is a checklist that can be used by construction companies to reduce the risk of workplace violence in their facilities and at their job sites. Most of these risk-reduction strategies grew out of the philosophy of Crime Reduction Through Environmental Design (CRTED).[8] CRTED has the following four major elements, to which I have added a fifth element (**administrative controls**):

✓	Identify high-risk areas and make them visible. Secluded areas invite violence.
✓	Install good lighting in parking lots.
✓	Minimize the handling of cash by employees and the amount of cash available on job sites.
✓	Install silent alarms and surveillance cameras where appropriate.
✓	Control access to all buildings and job sites (ensure use of employee badges, visitor check-in and check-out procedures, visitor passes).
✓	Discourage working alone, particularly late at night.
✓	Provide training in conflict resolution as part of a mandatory employee orientation.
✓	Conduct background checks before hiring new employees.
✓	Train employees on how to handle themselves and respond when a violent act occurs on the job.
✓	Develop policies that establish ground rules for employee behavior and responses in threatening or violent situations.
✓	Nurture a positive, harmonious work environment.
✓	Encourage employees to report suspicious individuals and activities or potentially threatening situations.
✓	Deal with allegations of harassment or threatened violence promptly before the situation escalates.
✓	Take threats seriously and act appropriately.
✓	Adopt a zero-tolerance policy toward threatening or violent behavior.
✓	Establish a crisis management team with responsibility for preventing and responding to violence.
✓	Establish an emergency response team to deal with the immediate trauma of workplace violence.

FIGURE 11–1 Checklist for reducing the risk of workplace violence.

- Natural surveillance
- Control of access
- Establishment of territoriality
- Activity support
- Administrative controls

Natural Surveillance

This strategy involves designing, arranging, and operating the workplace or job site in a way that minimizes secluded areas. Making all areas inside and outside of the facility and job sites easily observable allows **natural surveillance**.

Control of Access

One of the most common types of workplace violence involves an outsider entering the workplace or job site and harming employees. **Control of access** is the most effective way of stopping this type of incident from occurring at the workplace or job site. Channeling the flow of outsiders to an access-control station, requiring visitor's passes, issuing access badges to workers, and isolating pickup and delivery points can minimize the risk of violence perpetrated by outsiders.

Establishment of Territoriality

Establishment of territoriality involves giving employees control over the workplace or job site. With this approach, employees move freely within their established territory, but are restricted in other areas. Employees should come to know everyone who works in their territory and, as a result, immediately recognize anyone who should not be there.

Activity Support

Activity support involves organizing workflow and natural traffic patterns in ways that maximize the number of employees conducting natural surveillance. The more employees observing the activity in the workplace and on job sites, the better.

Administrative Controls

Administrative controls consist of management practices that can reduce the risk of workplace violence. These practices include establishing appropriate policies, conducting background checks, and providing training for employees.

CONTRIBUTING SOCIAL AND CULTURAL FACTORS

Another way to reduce the risk of workplace violence is to ensure that managers understand the social and cultural factors that can lead to it. These factors fall into two broad categories: individual and environmental.

Individual Factors Associated with Violence

The factors explained in this section can be predictors of the potential for violence. Employees and individuals who have one or more of the following characteristics may respond to anger, stress, or anxiety in a violent way.

1. *A record of violence.* Past violent behavior is typically an accurate predictor of future violent behavior. Consequently, thorough background checks should be a normal part of the employment process.
2. *Membership in a hate group.* Hate groups often promote violence against the subjects of their prejudice. Hate group membership on the part of an employee should raise a red flag in the eyes of management.
3. *Psychotic behavior.* Individuals who incessantly talk to themselves, express fears concerning conspiracies against them, say that they hear voices, or become increasingly disheveled over time may be prone to violence.
4. *Romantic obsessions.* Workplace violence is often the result of romantic entanglements or love interests gone awry. Employees who persist in making unwelcome advances may eventually respond to rejection with violence.
5. *Depression.* People who suffer from clinical depression are prone to hurt either themselves or someone else. An employee who becomes increasingly withdrawn or overly stressed may be suffering from depression.
6. *Finger pointers.* Refusal to accept responsibility is a factor often exhibited by perpetrators of workplace violence. An employee's tendency to blame others for his or her own shortcomings should raise the caution flag.
7. *Unusual frustration levels.* The workplace has become a competitive, stressful, and sometimes frustrating place. When frustration reaches the boiling point, the emotional explosion that results can manifest itself in violence.
8. *Obsession with weapons.* Violence in the workplace often involves a weapon (e.g., gun, knife, or explosive device). A normal interest in guns used for hunting or target practice need not raise concerns. However, an employee whose interest in weapons is unusually intense and focused is cause for concern.
9. *Drug dependence.* It is common for perpetrators of workplace violence to be drug abusers. Consequently, drug dependence should cause concern not only for all of the usual reasons, but also for its association with violence on the job.

Environmental Factors Associated with Violence

The environment in which employees work can contribute to workplace violence. An environment that produces stress, anger, frustration, feelings of powerlessness, resentment, and feelings of inadequacy can increase the potential for violent behavior. The following factors can result in such an environment.

1. *Dictatorial management.* Dictatorial, overly authoritative management that shuts employees out of the decision-making process can cause them to feel powerless, as if they have little or no control over their jobs. Some people respond to powerlessness by striking out violently—a response that gives them power, if only momentarily.
2. *Role ambiguity.* One of the principal causes of stress and frustration on the job is role ambiguity. Employees need to know what they are responsible for, how they will be held accountable, and how

much authority they have. When these issues are not clear, employees become stressed and frustrated—factors often associated with workplace violence.

3. *Partial, inconsistent supervision.* Supervisors who play favorites engender resentment in employees who are not the favorite. Supervisors who treat one employee differently than another or one group of employees differently than another group also cause resentment. Employees who feel that they are being treated unfairly or unequally may show their resentment in violent ways.
4. *Unattended hostility.* Supervisors who ignore hostile situations or threatening behavior are unwittingly giving them their tacit approval. An environment that accepts hostile behavior will have hostile behavior.
5. *No respect for privacy.* Supervisors and managers who go through the tool boxes and work areas of employees without first getting their permission can make workers feel invaded or even violated. Violent behavior is a possible response to these feelings.
6. *Insufficient training.* Holding employees accountable for performance on the job without providing the training they need to perform well can cause them to feel inadequate. People who feel inadequate can turn their frustration inward and become depressed or turn it outward and become violent, or both: first the depression, then the violence.

The overriding message in this section is twofold. First, construction professionals should endeavor to establish and maintain a positive work environment that builds employees up rather than tearing them down. Second, construction professionals should be aware of the individual characteristics that can contribute to violent behavior and respond promptly if employees with these characteristics show evidence of responding negatively.

COMPONENTS OF A VIOLENCE-PREVENTION PROGRAM

OSHA Guidelines

The Occupational Safety and Health Administration (OSHA) does not publish specific standards for workplace violence. However, under the "General Duty" clause of the Occupational Safety and Health Act (OSH Act), the following requirement is clearly set forth[9]:

> Each employer shall furnish to each of his [or her] employees employment and a place of employment which are free from recognized hazards that are causing or are likely to cause death or serious physical harm to his [or her] employees.

OSHA does publish voluntary guidelines that employers can use to help ensure a safe and healthy workplace free of violence.[10] The guidelines were not designed for construction companies, but many of them will work for any type of employer. The following are two key points to understand about these guidelines:

- The guidelines are *advisory* in nature and *informational* in content. The guidelines do not add to or enhance in any way the requirements of the General Duty clause of the OSH Act.
- The guidelines were developed with service establishments in mind. Consequently, they have a service-oriented emphasis. However, much of the advice contained in the guidelines can be adapted for use in construction settings.

✓	Management commitment and employee involvement
✓	Worksite analysis
✓	Hazard prevention and control
✓	Safety and health training

FIGURE 11–2 Elements of OSHA advisory guidelines on workplace violence that apply to a construction setting.

Figure 11–2 is a checklist of those elements of the specifications that have broader applications. Any management program related to safety and health in the workplace should include at least these four elements.

Management Commitment and Employee Involvement

Management commitment and employee involvement are fundamental to developing and implementing any safety program, but they are especially important when trying to prevent workplace violence. The effectiveness of a workplace violence prevention program may be a life-or-death proposition. Figure 11–3 is a checklist that explains what management commitment means in practical terms. Figure 11–4 describes the practical application of employee involvement. Figure 11–5 is a checklist that can be used to ensure that violence prevention

SAFETY FACTS & FINES

Workplace violence is an increasing problem. Ever more frequently, workers are resorting to violence when they feel slighted, mistreated, or powerless. A worker in Honolulu, Hawaii, returned to his former job site after being fired and took five former co-workers hostage. During the incident, which lasted six hours, the disgruntled worker shot and seriously wounded his former supervisor. The worker was eventually killed by police after he held a shotgun to the head of a former co-worker for several hours while negotiating with police. As the disgruntled worker counted down to pulling the trigger of the shotgun, the hostage grabbed the barrel, giving police an opportunity to fire.

- ✓ Hands-on involvement of executive management in developing and implementing prevention strategies
- ✓ Sincere, demonstrated concern for the protection of employees
- ✓ Balanced commitment to both employees and customers
- ✓ Inclusion of safety and health promotion and workplace violence-prevention criteria in the job descriptions of all executives, managers, supervisors, and employees
- ✓ Inclusion of safety and health promotion and workplace violence-prevention criteria in the performance evaluations of all executives, managers, supervisors, and employees
- ✓ Assignment of responsibility for providing coordination and leadership for safety and health promotion and workplace violence prevention to a management-level employee
- ✓ Provision of the resources needed to prevent workplace violence effectively
- ✓ Provision of or guaranteed access to appropriate medical counseling and trauma-related care for employees affected physically or emotionally by workplace violence
- ✓ Implementation, as appropriate, of the violence-prevention recommendations of committees, task forces, and safety professionals

FIGURE 11–3 Management commitment to preventing workplace violence.

- ✓ Staying informed concerning all aspects of the organization's safety, health, and workplace violence program
- ✓ Voluntarily complying—in both letter and spirit—with all aspects of the workplace violence-prevention program
- ✓ Making recommendations—through proper channels—concerning ways to prevent workplace violence and other hazardous conditions
- ✓ Accurate and immediate reporting of all violent or threatening incidents
- ✓ Voluntary participation on committees, task forces, or focus groups concerned with preventing workplace violence
- ✓ Voluntary participation in seminars, workshops, or other educational programs related to the prevention of workplace violence

FIGURE 11–4 Elements of employee involvement in preventing workplace violence.

becomes a standard component of organizational plans and operational practices. These three checklists can be used by construction companies to put the concepts of management commitment and employee involvement into operation.

- ✓ Include the prevention of workplace violence in the safety and health component of the organization's strategic plan.
- ✓ Adopt, disseminate, and implement a no-tolerance policy concerning workplace violence.
- ✓ Adopt, disseminate, and implement a policy that protects employees from reprisals when they report violent, threatening, or potentially threatening situations.
- ✓ Establish procedures for reporting violent and threatening incidents.
- ✓ Establish procedures for making recommendations for preventing workplace violence.
- ✓ Establish procedures for monitoring reports of workplace violence so that trends can be identified and incidents predicted and prevented.
- ✓ Develop a comprehensive workplace violence-prevention program that contains operational procedures and standard practices.
- ✓ Develop a workplace violence component to the organization's emergency response plan.
- ✓ Train all employees in the application of standard procedures related to workplace violence.
- ✓ Conduct periodic emergency response drills for employees.

FIGURE 11–5 Checklist for incorporating workplace violence prevention in strategic plans and operational practices.

Workplace or Job-Site Analysis

- ✓ Security analysis
- ✓ Employee input (surveys and focus groups)
- ✓ Trend monitoring and incident analysis
- ✓ Monitoring and tracking records

FIGURE 11–6 Essential components of workplace or job-site analysis.

Workplace Analysis

Workplace analysis is the same process that is used to identify potentially hazardous conditions that are unrelated to workplace violence. Workplace and job-site analysis should be ongoing and have at least four components (Figure 11–6). An effective way to conduct an ongoing program of workplace or job-site analysis is to establish a threat assessment team, with representatives from all departments, led by the organization's chief safety and health professional or a construction professional who is knowledgeable about safety and health principles.

Records Monitoring and Tracking

The purpose of **records monitoring and tracking** is to identify and chart all incidents of violence and threatening behavior that have occurred within a given time frame. Records to analyze include incident reports, police reports, employee evaluations, and letters of reprimand. Of course, the records of individual workers should be analyzed in confidence by the human resources member of the team. The type of information that is pertinent includes the following:

- Where specifically did the incident occur?
- What time of day or night did the incident occur?
- Was the victim an employee? Customer? Outsider? Contractor?
- Was the incident the result of a work-related grievance or a personal one?
- Was the assailant an employee? Customer? Outsider? Contractor?

Trend Monitoring and Incident Analysis

Trend monitoring and incident analysis may prove helpful in determining patterns of violence. If there have been enough incidents to create one or more graphs, the team should determine if the graphs suggest a trend or trends. If the organization has experienced only isolated incidents, the team may want to monitor national trends. By analyzing both local and national incidents, the team can generate information that will be helpful in predicting and, thereby, preventing workplace violence. The team should look for trends in severity, frequency, and type of incidents.

Employee Surveys and Focus Groups

Employees are one of the best sources of information concerning workplace hazards. This is also true when it comes to identifying vulnerabilities to workplace violence. Employee input should be solicited periodically through written **employee surveys** or **focus groups**, or both. Where are we vulnerable? What practices put our employees at risk? These are the types of questions that should be asked to employees. An effective strategy for use with focus groups is to give participants case studies of incidents that occurred in other organizations. Then ask questions such as "Could this happen here? Why or why not? How can we prevent such incidents from occurring here?"

Security Analysis

Is the workplace or job site secure, or could a disgruntled individual simply walk up and harm employees? It is important to ask this question. The team should periodically perform a **security analysis** of the workplace to identify conditions, situations, procedures, and practices that make employees vulnerable. The types of questions to ask include the following:

- Are there physical factors about the facility or job site that make employees vulnerable (e.g., isolated areas, poorly lighted, infrequently trafficked, or unobservable)?
- Is there a process for handling disgruntled people? Does it put employees at risk?
- Are the prevention strategies that have already been implemented working?
- Is the training provided to employees having a positive effect? Is more training needed? Who needs the training? What kind of training is needed?
- Are there situations in which employees have substantial amounts of money in their possession, onsite or offsite?
- Are there situations in which employees are responsible for highly valuable equipment or materials late at night or at isolated locations?

Hazard Prevention and Controls

Once hazardous conditions have been identified, the strategies and procedures necessary to eliminate them must be put in place. The two broad categories of prevention strategies are engineering controls and administrative controls, just as they are with other safety and health hazards. In addition to these, organizations should adopt **postincident response** strategies as a way to prevent future incidents.

Engineering Controls

Engineering controls related to the prevention of workplace violence serve the same purpose as engineering controls related to other hazards. Either they remove the hazard or create a barrier between it and employees. Engineering controls typically involve changes to the workplace and at job sites. Examples of engineering controls include the following:

- Installing devices and mechanisms that give employees a complete view of their surroundings (e.g., mirrors, glass, or clear plastic partitions)
- Installing surveillance cameras and television screens that allow for monitoring of the workplace or job site
- Installing adequate lighting, particularly in parking lots
- Pruning shrubbery and undergrowth outside and around the facility or job site
- Installing fencing so that routes of egress and ingress to company property and job sites can be channeled and, as a result, better controlled
- Arranging outdoor sheds, storage facilities, recycling bins, and other outside facilities for maximum visibility (including the use of lighting)

Administrative Controls

Whereas engineering controls involve making changes to the workplace, administrative controls involve making changes to how work is done. This amounts to changing

	Categories of Administrative Controls
✓	Proper work practices and procedures
✓	Monitoring and feedback
✓	Adjustments and modifications
✓	Enforcement

FIGURE 11–7 Administrative controls.

work procedures and practices. Administrative controls fall into four categories (Figure 11–7):

1. **Proper work practices** and procedures are those that minimize the vulnerability of employees. For example, if a driver has to pick up supplies in a high-crime area, the company might employ a security guard to go along or change suppliers, or both.
2. **Monitoring and feedback** ensure that proper work practices are being used and that they are having the desired effect. For example, a company established a controlled access system in which visitors must check in at a central location and receive a visitor's pass. Is the system being used? Are all employees sticking to specified procedures? Has unauthorized access to the workplace been eliminated?
3. **Adjustments and modifications** are made to violence-prevention practices if it becomes clear from monitoring and feedback that the practices are not working or that improvements are needed.
4. **Enforcement** involves applying meaningful sanctions when employees fail to follow the established and proper work practices. An employee who has been fully informed concerning a given administrative control and has received the training needed to practice it properly, but consciously decides not to follow the procedure, should be disciplined appropriately.

Postincident Response

Postincident response related to workplace violence is the same as postincident response related to any traumatic accident or incident. The first step is to provide immediate medical treatment for injured employees. The second step involves providing psychological treatment for traumatized employees. This step is ever more important for cases of workplace violence than for accidents. Employees who are present when a violent incident occurs in the workplace, even if they do not witness it, can experience the symptoms of psychological trauma shown in Figure 11–8. Employees experiencing any symptoms growing out of psychological trauma should be treated by professionals, such as psychologists, psychiatrists, clinical nurse specialists, or certified social workers. In addition to one-on-one counseling, employees may also be enrolled in support groups. The final aspect of

✓	Fear of returning to work
✓	Problems in relationships with fellow employees or family members
✓	Feelings of incompetence
✓	Guilt feelings
✓	Feelings of powerlessness
✓	Fear of criticism by fellow employees, supervisors, and managers

FIGURE 11–8 Symptoms of psychological trauma in employees exposed to workplace violence.

postincident response is the investigation, analysis, and report. In this step, construction professionals determine how the violent incident occurred and how future incidents may be prevented—just as postaccident investigations are handled.

Training and Education

Training and education are as fundamental to the prevention of workplace violence as they are to the prevention of workplace accidents and health-threatening incidents. A comprehensive safety and health training program should include a component covering all aspects of workplace violence (e.g., workplace analysis, hazard prevention, proper work practices, and emergency response). Such training should be provided on a mandatory basis for supervisors, managers, and employees.

Record Keeping and Evaluation

Accurate, comprehensive, up-to-date **record keeping** is just as important when dealing with violent incidents as it is when dealing with accidents and nonviolent incidents. **Evaluation of records** allows construction professionals to determine how effective their violence-prevention strategies are, where deficiencies exist, and what changes need to be made. Figure 11–9 shows the types of records that should be kept, and they are described here.

- *The OSHA log of injury and illness* must include any injury that requires more than first aid, is a lost-time injury, requires modified duty, or causes loss of consciousness, according to OSHA regulations. Of course, this applies only to establishments

	Record Keeping Checklist
✓	OSHA log of injury and illness
✓	Medical reports
✓	Incident reports
✓	Minutes of meetings
✓	Training records

FIGURE 11–9 Types of records that should be kept.

required to keep OSHA logs. Injuries caused by assaults, which are otherwise recordable, also must be included in the log. A fatality or catastrophe that results in the hospitalization of three or more employees must be reported to OSHA within eight hours. This includes those resulting from workplace violence and applies to all establishments.

- *Medical reports* of all work injuries should be maintained. These records should describe the type of assault (e.g., unprovoked sudden attack), the victim, and all other circumstances surrounding the incident. The records should include a description of the environment or location, potential or actual cost, lost time, and the nature of injuries sustained.
- *Incidents involving abuse,* verbal attacks, or aggressive behavior (e.g., pushing, shouting, or acts of aggression)—which may be threatening to the employee, but do not result in injury—should be routinely evaluated by the department affected.
- *Minutes of safety meetings,* records of hazard analyses, and details of corrective actions that were recommended and taken should be documented.
- *Records of all training programs,* attendees, and qualifications of trainers should be maintained.

As part of its overall program, a construction company should regularly evaluate its safety and security measures. Top management should review the program regularly, as well as each incident, to determine the program's effectiveness. Parties responsible (e.g., managers, supervisors) should collectively evaluate policies and procedures on a regular basis. Deficiencies should be identified and corrective action taken. An evaluation program should involve the following activities:

- Establishing a uniform violence reporting system and regular review of reports.
- Reviewing reports and minutes from staff meetings on safety and security issues.
- Analyzing trends and rates of illness and injury or fatalities caused by violence, relative to initial or baseline rates.
- Measuring improvements based on lowering the frequency and severity of workplace violence.
- Keeping up-to-date records of changes to administrative and work practices made to prevent workplace violence.
- Surveying employees before and after making job or workplace changes or installing security measures or new systems to determine the effectiveness of the measures.
- Keeping abreast of new strategies available to deal with violence as the strategies develop.
- Surveying employees who have been injured as the result of workplace violence about the medical treatment they received initially and, again, several weeks afterward and, then, several months later.
- Complying with OSHA and state requirements for recording and reporting deaths, injuries, and illnesses.
- Requesting periodic reviews of the workplace by law enforcement or outside consultants to generate recommendations on improving employee safety.

Management should share violence-prevention evaluation reports with all employees. Any changes in the program should be discussed at regular meetings of the safety committee, union representatives, or other employee groups.

CONFLICT RESOLUTION AND WORKPLACE VIOLENCE

When developing a violence-prevention program for a company, the natural tendency is to focus on protecting employees from outsiders. However, increasingly, the problem of workplace violence is internal. All too often in the modern workplace, conflict between employees is turning violent. Consequently, a violence-prevention program is not complete without the elements of **conflict management** and **anger management**.

Conflict Management Component

Disagreements on the job can generate counterproductive conflict. This is one of the reasons why supervisors and managers should do what is necessary to manage conflict properly. However, it is important to distinguish between conflicting opinions and counterproductive conflict. Not all conflict is bad. In fact, properly managed conflict that has the improvement of products, processes, people, or the work environment as its source is positive conflict.

Counterproductive conflict—the type associated with workplace violence—occurs when employees behave in ways that work against the interests of the company and its employees. This type of conflict is often characterized by deceitfulness, vindictiveness, personal rancor, and anger. Productive conflict occurs when right-minded, well-meaning people disagree—without being disagreeable—about the best way to achieve the company's goals. Conflict management has the following components:

- Establishing conflict guidelines.
- Helping employees develop conflict-prevention skills and resolution skills.
- Helping employees develop anger management skills.

Establishing Conflict Guidelines. **Conflict guidelines** establish ground rules for discussing differing points of view, differing ideas, and differing opinions

Middletown Construction Management (MCM)
Conflict Guidelines

MCM encourages discussion and debate among employees at all levels concerning better ways to continually improve the quality of our work processes and work environment. This type of interaction, if properly handled, will result in better ideas, policies, procedures, practices, and decisions. However, human nature is such that conflict can easily get out of hand, take on personal connotations, and become counterproductive. Consequently, to promote productive conflict, MCM has adopted the following guidelines. These guidelines are to be followed by all employees at all levels:

- The criteria to be applied when discussing or debating any point of contention is: Which recommendation is most likely to move our company closer to accomplishing its goals?
- Disagree, but don't be disagreeable. If the discussion becomes too hot, stop and give all parties an opportunity to cool down before continuing. Apply your conflict resolution and anger management skills. Remember, even when we disagree about how to get there, we are all trying to reach the same destination.
- Justify your point of view by tying it to our mission statement, and require others to do the same.
- In any discussion of differing points of view, ask yourself the question: "Am I just trying to win the debate just for the sake of winning [ego], or is my point of view really the most valid?"

FIGURE 11–10 Sample conflict guidelines.

concerning how best to accomplish the company's goals. Figure 11–10 is an example of a company's conflict guidelines. Guidelines such as these should be developed with a broad base of employee involvement from all levels in the company.

Develop Conflict-Prevention Skills and Resolution Skills. If managers are going to expect employees to disagree without being disagreeable, they are going to have to ensure that all employees are skilled in the art and science of conflict resolution. The first guideline in Figure 11–10 is an acknowledgment of human nature. It takes advanced human relation skills and constant effort to disagree without being disagreeable. Few people are born with this ability, but fortunately, it can be learned. The following strategies are based on a three-phase model developed by Tom Rusk and described in his book *The Power of Ethical Persuasion*.[11]

Explore the other person's viewpoint. Allow the other person to present his or her point of view, Figure 11–11. The following strategies help to make this phase of the discussion more positive and productive:

1. Establish that your goal at this point is mutual understanding.
2. Elicit the other person's complete point of view.
3. Listen nonjudgmentally and do not interrupt.

FIGURE 11–11 Both parties in conflict must be willing to listen.

Source: Chaloemphan/Fotolia

4. Ask for clarification, if necessary.
5. Paraphrase the other person's point of view, and restate it to show that you understand.
6. Ask the other person to correct your understanding if it appears to be incomplete.

Explain your viewpoint. After you accurately and fully understand the other person's point of view, present your own. The following strategies help to make this phase of the discussion more positive and productive:

1. Ask for the same type of fair hearing for your point of view that you gave the other party.
2. Describe how the person's point of view affects you. Do not point the finger of blame or be defensive. Explain your reactions objectively, keeping the discussion on a professional level.
3. Explain your point of view accurately and completely.
4. Ask the other party to paraphrase and restate what you have said.
5. Correct the other party's understanding, if necessary.
6. Review and compare the two positions (yours and that of the other party). Describe the fundamental differences between the two points of view, and ask the other party to do the same.

Agree on a resolution. Once both viewpoints have been explained and are understood, it is time to move to the resolution phase. This is the phase in which both parties attempt to come to an agreement. Agreeing to disagree—in an agreeable manner—is an acceptable solution. The following strategies help make this phase of the discussion more positive and productive:

1. Reaffirm the mutual understanding of the situation.
2. Confirm that both parties are ready and willing to consider options for coming to an acceptable solution.
3. If it appears that differences cannot be resolved to the satisfaction of both parties, try one or more of the following strategies:
   - Take time out to reflect and try again.
   - Agree to third-party arbitration or neutral mediation.
   - Agree to a compromise solution.
   - Take turns suggesting alternative solutions.
   - Yield (this time), once your position has been thoroughly stated and is understood. The eventual result may vindicate your position.
   - Agree to disagree while still respecting each other.

Develop Anger Management Skills

It is difficult, if not impossible, to keep conflict positive when anger enters the picture. If individuals in a company are going to be encouraged to question, discuss, debate, and even disagree, they must know how to manage their anger. Anger is an intense emotional reaction to conflict in which self-control may be lost, Figure 11–12. Anger is a major cause of workplace violence; it occurs when people feel that one or more of their fundamental needs are being threatened. These needs include the following:

FIGURE 11–12 Anger can lead to accidents and injuries.
Source: Auremar/Fotolia

1. Need for approval
2. Need to be valued
3. Need to be appreciated
4. Need to be in control
5. Need for self-esteem

An angry person can respond in one of four ways:

1. *Attacking.* In this response, the source of the threat is attacked. This response often leads to violence or, at least, verbal abuse.
2. *Retaliating.* In this response, the person receiving the anger fights fire with fire, so to speak. Whatever is given, the person gives back. For example, if someone calls a suggestion ridiculous (threatens the need to be valued), it is natural to feel like retaliating by calling his or her suggestion dumb. However, retaliation can escalate into violence.
3. *Isolating.* This response is the opposite of venting. With the isolation response, the person receiving the anger internalizes the anger, finds a place to be alone, and simmers. The childhood version of this response for an individual was to go to his room and pout. For example, when someone fails to even acknowledge a suggestion (i.e., threatens the need to be appreciated), one response is to swallow the resulting anger, go somewhere, and boil over in private.
4. *Coping.* This is the only positive response to anger. Coping does not mean that a person does not become angry. It means that, even when angry, people should control their emotions, rather than letting the emotions control them. A person who copes well with anger is a person who, in spite of his or her anger, stays in control. The following strategies help employees manage their anger by becoming better at coping:
   - Avoid the use of anger-inducing words and phrases, including the following: *but, you should, you made me, always, never, I can't, you can't.*
   - Admit that others do not make someone angry; a person allows himself to become angry. People are responsible for their emotions and their responses to them.
   - Do not let pride get in the way of progress. Nobody has to be right every time.
   - Drop defenses when dealing with people. Be open and honest.
   - Relate to other people as equals. Regardless of position or rank, people are equal as human beings.

- Avoid the human tendency to rationalize angry responses. People are responsible and accountable for their behavior.

If employees in a company can learn to manage conflict properly and to deal with anger positively, the potential for workplace violence will be substantially diminished. Conflict and anger management do not prevent violent acts from outsiders. There are other methods for dealing with outsiders. However, properly managing conflict and anger can protect employees from each other.

GUIDELINES FOR SUPERVISORS

Supervisors can play a pivotal role in the prevention of workplace violence, Figure 11–13. The following are some rules of thumb that enhance the effectiveness of supervisors:[12]

- *Do not* try to diagnose the personal, emotional, or psychological problems of employees.
- *Do not* discuss an employee's drinking unless it occurs on the job. Restrict comments to performance.
- *Do not* preach to employees. Counsel employees about attendance, tardiness, and job performance, not about how they should live their lives.
- *Do not* cover up for employees or make excuses for inappropriate behavior. Misguided kindness may allow problems to escalate and get out of hand.
- *Do not* create jobs to get problem employees out of the way. Stockpiling an employee simply gives him or her more time to brood and allows resentment to build.
- *Do not* ignore the warning signs explained earlier in this chapter. The problems they represent do not simply go away. Sooner or later, these problems have to be handled; sooner is better.
- *Do* remember that chemical dependence and emotional problems tend to be progressive. Left untreated, they get worse, not better, and can lead to violent behavior.
- *Do* refer problem employees to the employee assistance program or to mental health service providers.
- *Do* make it clear to employees that job performance is the key issue. They are expected to do what is necessary to maintain and improve their performance.
- *Do* make it clear that inappropriate behavior is not tolerated.

COMMUNICATING WITH WORKERS IN THE AFTERMATH OF A VIOLENT INCIDENT

Communicating with workers is always an important responsibility of construction professionals, but it becomes even more important in the aftermath of a violent incident on the job. When workers are distracted

FIGURE 11–13 Close supervision can prevent workplace violence.
Source: Espion/Fotolia

by their feelings of uncertainty and even fear in the aftermath of a violent incident on the job, the potential for accidents and injuries increases. To prevent this, construction professionals should communicate openly and often about the incident making sure that it does not become the proverbial elephant in the living room everyone just walks around but is afraid to talk about. The following strategies will help workers cope with the emotions, fear, and feelings of uncertainty they might feel following a violent incident:

- Make sure workers know that you are concerned and will do whatever is necessary to help them adjust. Give workers plenty of opportunities to discuss their feelings with you in private as well as in group meetings. Encourage them to talk about what they are feeling.
- Let workers know that feelings of fear, grief, shock, and anger are normal. Explain that you feel these things too.
- Share all relevant information with workers as soon as it is available. Do not let them get distracted wondering what is happening.
- Refer workers who appear to need help to your company's Employee Assistance Program (EAP) or to the Human Resources Department. Do not try to be an amateur psychologist.
- Encourage workers to look out for each other.
- Get workers back into the routine of work as soon as possible. There is comfort and healing in a familiar routine.

EMERGENCY PREPAREDNESS PLAN

To be prepared to properly handle a violent incident in the workplace, the employer should form a crisis management team.[13] The team should have only one mission: immediate response to traumatic acts on the job. Team members should receive special training and be updated regularly. The team's responsibilities should be as follows:

- Undergo trauma response training.
- Handle media interaction.
- Establish telephone and communication teams to operate emergency communication equipment.
- Develop and implement, as necessary, an emergency evacuation plan.
- Establish a backup communication system.
- Calm personnel after an incident.
- Debrief witnesses after an incident.
- Ensure that proper security procedures are established, kept up-to-date, and enforced.
- Help employees deal with posttraumatic stress.
- Keep employees informed about workplace violence, how to respond when it occurs, and how to help prevent it.

Summary

Dealing with workplace violence involves such activities as hazard analysis, record analysis and tracking, trend monitoring, and incident analysis. Although it is important that employers consider the perpetrator's rights when dealing with workplace violence, it is equally important that they act prudently to prevent harm to other employees and customers. The exclusivity provision of workers' compensation laws provides employers with some protection from liability in cases of workplace violence, provided the incident is work related. When this is the case, workers' compensation is the injured employee's exclusive remedy. A violent act can be considered an on-the-job incident, even if it is committed away from the workplace. Specific guidelines have been established by NIOSH for determining whether a violent act can be classified as an on-the-job incident.

The concept of crime prevention through environmental design (CRTED) has four major elements: natural surveillance, control of access, establishment of territoriality, and activity support. I have added another: administrative controls.

OSHA has produced voluntary advisory guidelines related to workplace violence. Although the guidelines are aimed specifically at the service sector, they provide an excellent framework that can be used in the construction industry. The framework has the following broad elements: management commitment and employee involvement, worksite analysis, hazard prevention and control, and safety and health training.

Key Terms and Concepts

Activity support
Adjustments and modifications
Administrative controls
Anger management
Conflict guidelines
Conflict management
Control of access
Customer
Employee
Employee-related outsider
Employee surveys
Enforcement
Engineering controls
Establishment of territoriality
Evaluation of records
Exclusivity provision
Focus groups
Management commitment and employee involvement
Monitoring and feedback
Natural surveillance
Occupational violent crime (OVC)
Outsider
Postincident response
Proper work practices
Record keeping
Records monitoring and tracking

Security analysis
Trend monitoring and incident analysis
Workplace analysis
Workplace violence

Review Questions

1. Define the following terms as they relate to violence in the workplace: occupational violent crime, employee, and outsider.
2. Approximately how many people are direct victims of workplace violence annually?
3. Defend or refute the following statement: Employees who commit violent acts forfeit their rights and can be dealt with accordingly.
4. What is the "exclusivity provision" of workers' compensation laws? Why is this provision significant?
5. Defend or refute the following statement: A violent act that occurs away from the employer's premises cannot be considered work related.
6. Explain the concept of crime reduction through environmental design (CRTED).
7. Defend or refute the following statement: A contractor must comply with the OSHA guidelines on workplace violence.
8. What elements of the OSHA guidelines on workplace violence can be adapted for use in construction companies?
9. What are the primary causes of conflict on the job?
10. Describe the five ways in which an angry person may respond in a work setting.

Critical Thinking and Discussion Activities

1. Mark Raflin, a supervisor for Canton Construction Company, wants to prevent a problem before it happens. Concerned that one of his new employees, Jeff Spencer, might have violent tendencies, he asked for an appointment with the human resources (HR) director. Here is how the conversation went:

   Raflin: "We need to get rid of Jeff Spencer right now. I've seen him in action in another setting, and I'm telling you, he blew up. He can't control his temper."

   HR director: "But you say he has done nothing wrong, nor has he lost his temper, at least not yet. We can't fire Spencer just because you think he might get violent." Who is right in this case, Raflin or the HR director? How should this situation be handled?
2. Employees of Bearden Construction Company have begun to complain regularly about Josh Randall. They say that his behavior has become unpredictable and that he blows up at the slightest provocation. One employee even said, "This guy is going to hurt somebody." To each complaint, the supervisor's response is the same: "Be patient. Josh is a good man. He's just dealing with some personal problems right now. He'll get over it." What do you think of the supervisor's response to employee concerns? How do you think this situation should be handled?

Application Activities

1. Develop a comprehensive risk-reduction plan, based on OSHA's voluntary guidelines, for preventing workplace violence. Identify a construction company that will cooperate with you in completing this activity, or create a fictitious company.
2. Develop a set of conflict guidelines for the company. Identify a construction company that will cooperate with you in completing this activity, or create a fictitious company.
3. Develop an emergency preparedness plan for the company. Identify a construction company that will cooperate with you in completing this activity, or create a fictitious company.

Endnotes

1. ASIS/SHRM. *Wokplace Violence Prevention and Intervention.* Retrieved from http://www.shrm.org/hrstandards/documents/wvpi%20std.pdf on January 9, 2016.
2. Ibid., 2.
3. Ibid., 7.
4. Ibid., 12.
5. Ibid., 26.
6. Ibid., 27.
7. Ibid., 29.
8. Ibid., 31.
9. Public Law 91–596 (General Duty Clause) Section 5(a)(1).
10. U.S. Department of Labor, Occupational Safety and Health Administration. *Guidelines for Workplace Violence Prevention Programs for Night Retail Establishments* (May 1997).
11. Rusk, Tom, *The Power of Ethical Persuasion* (New York: Penguin Books, 1994), xv–xvii.
12. Illinois State Police, "Dos and Don'ts for the Supervisor," retrieved January 8, 2006 from http://www.state.il.us/isp/viowkplc/vwpp6c.htm
13. Ibid.

CHAPTER **TWELVE**

PROMOTING SAFETY, SAFETY TRAINING, AND MENTORING

LEARNING OBJECTIVES

- Describe the role of a company safety policy in construction safety.
- List the guidelines for developing safety rules and regulations.
- Explain the importance of employee participation in promoting safety.
- Summarize how training can be used to promote/ensure safety.
- Explain how to use suggestion systems to promote safety.
- Describe the role visual awareness can play in promoting safety.
- Summarize how to make the safety committee an effective part of a company's safety program.
- Explain how to use employees' signatures for gaining their personal commitment to safety.
- Describe how to gain the commitment of management and labor to safety.
- Explain the potential benefits and problems associated with using incentives to promote safety.
- Explain how to use competition to promote safety without it causing safety problems.
- Describe how to use company-sponsored **wellness programs** to promote safety.
- Summarize the teamwork approach to promoting safety.
- Explain how to use mentoring to promote safety.
- Explain how to build safety into construction contracts.

The purpose of safety promotion is to keep employees focused on doing their work the safe way, every day. This chapter provides prospective and practicing construction professionals with information that will enable them to promote safety effectively.

COMPANY SAFETY POLICY

Promoting safety begins with having a published company **safety policy**. The policy should make it clear that safe work practices are expected of all employees at all levels at all times. The safety policy serves as the foundation upon which all other promotional efforts are built.

Baker Construction Company
519 Baker Highway
Crestview, Florida 36710

Safety Policy

It is the policy of this company and its top management to ensure a safe and healthy workplace for employees; safe and healthy buildings for customers; and a safe and healthy environment for the community. BCC is committed to safety on the job and off. Employees are expected to perform their duties with this commitment in mind.

FIGURE 12–1 Sample company safety policy.

Figure 12–1 is an example of a company safety policy. This policy succinctly expresses the company's **commitment** to safety. It also indicates clearly that employees are expected to perform their duties with safety foremost in their minds. With such a policy in place and clearly communicated to all employees, other efforts to promote safety will have solid backing.

A company's safety policy need not be long and elaborate. In fact, a short and simple policy is better. Regardless of its length or format, a safety policy should convey at least the following messages:

- The company and its top managers are committed to safety and health.
- Employees are expected to perform their duties in a safe and healthy manner.
- The company's commitment extends beyond the walls of its plant to include customers and the community.

Promoting Safety by Example

After a safety policy has been implemented, its credibility with employees will be determined by the example set by management—from supervisors through executives. It is critical that managers follow the company safety policy in both letter and spirit. Managers who set a poor example undermine all of the company's efforts to promote safety. The "Do as I say, not as I do" approach will not work with employees today.

Positive or good examples of safe work behaviors tend to break down most frequently under the pressure of deadlines. To meet a deadline, supervisors may encourage their team members to take shortcuts, or, at least, look the other way when they do. This type of behavior conveys the message that safety is not really important; it is something we talk about, but not something we believe in. This issue of setting a positive example is discussed further in the next section.

SAFETY RULES AND REGULATIONS

A company's safety policy is translated into everyday action and behavior by rules and regulations. Rules and regulations define behavior that is acceptable and unacceptable from a safety and health perspective. From a legal point of view, an employer's **obligations regarding safety rules** can be summarized as follows:

- Employers must have rules that ensure a safe and healthy workplace.
- Employers must ensure that all employees know about the rules.
- Employers must ensure that safety rules are enforced objectively and consistently.

The law tends to view employers who do not meet these three criteria as being *negligent*. Having the rules is not enough. Having rules and making employees aware of them is not enough. Employers must develop appropriate rules, familiarize all employees with them, and enforce the rules. It is this final step—enforcement—from which most negligence charges arise.

Although it is acceptable to prioritize rules and assign different levels of punishment for failing to observe them, it is unacceptable to ignore rules. If the punishment for failure to observe Rule X is a letter of reprimand, then every person who fails to observe Rule X should receive such a letter every time. Of course, consequences for repeated violations should be more than a letter of reprimand.

Objectivity and consistency are critical when enforcing rules. **Objectivity** means that rules are enforced without bias. **Consistency** means that the rules are enforced in the same manner every time with no regard to any outside factors. This means that the same punishment is assigned regardless of who commits the infraction. Objectivity and consistency are similar, yet different, concepts in that one can be consistent without being objective. For example, one could be consistently biased. Failure to be objective and consistent can undermine the credibility and effectiveness of a company's efforts to promote safety.

Figure 12–2 contains guidelines to follow when developing safety rules. These guidelines help ensure a safe and healthy workplace without unduly inhibiting workers in the performance of their jobs. This is an important point for construction professionals to understand. Fear of negligence charges can influence an employer in such a way that the book of safety rules becomes a multivolume nightmare that is beyond the comprehension of most employees.

- Minimize the number of rules to the extent possible. Too many rules can result in *rule overload*.
- Write rules in clear and simple language. Be brief and to the point, avoiding ambiguous or overly technical language.
- Write only the rules that are necessary to ensure a safe and healthy workplace. *Do not nit-pick.*
- Involve employees in the development of rules that apply to their specific areas of operation.
- Develop only rules that can and will be enforced.
- Use common sense in developing rules.

FIGURE 12–2 Guidelines for developing safety rules and regulations.

Such attempts to avoid costly litigation, penalties, or fines by regulating every move that employees make and every breath that they take are likely to backfire. Remember, employers must do more than just write rules. They must also familiarize all employees with them and enforce them. This is not possible if the rulebook is as thick as an unabridged dictionary. Apply common sense when writing safety rules.

EMPLOYEE PARTICIPATION IN PROMOTING SAFETY

One of the keys to promoting safety successfully is to involve employees. This is known as **employee participation**. Employees usually know better than anyone where hazards exist. In addition, they are the ones who must follow safety rules. A fundamental rule of management is "If you want employees to make a commitment, involve them from the start." One of the most effective strategies for getting employees to commit to the safety program is to involve them in the development of it. Employees should also be involved in the implementation, monitoring, and follow-up. In all phases, employees should be empowered to take action to improve safety. The most effective safety program is one that employees view as *their* program.

SAFETY TRAINING

One of the best ways to promote safety in the workplace is to provide all employees with ongoing **safety training**. Initial safety training should be part of the orientation process for new employees. Subsequent safety training should be aimed at developing new, more specific, and more in-depth knowledge and at renewing and updating existing knowledge.

Training serves a dual purpose in the promotion of safety. First, it ensures that employees know how to work safely and why doing so is important. Second, it shows that management is committed to safety.

OSHA's Safety Training Rule

OSHA's safety training rule (29 CFR 1926.21) applies to most construction situations and should, therefore, be familiar to construction professionals. In this rule OSHA requires that employers provide instruction to employees in the recognition and avoidance of hazardous working conditions and all applicable regulations relating to the work environment in question. The goal is to ensure that employees understand how to control or eliminate hazards and any other unsafe working conditions that might exist in their work environment so as to prevent accidents and injuries.

OSHA's safety training rule requires employers to provide general safety training and specific training concerning how to deal safely with the following situations as a minimum: (1) handling and safe use of harmful/toxic substances, (2) avoiding hazards relating to potentially dangerous plants and animals at the construction site, (3) handling and safe use of flammable liquids, gases, and other potentially harmful substances, and (4) entry into and working in confined spaces. A complete list of OSHA's training related requirements is available in OSHA Publication 2254 (*Training Requirements in OSHA Standards and Training Guidelines)* at www.osha.gov/publications/osha2254.pdf.

OSHA's Outreach Training Program

This program provides construction safety training for employers and workers on the recognition, avoidance, mitigation, and prevention of workplace hazards. The program also covers workers' rights, employer responsibilities, and how to file a complaint about hazardous working conditions. OSHA's 10- and 30-hour training courses fall under this program. The 10-hour course is for entry-level construction workers. The 30-hour course is for supervisors and workers who have specific responsibilities for safety. The Outreach Program also has training for construction professionals who would like to become OSHA certified safety trainers. Complete information about OSHA's Outreach Training Program is available at www.osha.gov/dte/outreach/construction.

OSHA's 10-Hour Construction Safety Course This course is for workers who have not had safety training and are not specifically responsible for safety. It teaches participants how to recognize, avoid, mitigate, and prevent workplace hazards. Major topics covered include worker rights, employer responsibilities, fall hazards, caught-in-or-between hazards, struck-by hazards, electrocution, selection and proper use of PPE and lifesaving equipment, major health hazards in construction, material handling hazards, hand and power tool hazards, role of the workforce in improving the current culture relating to safety, and role of management in improving the current culture relating to safety.

OSHA's 30-Hour Construction Safety Course This course is for foremen, supervisors, personnel responsible for construction-related projects, and workers with specific responsibilities for safety. It teaches participants how to recognize, avoid, mitigate, and prevent workplace hazards. Major topics covered include: worker rights, employer responsibilities, the competent person, fall hazards, caught-in-or-between hazards, struck-by hazards, electrocution, selection and proper use of PPE and lifesaving equipment, major health hazards in the construction industry, material handling hazards, hand and power tool hazards, role of the workforce in improving the current culture relating to safety, role of management in improving the current culture relating to safety, crane hazards, motorized vehicle/mechanized equipment/marine operations hazards, cranes and rigging, excavations, work zone traffic control, forklift hazards, fire hazards, concrete and masonry hazards, steel erection hazards, welding and cutting hazards, confined space hazards, MSD and RMD hazards, providing a comprehensive safety and health program, supervisor's requirements in construction safety, legal responsibilities of supervisors relating to safety, incident investigations, and arc flash hazards.

OSHA's On-Line Training

OSHA provides approximately 40 different safety training courses in an on-line format. All of these courses come with a printable *certificate of completion.* The course ranges from a 10-hour construction safety orientation to specific courses on a full range of safety topics including bloodborne pathogens, drug and alcohol abuse, HAZWOPER, hazardous materials, ergonomics in construction, asbestos in the workplace, materials handling and storage, respiratory protection, scaffold safety, use of explosives, welding and cutting, workplace violence prevention, and many others. The complete list of OSHA's on-line training courses is available at www.osha.com/courses/construction.html (notice the .com in the address rather than the usual .gov).

Training Topics

The training topics provided by OSHA in its 10- and 30-hour courses as well as what is available on-line from OSHA are comprehensive enough to meet the needs of almost any safety situation a construction professional is likely to confront. However, it is a good idea for construction professionals to be familiar with the types of training that are most frequently requested in the construction industry. These topics—provided in alphabetical not priority order because needs vary according to the type of construction in question—are as follows:

- Asbestos awareness and abatement
- Back safety and lifting
- Bilingual safety programs
- Bloodborne pathogens
- Chemical labels
- Cold weather work
- Emergency action plans
- Emergency preparedness
- Ergonomics
- Fall protection
- Forklift safety
- Hand, foot, eye, and head protection
- Housekeeping
- Lockout/tagout
- Machine guards
- Safety data sheets
- Power tools
- Respirators
- Sexual harassment and safety
- Stress
- Violence in the workplace

Toolbox Talks for Construction Safety Training

Gathering workers around the toolbox to discuss safety issues and concerns is an excellent training method. Toolbox talks can be informal and open-ended or they can focus on specific safety issues. Regardless of the topics dealt with in toolbox talks by construction professionals, the following guidelines will help ensure the effectiveness of the talks:

- Choose a place to hold toolbox talks that is conducive to discussion (i.e., free of noise and other distractions and impediments).
- Make the talk a two-way conversation. Present the material to be covered, but give participants plenty of time and opportunities to make comments, ask questions for clarification, and discuss the issue at hand.
- Whenever appropriate, use props and other devices that will help hold the attention of participants. Perhaps the best-known example of this strategy is using a watermelon to demonstrate what happens when a heavy tool such as a hammer is dropped on to it from 10 to 15 feet above. With the first drop of the hammer, a hard hat covers the melon. With the second the hard hat is removed. What happens to the melon (i.e., the worker's head) with the second drop of the hammer drives home the importance of always wearing a hard hat on a construction site.
- Make sure that as a trainer you always set a good example of doing what you recommend employees do during toolbox talks.

Other Approaches to Training

How training is presented does matter. Different approaches are likely to yield different results; although the teaching talent of the trainer is an important factor regardless of the approach. What follows are some approaches that typically yield good results when providing construction training:

- *Blended learning.* Blended learning involves combining more than one teaching approach (e.g., combining traditional classroom instruction and on-line learning). With blended learning participants study the basic material in one format—say on-line—and receive clarification, expansion, and discussion of the material in a more traditional format.
- *Case studies.* Case studies provide real situations that have occurred and allow trainers to explain the who, what, where, when, why, and how of the case. They also allow participants to discuss and debate these aspects of the case. Well-chosen case studies can be effective in bringing safety training to life for participants and making it more real and relevant.
- *Computer-based/on-line learning.* The benefits of this approach include that is self-paced, individualized, and interactive. Provided the computer-based and on-line materials are of sufficient quality, this can be an especially effective approach. Of course, this approach does not provide face-to-face interaction with the trainer, which is why it works best as a component of the blended approach.
- *Lecture method.* When using this method—sometimes called the traditional classroom method—it is important to supplement lectures with materials that make the trainer's words visual (e.g., handouts and PowerPoints). An effective way to improve this method is to combine it in a blended format with computer-based instruction.

Multilingual Safety Training

Construction professionals should be prepared to provide training for a multilingual audience. The percentage of immigrants in the workforce now approaches 15 percent and is highest in the construction industry. Nearly half of the immigrants working in construction are considered to have a limited command of English. For this reason it may be necessary to provide safety training in another language—most commonly Spanish in the construction industry. Fortunately, OSHA provides many of its training programs in Spanish. English-as-a-second-language programs in the community can also provide assistance to construction companies in this area. Having a bi-lingual employee become an

OSHA certified trainer is another approach for providing training in a language other than English.

SUGGESTION PROGRAMS

Suggestion programs, if properly handled, promote safety and health. Well-run suggestion programs offer two advantages: (1) They solicit input from the people most likely to know where hazards exist and (2) they involve and empower employees, which in turn gives them **ownership** of the safety program.

Suggestion programs must meet certain criteria to be effective:

- All suggestions must receive a **formal response**.
- All suggestions must receive an **immediate response**.
- Management must monitor the performance of each department in generating and responding to suggestions.
- System costs and savings must be reported.
- Recognition and awards must be handled promptly.
- Good ideas must be implemented.
- Personality conflicts must be minimized.[1]

Suggestion programs that meet these criteria are more likely to be successful than those that do not. Note that all of the following must be recorded:

- The date that the suggestion was submitted.
- The date that the suggestion was logged in.
- The date that the employee received a response.

This form satisfies the formal response and immediate response criteria. It also makes it easier to monitor responses. Walton Contractors (in Figure 12–3) publishes system costs and savings in its monthly newsletter for employees, implements good ideas, and recognizes employees through a variety of awards—ranging from certificates to cash—at a monthly recognition ceremony. This company's suggestion program is an example of one that promotes not just safety, but continual improvements in quality, productivity, and competitiveness.

Walton Contractors, Inc.
Highway 331
DeFuniak Springs, Florida 32614

Suggestion Form

Name of employee: ______________________

Date of suggestion: ______________________

Department: ______________________

Suggested improvement: ______________________

____

____

Date logged in: ____________ Time: ____________

Action taken: ______________________

____

Current status: ______________________

Date of response to employee: ______________________

Person responding: ______________________

(Signature)

FIGURE 12–3 Sample safety suggestion form.

VISUAL AWARENESS

We tend to be a visual society. This is why television and billboards are so effective in marketing promotions. Consequently, **visual awareness** is important. Making a safety and health message *visual* can be an effective way to get the message across. Figure 12–4 is a sign that gives workers a visual reminder to wear PPE.

Figure 12–5 may be placed on the door leading into the hard hat area or on a stand placed prominently at the main point of entry if there is no door. Such a sign helps prevent inadvertent slipups when employees are in a hurry or are thinking about something else. Figures 12–6, 12–7, and 12–8 are additional examples of signs and posters that make a safety message visual.

Several rules of thumb can help ensure the effectiveness of efforts to make safety visual:

- Change signs, posters, and other visual aids periodically. Visual aids left up too long begin to blend into the background and are no longer noticed.

FIGURE 12–4 Sample safety reminder sign.
Source: Barry Barnes/Fotolia

CAUTION

HARD HAT AREA

FIGURE 12–5 Sample safety reminder sign.

FIGURE 12–6 Sample safety reminder sign.

CAUTION

WEAR EYE PROTECTION
WHILE OPERATING

FIGURE 12–7 Sample safety reminder sign.

NOTICE
SAFETY SHOES ARE
REQUIRED IN THIS
AREA

FIGURE 12–8 Sample safety reminder sign.

- Involve employees in developing the messages that are displayed on signs and posters. Employees are more likely to notice and heed their own messages than those of others.
- Keep visual aids simple and the message brief.
- Make visual aids large enough to be seen easily from a reasonable distance.
- Locate visual aids for maximum effect. For example, the sign in Figure 12–4 should be located on the machine in question, preferably near the on-off switch, so that the operator cannot activate the machine without seeing it.
- Use color whenever possible to attract attention to the visual aid, but be sure to follow OSHA's color standards where applicable.

SAFETY COMMITTEES

Another way to promote safety through **employee involvement** is the safety committee. **Safety committees** provide a formal structure through which employees and management can funnel concerns and suggestions about safety and health issues. The composition of the safety committee can be a major factor in the committee's success or failure.

The most effective committees are those with a broad cross section of workers representing all departments. This offers two advantages: (1) It gives each member of the committee a constituent group for which he or she is responsible and (2) it gives all employees a representative voice on the committee.

There is disagreement over whether an executive-level manager should serve on the safety committee. On the one hand, an executive-level participant can give the committee credibility, visibility, and access. On the other hand, the presence of an executive manager can inhibit the free flow of ideas and concerns. The key to whether an executive manager's participation will be positive or negative lies in the personality and management skills of the executive.

An executive who knows how to put employees at ease, interact in a nonthreatening manner, and draw people out will add to the effectiveness of the committee. An executive with a threatening attitude will render the committee useless. Consequently, I recommend the involvement of a carefully selected executive manager on the safety committee.

The construction professional responsible for safety should be a member of the committee, serving as an advisor, a facilitator, and a catalyst. Committee members should select a chairperson from the membership and a recording secretary for taking minutes and maintaining committee records. Neither the executive manager nor the safety director should serve as chairperson, but either can serve as recording secretary. Excluding executive managers and safety professionals from the chair gives employees more ownership in the committee.

Safety committees work only if members are truly empowered to identify hazards and take steps to eliminate them.

Safety Committee Meetings

Every employee who serves on a safety committee has another job to do. Consequently, it is important to have meetings that are both efficient and effective. This is accomplished by having an agenda that gives both structure and direction to meetings.

A typical meeting of the safety committee should proceed as follows:

- Call to order.
- Record attendance.
- Review and approve previous minutes.

- Discuss old business.
- Discuss new business.
- Discuss new accidents.
- Discuss near misses.
- Report on inspections, subcommittee work, special assignments, and safety programs.
- Make special presentations (guests, videos, demonstrations).
- Make announcements.
- Adjourn.

Guidelines for a Safety Committee

For both management and employees to participate willingly and contribute effectively, safety committees must work well. In addition to having an agenda for meetings and sticking to it, the following rules should be observed[2]:

Do's

- Suggest strategies and options for management to improve health and safety performance.
- Train committee members so that they can successfully carry out their responsibilities.
- Give the safety committee authority commensurate with its responsibilities.
- Have goals and objectives, and measure against them to track progress.
- Encourage employee involvement by actively creating an atmosphere of trust, teamwork, respect, and partnership.
- Be patient. Be reasonable in allowing enough time for the committee to work.
- Reward progress, participation, and leadership.
- Train management on its responsibility for safety and on the support role of the safety committee.
- Stagger committee memberships to maintain a mixture of experience levels.

Dont's

- Allow the safety committee to function as "safety cops." Keep management responsible for decisions and enforcements.
- Discuss topics unrelated to health and safety at safety committee meetings. Stay away from labor and personnel issues.
- Rotate members too quickly. A one-year minimum membership is the norm.
- Let any one member dominate safety committee meetings. Encourage and maintain equal participation.
- Allow safety committee members to bring just problems to the meetings. Have them bring solutions as well.
- Allow the safety committee to become a scapegoat when something goes wrong. Management is responsible for safety performance.
- Punish. It creates fear, which inhibits communication and partnership. For example, blaming an injured employee can obscure other contributing causes and encourage underreporting. Do hold persons accountable for their responsibilities (which differs from punishment).

GAINING A PERSONAL COMMITMENT

If every employee is committed to working safely every day, workplace safety will take care of itself. But how does a company gain this type of **personal commitment** from its employees? One way is to have employees commit themselves to safety by signing on the bottom line.[3]

This approach to gaining a personal commitment from employees has merit. Ours is a society that revolves around the written signature. We sign countless documents in our lives from credit statements to bank loans to home mortgages to college registration forms. In each of these cases, our signature is a written pledge of our commitment to meet certain responsibilities.

Companies gain the following advantages from making signing on the dotted line part of their program to promote safety:

1. By their signatures, employees make a personal commitment.
2. By their signatures, employees promise to interact positively with fellow workers when they see them ignoring safety precautions.
3. By their signatures, employees give fellow workers permission to correct them when they ignore safety precautions.[4]

EMPLOYEE AND MANAGEMENT PARTICIPATION

An excellent way to promote safety is to secure the cooperation of management and labor. For a company's safety program to succeed, **employee and management participation** and support are critical. Fortunately, employee and management agreement on workplace safety is common.

When disagreement over a safety procedure does surface, the issue at the heart is usually money. Employees are likely to favor procedures that enhance workplace safety regardless of cost. Management, on the other hand, is likely to want to weigh the cost versus the benefits of safety improvement strategies. However, sometimes employees, rather than managers, question safety strategies.

An example of an employee's questioning a safety enhancement strategy is the "Sign Up for Safety" campaign conducted by a company in Muskegon, Michigan.[5] In an attempt to gain a personal commitment to safety, Fettig asked employees to sign a declaration that they will work in a safe manner. This is a technique that has met with a great deal of success.

However, in attempting to sell employees on the strategy, Fettig ran into resistance from an employee who refused to sign his name, claiming that the company's management team might use it against him.

The employee had to be persuaded the strategy was not a management trick. The employee's eventual willingness to sign the safety declaration in this case is what made the program work. With management and employees on the same team, the safety program is much more likely to succeed.[6]

SAFETY INCENTIVES: BENEFITS AND PROBLEMS

Some construction companies use incentives to promote workplace safety. However, safety professionals need to know that OSHA is skeptical of this approach.[7] OSHA is not against the use of safety incentives per se. Rather, OSHA is concerned that if safety incentives are not used properly and monitored carefully they might lead to underreporting of accidents, incidents, and near-misses. One way to ensure that incentives are forthcoming is to cover up safety problems rather than report them. Construction professionals should always consider human nature in all decisions including, of course, whether or not to use safety incentives.

Consequently, if safety incentives are used it is important to monitor workers even more closely to ensure that problems are not being covered up or simply ignored. In addition to close monitoring, the following strategies will help improve the effectiveness of safety incentives:

- *Define objectives.* Begin by deciding what is supposed to be accomplished by the incentive program.
- *Develop specific criteria.* On what basis will the incentives be awarded? This question should be answered during the development of the program. Specific criteria define the type of behavior and level of performance that is to be rewarded as well as guidelines for measuring success.
- *Make rewards meaningful.* For an incentive program to be effective, the rewards must be meaningful to the recipients. Giving an employee a reward that he or she does not value will not produce the desired results. To determine what types of rewards will be meaningful, it is necessary to involve employees.
- *Recognize that only employees who will participate in an incentive program know what incentives will motivate them.* In addition, employees must feel it is their program. This means that employees should be involved in the planning, implementation, and evaluation of the incentive program.
- *Keep communications clear.* It is important for employees to fully understand the incentive program and all of its aspects. Communicate with employees about the program, ask for continual feedback, listen to the feedback, and act on it.
- *Reward teams.* Rewarding teams can be more effective than rewarding individuals. This is because work in the modern industrial setting is more likely to be accomplished by a team than an individual. When this is the case, other team members may resent the recognition given to an individual member. Such a situation can cause the incentive program to backfire.[8]

COMPETITION

Competition is another strategy that can be used to promote safety. However, if this approach is not used wisely, it can backfire and do more harm than good. To a degree, most people are competitive. A child's competitive instinct is nurtured through play and reinforced by sports and school activities. Construction professionals can use the adult's competitive instinct when trying to motivate employees, but competition on the job should be carefully organized, closely monitored, and strictly controlled. Competition that is allowed to get out of hand can lead to cheating and hard feelings among co-workers.

Competition can be organized among teams, shifts, divisions, or even job sites. Here are some tips that will help safety and health professionals use competition

SAFETY FACTS & FINES

Complaints from nonoccupants of a building can result in fines just as large as those resulting from employee complaints. After a parent complained, the Environmental Protection Agency (EPA) investigated all 263 buildings in a Michigan school district and fined the district $1.4 million. The EPA found that in 256 of the buildings, the district had failed to properly monitor asbestos. The EPA could not tell whether the other schools had the same problems because the district had not kept records.

in a positive way while ensuring that it does not get out of hand:

- Involve the employees who will compete in planning programs of competition.
- Where possible, encourage competition among groups rather than individuals, while simultaneously promoting individual initiative with groups.
- Make sure that the competition is fair by ensuring that the resources available to competing teams are equitably distributed and that human talent is as appropriately spread among the teams as possible.

The main problem with using competition to promote safety is that it can induce competing teams to cover up or fail to report accidents in order to win. Construction professionals should be particularly attentive to this situation and should watch carefully for evidence that accidents are going unreported. If this occurs, the best approach is to confront the situation openly and frankly. Employees should be reminded that improved safety is the first priority, and winning the competition is second. Failing to report an accident should be grounds for eliminating a team from competition.

TEAMWORK APPROACH TO PROMOTING SAFETY

Increasingly, teamwork is stressed as the best way to get work done in the contemporary workplace. Consequently, it follows that the teamwork approach is an excellent way to promote safety.

Characteristics of Effective Teams

Effective teams share several common characteristics: supportive environment, team-player skills, role clarity, clear direction, team-oriented rewards, and accountability.

Supportive Environment

The characteristics of a team-supportive environment are well known. These characteristics are as follows:

- Open communication
- Constructive, nonhostile interaction
- Mutually supportive approach to work
- Positive, respectful climate

Team-Player Skills

Team-player skills are personal characteristics of individuals that make them good team players. They include the following:

- Honesty
- Selflessness
- Initiative
- Patience
- Resourcefulness
- Punctuality
- Tolerance
- Perseverance

Role Clarity

On any team, different members play different roles. Consider the example of a football team. When the offensive team is on the field, each of the 11 team members has a specific role to play. The quarterback plays one role; the running backs, another; the receivers, another; the center, another; and the linemen, another. Each of these roles is different but important to the team. When each of these players executes his role effectively, the team performs well.

But what would happen if the center suddenly decided he wanted to pass the ball? What would happen if one of the linemen suddenly decided that he wanted to run the ball? Of course, chaos would ensue. A team cannot function if the team members try to play roles that are assigned to other team members. **Role clarity** means that all members understand their respective roles on the team and play those roles.

Clear Direction

What is the team's purpose? What is the team supposed to do? What are the team's responsibilities? These are the types of questions that people ask when they are assigned to teams. The team's charter should answer such questions. The various components of a team's charter are as follows:

1. *Mission.* The team's **mission** statement defines its purpose and how the team fits into the larger organization. In the case of a safety promotion team, it explains the team's role in the organization's overall safety programs.
2. *Objectives.* The team's **objectives** spell out exactly what the team is supposed to accomplish in terms of the safety programs.
3. *Accountability measures.* The team's accountability measures spell out how the team's performance will be evaluated.

Figure 12–9 is an example of a team charter for the safety promotion team in a construction company. This charter clearly defines the team's purpose, where it fits into the overall organization, what it is supposed to accomplish, and how the team's success will be measured.

Team Charter
Safety Promotion Team
MTC Construction

Mission

The mission of the Safety Promotion Team at MTC Construction is to make all employees at all levels of the company aware of the importance of safety and health and, having made them aware, to keep them aware.

Objectives

1. Identify innovative, interesting ways to communicate the company's safety rules and regulations to employees.
2. Develop a company-wide suggestion system to solicit safety-related input from employees.
3. Identify eye-catching approaches for making safety a visible issue.
4. Develop appropriate safety competition activities.

Accountability Measures

The quality of participation in all of the activities of this team will be assessed by the team leader and included in the annual performance appraisal of each team member. Team members are expected to be consistent in their attendance, punctual, cooperative, and mutually supportive.

FIGURE 12–9 Sample team charter.

Team-Oriented Rewards

One of the most commonly made mistakes in organizations is attempting to establish a teamwork culture while maintaining an individual-based reward system. If teams are to function fully, the organization must adopt team-oriented rewards, incentives, and recognition strategies. For example, a team functions best when the financial rewards of its members are tied at least partially to team performance. Performance appraisals that contain criteria related to team performance promote teamwork in addition to individual performance. The same concept applies to recognition activities.

Accountability

There is a rule of thumb in management that says "If you want to improve performance, measure it." **Accountability** is about being held responsible for accomplishing specific objectives or undertaking specific actions. The most effective teams know what their responsibilities are and how their success will be measured.

Potential Benefits of Teamwork in Promoting Safety

Teamwork can have both direct and indirect benefits for an organization. Through teamwork, counterproductive internal competition and internal politics are replaced by collaboration, Figure 12–10. When this happens, the following types of benefits typically accrue:

- Better understanding of safety rules and regulations.
- Visibility for safety.

FIGURE 12–10 Teamwork promotes safety.
Source: Andres Rodriguez/Fotolia

- Greater employee awareness.
- Positive, productive competition.
- Continual improvement.
- Broader employee input and acceptance.

Potential Problems with Teams

Teamwork can yield important benefits, but as with any concept, there are potential problems. The most pronounced potential problems with teams are as follows:

- It can take a concerted effort over an extended period of time to mold a group into an effective team, but a team can fall apart quickly.
- Personnel changes are common in organizations, but personnel changes can disrupt a team and break down team cohesiveness.
- Participative decision making is inherent in teamwork. However, this approach to decision making takes time, and time is often in short supply.
- Poorly motivated and lazy employees can use a team to blend into the crowd and to avoid participation. If one team member sees another slacking, he or she may respond in kind.

These potential problems can be prevented, of course. The first step in doing so is recognizing them. The next step is ensuring that all team members fulfill their responsibilities to the team and to one another.

Responsibilities of Team Members

Accountability in teamwork amounts to team members fulfilling their individual responsibilities to the team and to one another. These responsibilities are as follows:

- Actively participating in all team activities.
- Being punctual in attendance of meetings.
- Being honest and open toward fellow team members.
- Making a concerted effort to work well with team members.

- Being a good listener for other team members.
- Being open to the ideas of others.

If individual team members fulfill these responsibilities to one another and the team, the potential problems with teams can be overcome and the benefits of teamwork can be fully realized. It is important for members of the safety team to understand these responsibilities, accept them, and set an example of fulfilling them. If this happens, the benefits to the organization will go well beyond just safety and health.

MENTORING FOR SAFETY

Because safety is an essential ingredient in organizational excellence, construction professionals have every reason to engage in on-going employee development. One of the most effective forms of employee development is mentoring. Effective mentoring can improve the performance of the employee being mentored with regard to safety as well as overall performance, Figure 12–11.

Mentoring Defined

Mentoring for safety is a process in which a more experienced individual helps one who is less experienced develop the knowledge, skills, and attitudes needed to work safely and help other do the same. Being a mentor involves being a role model, but the two concepts are not the same. A role model can be anyone who exemplifies admirable qualities that are worthy of being emulated. It is not necessary to know or even meet a role model. Role models can be found in books, movies, television programs, or any other medium that can convey their examples. Employees can have a face-to-face relationship with a role model, but it is not necessary.

Mentoring is different. Mentoring requires a face-to-face, one-on-one relationship that is reinforced by frequent contact and effective communication. Mentors invest time, energy, expertise, and caring in helping their protégés learn, grow, and develop. An effective mentoring relationship will help prepare the protégé to work safely in all situations. When mentoring is done well, both the protégé and the mentor benefit from the experience.

FIGURE 12–11 Effective mentoring can improve safety.
Source: Viappy/Fotolia

Responsibilities of Mentors

In order to ensure positive results from mentoring, it is necessary for construction professionals to understand the responsibilities of mentors. In broad terms, mentors must be willing to give of their time, remain open-minded, give feedback that is constructive and tactful, and listen well. In addition to these things, they must care about the development of their protégés as individuals. In addition to these broad responsibilities, there are several specific responsibilities mentors must be willing to accept:

- Communicating openly, frankly, tactfully, and frequently with protégés.
- Serving as a sounding board and patient, attentive listener for protégés.
- Providing encouragement, recognition, and support for protégés.
- Providing a steady flow of accurate, up-to-date information about opportunities, issues, problems, and options for protégés.
- Being a consistent role model of a safety-first attitude and behaviors for protégés.
- Helping protégés set goals and realistic timetables for achieving them.
- Helping protégés develop effective strategies for achieving goals.
- Helping protégés understand the organization's culture overall as well as specifically relating to safety.
- Introducing protégés to useful contacts who are good role models of working safely.
- Helping protégés develop specific knowledge and skills relating to safety.
- Helping protégés become mentors to their fellow workers: parallel mentoring.

How Employees Can Mentor Their Fellow Workers: Parallel Mentoring?

Traditional mentoring involves more experienced personnel helping less experienced individuals develop, grow, and succeed. But there is another side to mentoring that gets very little attention—employees mentoring their fellow workers in selected areas—areas such as safety. The author refers to this kind of mentoring as *parallel mentoring* or mentoring provided by workers of equal rank and status.

Parallel Mentoring Strategies

Workers who have been mentored can be valuable assets in spreading the word among fellow workers about what they have learned. Consequently, construction professionals should teach their employees the following strategies for helping enhance the safety-related performance of their fellow workers: (1) be a second pair of eyes and ears for them, (2) work quietly to enhance their safety-related performance, and (3) point out hazardous conditions and/or behaviors in real time.

- *Be a second pair of eyes and ears.* Every worker has different experiences, knowledge, and levels of skill. Consequently, one way employees can help their fellow workers is to serve as a second pair of eyes and ears for them. Safety professionals should teach their protégés to spread the word about what they have learned. If there is something relating to safety that a fellow worker does not appear to know, these protégés should step forward and teach them or show them the safe way. Parallel mentoring multiplies the value of the mentoring process.
- *Work quietly to enhance their safety-related performance.* To multiply the value of the mentoring process, construction professionals should encourage their protégés to share what they have learned with their fellow workers, but not in a bossy or high-handed way. Rather, they should quietly work with their colleagues to show them the safe way.
- *Point out hazardous conditions and/or behaviors in real time.* Workers who have been mentored may be better able to identify hazardous conditions than can fellow workers who have not been mentored. They will also recognize unsafe behaviors when they observe them. By pointing out hazardous conditions before they cause accidents and injuries, workers who have been mentored can multiply the value of the mentoring process. By quietly pointing out the fact when their fellow workers behave in an unsafe manner, workers who have been mentored can exert positive peer pressure in favor of safety.

SAFETY IN CONSTRUCTION CONTRACTS

An effective way to promote safety in construction is to build safety requirements into construction contracts. In this way, a general contractor or an owner can ensure that it is working competent contractors and subcontractors that give safety an appropriate priority. Issues to consider building into construction contracts include the following:

- Required safety meetings.
- Required monthly safety reports on specified criteria.
- Required safety training.
- Mechanisms for ensuring that all subcontractors meet the safety requirements.
- An approved safety plan that includes at least the following elements: (1) major incident procedures, (2) first aid facilities and provisions, (3) safety inspection procedures, (4) personal protective equipment requirements, (5) housekeeping requirements, (6) fire precaution requirements, (7) site access and security, (8) public protection, (9) waste clearance and disposal, (10) material storage, (11) lighting requirements, and (12) job-specific safety procedures for such concerns as cranes and hoisting equipment, noise, fall protection, scaffolding, etc.

Building safety requirements and plans into construction contracts serves notice to all stakeholders that safety is an important concern on the construction project in question. It also provides both a practical and a legal basis for effectively monitoring job-site safety.

Summary

A company's safety policy should convey the following messages: (1) a company-wide commitment, (2) the expectation that employees will perform their duties in a safe manner, and (3) the company's commitment includes customers and the community.

From a legal perspective, an employer's obligations regarding safety rules can be summarized as follows: (1) employers must have rules that ensure a safe and healthy workplace, (2) employers must ensure that all employees are knowledgeable about the rules, and (3) employers must ensure that safety rules are enforced objectively and consistently.

A fundamental rule of management is "If you want employees to make a commitment, involve them from the start." This is especially important when formulating safety rules.

Safety training should be a fundamental part of any effort to promote safety. Safety training ensures that employees know how to work safely, and it shows that management is committed to safety.

Well-run suggestion programs promote safety by: (1) soliciting input from the people who are most likely to know where hazards exist, and (2) involving employees in a way that lets them feel ownership in the safety program.

Safety committees can help promote safety if they are properly structured. The composition of the committee can be a major factor in the committee's success. The most effective committees are composed of a broad cross section of workers representing all departments.

Employee and management agreement is important in promoting safety. Fortunately, safety is an issue on which employees and management can usually agree.

Incentives can promote safety if they are properly applied. To enhance the effectiveness of incentives, the following steps should be followed: (1) define objective, (2) develop specific criteria, (3) make rewards meaningful, (4) keep communications clear, (5) involve employees in planning the incentives, and (6) reward teams.

Competition can promote safety, but it can also get out of hand and do more harm than good. To keep competition positive, involve employees in planning programs of competition, and encourage competition between teams rather than individuals.

OSHA's safety training rule (29 CFR 1926.21) applies to most construction situations and, therefore, should be familiar to construction professionals. OSHA's safety training rule requires employers to provide general safety training plus specific training covering the following situations: (1) handling/use of harmful/toxic substances, (2) avoiding plant and animal hazards at job sites, (3) handling and use of flammable liquids, gases, and other potentially harmful substances, and (4) confined spaces. **OSHA's Outreach Training Program** provides construction safety training for employers and workers on the recognition, avoidance, mitigation, and prevention of workplace hazards. The program also covers workers' rights, employers' responsibilities, and how to file complaints about hazardous conditions.

Key Terms and Concepts

- Accountability
- Commitment
- Competition
- Consistency
- Employee and management participation
- Employee involvement
- Employee participation
- Formal response
- Immediate response
- Incentives
- Mission
- Objectives
- Objectivity
- OSHA's Outreach Training Program
- Ownership
- Personal commitment
- Role clarity
- Safety committee
- Safety policy
- Safety rules
- Safety training
- Suggestion programs
- Visual awareness
- Wellness programs

Review Questions

1. What messages should a company's safety policy convey?
2. Explain why promoting safety by example is so important.
3. What are the employer's obligations regarding safety rules and regulations?
4. Explain the concept of negligence as it relates to a company's safety.
5. What is the significance of objectivity and consistency when enforcing safety rules?
6. Why is employee participation/involvement so critical in the promotion of safety?
7. If your task was to establish a safety committee, who would you ask to serve on it?
8. List the three benefits that companies gain from asking employees to sign a declaration of safety.
9. What are the steps for ensuring that incentives actually promote safety?
10. What are the characteristics of effective teams?
11. What are the requirements of OSHA's safety training rule (29 CFR 1926.21)?
12. Describe what OSHA's Outreach Training Program provides for employers and workers.

Critical Thinking and Discussion Activities

1. Construction Management, Inc. (CMI) has a safety awareness program, but it does not seem to be working. CMI's president and his two vice presidents developed the program themselves and implemented it company wide six months ago. Unfortunately, they have seen no decrease in the number of accidents and incidents. What possible problems do you see with CMI's safety awareness program? What changes or improvements would you recommend?
2. "I'm not going to sign anything!" The employee was adamant about not signing a commitment-to-work-safely form. "This is nothing but a sneaky way to get the documentation necessary to fire us." The safety director really wants 100 percent commitment from all employees. How would you recommend he handle this situation? What would you say to this employee?

Application Activities

1. Develop a safety policy and a corresponding set of safety rules that could be used by a small construction company.
2. Develop an incentive program that could be used by a large construction company to promote safety.
3. Develop a safety training program for a medium-sized construction company.

Endnotes

1. McDermott, B., "Employees Are Best Source of Ideas for Constant Improvement," *Total Quality Newsletter* 1, no. 4: 5.
2. Maurer, R. "Making Safety Committees Work," http://www.shrm.org/hrdisciplines/safetysecurity/articles/pages/workplace-safety-committees.aspx, May 3, 2016.
3. Fettig, A., "Sign Up for Safety," *Safety & Health* 144, no. 1: 26.
4. Fettig, "Sign Up for Safety," 26.
5. Ibid., 27.
6. Ibid.
7. http:// www.osha.gov/as/opa/whistleblowermemo.html Retrieved on January 19, 2016.
8. Ibid.

GUIDELINES FOR CONDUCTING ON-THE-SPOT ACCIDENT INVESTIGATIONS

Checklist for On-the-Spot Accident Investigations

- What was the injured employee doing at the time of the accident?
- Had the injured employee received proper training in the task before being asked to perform it?
- Was the injured employee authorized to use the equipment or perform the task involved in the accident?
- Were other employees present at the time of the accident? If so, who are they and what were they doing? (Interview them *separately* as soon as possible after the accident.)
- Was the task in question being performed according to properly approved procedures?
- Was proper personal protective equipment being used and were work procedures being followed at the time of the accident?
- Was the injured employee new to the job in question?
- Was the process, equipment or system that was involved new? Old? Properly maintained?
- Was the injured person being supervised at the time of the accident?
- Has a similar accident occurred before? If so, were corrective measures recommended? Were they implemented?
- Are there obvious factors that led to the accident or that could have prevented the accident?

CHAPTER THIRTEEN

SUBPARTS A THROUGH E AND RELATED SAFETY PRACTICES

LEARNING OBJECTIVES

- Summarize the general provisions of Subpart A.
- Summarize the general provisions of Subpart B.
- Summarize the general provisions of Subpart C.
- List/explain the specific requirements set forth in Subpart D.
- List/explain the specific requirements set forth in Subpart E.

The OSHA standard that applies specifically to construction is 29 CFR 1926. It is divided into subparts lettered from A to Z. Each subpart covers a specific area of concern. This chapter covers Subparts A through E. In addition, it explains how construction companies can adopt safety practices that will help them comply with the standard and make them more competitive in the marketplace. Although Subparts A, B, and C are included here, most compliance requirements occur beginning in Subpart D. Consequently, Subparts D through Z are covered in greater detail than Subparts A through C.

SUBPART A: GENERAL REQUIREMENTS

Subpart A covers purpose, scope, variances from safety and health standards, and enforcement of the standards. A major focus of this subpart is the right of the U.S. secretary of labor or the designated representative of the secretary to enter any job site that falls under the jurisdiction of the Contract Work Hours and Safety Standards Acts (Section 107). The principal message contained in this subpart is that the secretary has the right to conduct onsite inspections to determine if a construction company is complying with applicable standards. There is also a detailed explanation of the rules governing administrative adjudication of cases when violations are noted by an inspector in the process of enforcing Occupational Safety and Health Administration (OSHA) standards. Subpart A contains the following sections:

- 1926.1 Purpose and scope
- 1926.2 Variations from safety and health standards
- 1926.3 Inspections: right of entry
- 1926.4 Rule of practice for administrative adjudications for enforcement of safety and health standards
- 1926.5 OMB control numbers under the Paperwork Reduction Act
- 1926.6 Incorporation by reference

SUBPART B: GENERAL INTERPRETATIONS

The goal of Subpart B is to ensure that contractors provide workers with a safe and healthy work environment. The primary focus of this subpart is sanitation (toilet facilities) and first aid. The point is made in this subpart that subcontractors are not absolved of their responsibility to comply with all applicable regulations just because they have an agreement with a prime contractor to satisfy certain requirements. Liability can still be assigned in both directions if either the subcontractor or primary contractor fails to perform properly.

Subpart B applies to work performed under contract to a federal government agency. Consequently, all labor standards and statutes that apply when working on a government contract apply here. This includes such federal laws as the Davis–Bacon Act, Federal Aid Highway Acts, National Housing Act, and Walsh–Healey Public Contracts Act. Subpart B contains the following sections:

- 1926.10 Scope of the subpart
- 1926.11 Coverage under Section 103 of the act distinguished
- 1926.12 Reorganization Plan Number 14 of 1950
- 1926.13 Interpretation of statutory terms
- 1926.14 Federal contracts for "mixed" types of performance
- 1926.15 Relationship to the Service Contract Act, Walsh–Healey Public Contracts Act
- 1926.16 Rules of construction

SUBPART C: GENERAL SAFETY AND HEALTH PROVISIONS

Subpart C covers a variety of general safety and health provisions, most of which are covered in greater depth in later subparts of the standard. Of particular importance in this subpart are the requirements that construction companies implement accident-prevention programs, require onsite inspections by designated competent persons, and provide safety and health training for workers. Also included are the general requirements for first aid, housekeeping, illumination, fire protection, sanitation, means of egress, **personal protective equipment (PPE)**, and emergency plans. These provisions are covered when specific requirements related to them come up in discussion of later subparts. Subpart C contains the following sections:

- 1926.20 General safety and health provisions
- 1926.21 Safety training and education
- 1926.22 Recording and reporting of injuries (Reserved)
- 1926.23 First aid and medical attention
- 1926.24 Fire protection and prevention
- 1926.25 Housekeeping
- 1926.26 Illumination
- 1926.27 Sanitation
- 1926.28 Personal protective equipment
- 1926.29 Acceptable certifications
- 1926.30 Shipbuilding and ship repairing
- 1926.31 Incorporation by reference
- 1926.32 Definitions
- 1926.33 Access to employee exposure and medical records
- 1926.34 Means of egress
- 1926.35 Employee emergency action plans

SUBPART D: OCCUPATIONAL HEALTH AND ENVIRONMENTAL CONTROLS

Subpart D covers numerous health and environmental concerns. The requirements for providing medical services and first aid are explained. Sanitation requirements related to the provision of toilets, based upon the number of workers at a job site, are also explained. Concerns about a variety of hazards that are often grouped under the generic heading of "industrial hygiene" are covered, including concerns related to noise, radiation, and a variety of physical and chemical hazards. Subpart D contains the following sections:

- 1926.50 Medical services and first aid
- 1926.51 Sanitation
- 1926.52 Occupational noise exposure
- 1926.53 Ionizing radiation
- 1926.54 Nonionizing radiation
- 1926.55 Gases, vapors, fumes, dusts, and mists
- 1926.56 Illumination
- 1926.57 Ventilation
- 1926.58 (Reserved)
- 1926.59 Hazard communications
- 1926.60 Methylenedianiline
- 1926.61 Retention of DOT markings
- 1926.62 Lead
- 1926.64 Process safety management of highly hazardous chemicals
- 1926.65 Hazardous waste operations and emergency response
- 1926.66 Criteria for design and construction of spray booths

Several of the safety and health concerns that grow out of Subpart D are important enough to construction professionals that they warrant more in-depth treatment. These concerns include the following: medical services and first aid; occupational noise exposure; ionizing radiation; nonionizing radiation; **airborne contaminants** (gases, vapors, fumes, dusts, and mists); hazard communication; hazardous waste operations; and lead. Detailed information on safety practices related to these aspects is contained in the sections that follow.

MEDICAL SERVICES AND FIRST AID

Workplace emergencies often require a medical response. The immediate response is usually first aid. First aid consists of lifesaving measures taken to assist an injured person until medical help arrives. Since there is no way to predict when first aid may be needed, providing first aid training to employees should be part of preparing for emergencies. In fact, OSHA requires that companies provide for

Basic first aid
Cardiopulmonary resuscitation
Severe bleeding
Broken bones and fractures
Burns
Choking on an obstruction
Head injuries and concussion
Cuts and abrasions
Electrical shock
Heart attack
Stroke
Moving an injured person
Drug overdose
Unconscious victim
Eye injuries
Chemical burns
Rescue

FIGURE 13–1 Sample course outline for first aid class.

medical services and first aid (29 CFR 1926.50). Figure 13–1 contains a list of the topics that may be covered in a first aid class for construction workers.

First Aid Training Program

First aid training programs are available in most communities. The continuing education departments of community colleges and universities typically offer first aid training. Classes can often be provided onsite and customized to meet the specific needs of individual companies.

The American Red Cross provides training programs in first aid specifically geared toward the workplace. For more information about these programs, construction professionals may contact the national office of the American Red Cross at (202) 639-3200.

The National Safety Council (NSC) also provides first aid training materials. The First Aid and Emergency Care Teaching Package contains a slide presentation, overhead transparencies, a test bank, and an instructor's guide. The NSC also produces a book entitled *First Aid Essentials*. For more information about these materials, construction professionals may contact the publisher for the NSC at (800) 843-0034 or http://www.nsc.org.

Beyond Training

Training employees in first aid techniques is an important part of preparing for emergencies. However, there is more to being prepared to administer first aid than just training. In addition, it is important to do the following:

1. *Have well-stocked first aid kits available.* First aid kits should be placed at all job sites in clearly visible, easily accessible locations. They should be properly and fully stocked and periodically checked to ensure that they stay fully stocked. Figure 13–2 lists the minimum recommended contents for a workplace first aid kit.
2. *Have appropriate personal protective devices available.* With the concerns about AIDS and hepatitis, administering first aid has become more complicated than in the past. The main concern is contact with blood and other body fluids. Consequently, a properly stocked first aid kit should contain rubber surgical gloves and face masks or mouthpieces for cardiopulmonary resuscitation (CPR).
3. *Postemergency telephone numbers.* The advent of 911 service has simplified the process of calling for medical care, police, or firefighting assistance. If 911 services are not available, emergency numbers for ambulance, hospital, police, fire department, and appropriate internal personnel should be posted at clearly visible locations near all telephones at all job sites.
4. *Keep all employees informed.* Some companies require all employees to undergo first aid training; others choose to train one or more employees in each department. Regardless of the approach used, all employees must be provided and kept up-to-date about basic first aid information. Figures 13–3 and 13–4 are first aid fact sheets of the type used to keep employees informed.

Absorbent cotton
Adhesive tape
Aromatic spirits of ammonia
Assorted adhesive bandages
Baking soda (bicarbonate of soda)
Face masks or mouthpieces
Flashlight
Hot-water bottle
Ipecac syrup (to induce vomiting)
Measuring cup
Medicine dropper (eye dropper) (sterile)
Mild antiseptic for minor wounds
Oral thermometer
Petroleum jelly
Powdered activated charcoal (to absorb swallowed poisons)
Rectal thermometer
Roll of gauze bandages
Rubber surgical gloves
Scissors
Sharp knife or stiff-backed razor blades
Sterile gauze dressings (individually wrapped)
Sterile saline solution
Triangular bandages
Tweezers
Wooden safety matches

FIGURE 13–2 Minimum recommended contents of workplace first aid kits.

OCCUPATIONAL NOISE

The modern construction site can be a noisy place. This poses two safety- and health-related problems. First, there is the problem of distraction. Noise can distract workers and disrupt their concentration, which can lead to accidents. Second, there is the problem of hearing loss. Exposure to noise that exceeds prescribed levels can result in permanent hearing loss. These levels are set forth in 29 CFR 1926.52. Figure 13–5 is a table showing decibel levels of selected sounds.

OSHA's Hearing Conservation Program

OSHA's 29 CFR 1926.52 establishes limits on the sound levels that workers can be exposed to. Construction professionals need to be familiar with these limits and how to prevent over exposure to occupational noise. The permissible levels established by OSHA are as follows:

- 90 decibels for 8 hours
- 92 decibels for 6 hours
- 95 decibels for 4 hours

Carson Contracting, Inc.
301 N. Birch Street, Fort Walton Beach, Florida 32547

First Aid Fact Sheet No. 3
Moving an Injured Person

If a worker has a neck or back injury, do not move him or her unless it must be done to prevent additional injuries. If it is absolutely essential to move the victim, remember the following rules:

1. Call for professional medical help.
2. Always pull the body lengthwise, never sideways.
3. If there is time, slip a blanket under the victim and use it to pull him or her to safety.
4. If the victim must be lifted, support all parts of the body so that it does not bend or jackknife.
5. Always use a back support board and a neck stabilizer, if available.

FIGURE 13–3 First aid fact sheet.

Carson Contracting, Inc.
301 N. Birch Street, Fort Walton Beach, Florida 32547

First Aid Fact Sheet No. 2
ABCs of First Aid

If a worker is injured and you are the first person to respond, remember the ABCs of first aid:

A = Airway
Is the airway blocked? If so, clear it quickly.

B = Breathing
Is the victim breathing? If not, begin artificial respiration.

C = Circulation
Is the victim bleeding severely? If so, stop the bleeding. Is there a pulse? If not, begin CPR.

FIGURE 13–4 First aid fact sheet.

Source	Decibels (dB)
Whisper	20
Quiet office	50
Normal conversation	60
Noisy office	80
Power saw	90
Chain saw	90
Grinding operations	100
Passing truck	100
Jet aircraft	150

FIGURE 13–5 Selected sounds and their decibel levels.

- 97 decibels for 3 hours
- 100 decibels for 2 hours
- 102 decibels for 1.5 hours
- 105 decibels for 1 hour
- 110 decibels for 0.5 hour
- 115 decibels for 0.25 hour or less

If workers are exposed to noise levels that exceed the sound levels and duration limits, the construction company must have a continual and effective hearing conservation program in place. The program is required to encompass all workers whose work brings them—even intermittently—into areas where the sound level exceed the limits listed above. The recommended elements of the hearing conservation program are: *comprehensive monitoring, training, hearing protection that effectively reduces exposure to within the prescribed limits,* and *periodic audiometric testing.*

Construction professionals need to understand the hazards associated with noise and vibration, how to identify and assess these hazards, and how to prevent injuries related to them.

Hearing Loss Terms

There are certain terms common to hearing loss prevention that must be understood by construction professionals concerned with safety and health. The reader may find the definitions in this section helpful when trying to understand hearing-related issues.

- *Attenuation.* Estimated degree of sound protection that is provided by hearing protective devices as worn by employees in real-world environments is called **attenuation**.
- *Baseline audiogram.* A valid audiogram (hearing sensitivity test) against which subsequent audiograms are compared to determine if hearing thresholds have changed. The baseline audiogram is preceded by a quiet period to obtain the best estimate of the person's hearing at that time.
- *Continuous noise.* Noise of a constant level measured over at least one second, using the "slow" setting on a sound level meter. Note that an intermittent noise (e.g., on for over a second and then off for a period) is both variable *and* continuous.
- *Decibel.* The unit used to express the intensity of sound. The **decibel (dB)** was named after Alexander Graham Bell. The decibel scale is a logarithmic scale, in which 0 dB approximates the threshold of hearing in the mid-frequencies for young adults, the threshold of discomfort is between 85 and 95 dB, and the threshold for pain is between 120 and 140 dB.
- *Dosimeter.* When applied to noise, the instrument that measures sound levels over a specified interval, stores the measures, and calculates the sound as a function of sound level and sound duration. The **dosimeter** displays the results in terms of dose, time-weighted average, and other parameters, such as peak level, equivalent sound level, and sound exposure level.
- *Exchange rate.* The relationship between intensity and dose. OSHA uses a 5-dB exchange rate. Thus, if the intensity of an exposure increases by 5 dB, the dose doubles. This may also be referred to as the *doubling rate*. The U.S. Navy uses a 4-dB exchange rate; the U.S. Army and U.S. Air Force use a 3-dB exchange rate.
- *Hazardous noise.* Any sound for which any combination of frequency, intensity, or duration is capable of causing permanent hearing loss in a specified population. Hearing loss is often characterized by the area of the auditory system responsible for the loss. For example, when injury or a medical condition affects the *outer ear* or *middle ear* (i.e., from the pinna, ear canal, and eardrum to the cavity behind the eardrum, which includes the ossicles), the resulting hearing loss is referred to as a *conductive* loss. When an injury or medical condition affects the *inner ear* or the auditory nerve that connects the inner ear to the brain (i.e., the cochlea and cranial nerve VIII), the resulting hearing loss is referred to as a *sensorineural* loss. Thus, a welder's spark that damages the eardrum causes a conductive hearing loss. Because noise can damage the tiny hair cells located in the cochlea, **hazardous noise** causes a sensorineural hearing loss.
- *Hearing threshold level (HTL).* The hearing level, above a reference value, at which a specified sound or tone is heard by an ear in a specified fraction of the trials. **Hearing threshold levels (HTLs)** have been established so that the given HTL, in decibels, reflects the best hearing of a group of persons.
- *Hertz (Hz).* The unit of measurement for audio frequencies. The frequency range for human hearing

lies between 20 and approximately 20,000 Hz. The sensitivity of the human ear drops off sharply below about 500 Hz and above 4,000 Hz.

- *Impulsive noise.* Generally used to characterize impact or impulse noise typified by a sound that rapidly rises to a sharp peak and then quickly fades. The sound may or may not have a "ringing" quality (such as striking a hammer on a metal plate or a gunshot in a reverberant room). **Impulsive noise** may be repetitive or may be a single event (as with a sonic boom). If impulses occur in very rapid succession (e.g., some jackhammers), the noise is not described as impulsive.
- *Material hearing impairment.* As defined by OSHA, a material hearing impairment is an average hearing threshold level of 25 dB HTL at the frequencies of 1,000, 2,000, and 3,000 Hz.
- *Noise.* Any unwanted sound.
- *Noise dose.* The noise exposure expressed as a percentage of the allowable daily exposure. For OSHA, a 100-percent dose equals an eight-hour exposure to a continuous 90-dBA noise; a 50-percent dose equals an eight-hour exposure to an 85-dBA noise or a four-hour exposure to a 90-dBA noise. If 85 dBA is the maximum permissible level, an eight-hour exposure to a continuous 85-dBA noise equals a 100-percent dose. If a 3-dBA exchange rate is used in conjunction with an 85-dBA maximum permissible level, a 50-percent dose equals a two-hour exposure to 88 dBA or an eight-hour exposure to 82 dBA.
- *Noise-induced hearing loss.* A sensorineural hearing loss that is attributed to noise and for which no other cause can be determined.
- *Standard threshold shift (STS).* OSHA uses this term to describe a change in hearing threshold relative to the baseline audiogram of an average of 10 dB or more at 2,000, 3,000, and 4,000 Hz in either ear. Used by OSHA to trigger additional audiometric testing and related follow-up.
- ***Significant threshold shift (STS).*** The National Institute for Occupational Safety and Health (NIOSH) uses this term to describe a change of 15 dB or more, at any frequency between 400 and 6,000 Hz, from baseline levels, which is present on a retest in the same ear and at the same frequency. NIOSH recommends a confirmation audiogram within 30 days, with the confirmation audiogram preceded by a quiet period of at least 14 hours.
- ***Time-weighted average (TWA).*** A value, expressed in dBA, computed so that the resulting average is equivalent to an exposure to a constant noise level over an eight-hour period.

Hazard Levels and Risks

The fundamental hazard associated with excessive noise is hearing loss. Exposure to excessive noise levels for an extended period of time can damage the inner ear, so that the ability to hear high-frequency sound is diminished or lost altogether. Additional exposure can increase the damage until even lower frequency sound cannot be heard.

A number of factors affect the risk of hearing loss associated with exposure to excessive noise. The following are the most important of these:

- Intensity of the noise (sound pressure level)
- Type/frequency of noise (wide band, narrow band, or impulse)
- Duration of daily exposure
- Total duration of exposure (number of years)
- Age of the individual
- Coexisting hearing disease
- Nature of environment in which exposure occurs
- Distance of the individual from the source of the noise (distribution)
- Position of the ears relative to the sound waves

Of these factors, the most critical are the sound level, frequency, duration, and distribution of noise. The unprotected human ear is at risk when exposed to sound levels exceeding 115 dBA. Exposure to sound levels below 80 dBA is generally considered safe. Prolonged exposure to noise levels higher than 80 dBA should be minimized through the use of appropriate personal protective devices.

To decrease the risk of hearing loss, exposure to noise should be limited to a maximum eight-hour TWA of 85 dBA. The following general rules should be applied when dealing with noise in the workplace:

- Exposures of less than 80 dBA may be considered safe for the purpose of risk assessment.
- A TWA (threshold) of 85 dBA should be considered the maximum limit of continuous exposure without protection during eight-hour workdays.[1]

OSHA Regulations

OSHA's construction regulations related to occupational noise exposure and hearing conservation are found in 29 CFR 1926.52. The basic requirements generated from this standard for hearing conservation programs are as follows:

- Monitoring hearing hazards
- Engineering and administrative controls
- Personal hearing protection devices
- Education and motivation

- Record keeping
- Program evaluation including audiometric testing

Monitoring Hearing Hazards. As with any health hazard, it is important to determine accurately the nature of the hearing hazard and to identify the affected employees. Those responsible for **monitoring hearing hazards** must ensure that the exposures of all employees are properly evaluated and that reevaluations are conducted when changes in equipment or operations significantly alter working conditions. Recent evidence has indicated that exposure to aromatic solvents, metals, and petrochemicals may be associated with occupational hearing loss. Although studies are exploring the relationship between hearing loss and chemical exposures, there is insufficient information about this relationship to speculate on potential risk factors. Therefore, this section focuses on monitoring noise exposure, which is the major factor associated with occupational hearing losses. Hearing hazard exposure monitoring is conducted for various purposes, including the following:

- Determining whether hazards to hearing exist
- Determining whether noise presents a safety hazard by interfering with speech communication or the recognition of audible warning signals
- Identifying workers for inclusion in the hearing loss prevention program
- Classifying workers' noise exposures for prioritizing noise-control efforts and defining and establishing hearing protection practices
- Evaluating specific noise sources for noise-control purposes
- Evaluating the success of noise-control efforts

Various kinds of incrementation and measurement methods may be used, depending on the type of measurements being conducted. The most common measurements are area surveys, dosimetry, and engineering surveys.

In an area survey, environmental noise levels are measured using a sound level meter to identify work areas where exposures are above or below hazardous levels and where more thorough exposure monitoring may be needed. The result is often plotted in the form of a "noise map," showing noise level measurements for the different areas of the workplace.

Dosimetry involves the use of instruments (dosimeters) worn on the body to monitor a worker's noise exposure over the work shift. The monitoring results for one worker can be used to represent the exposures of other workers in the area with similar noise exposures. It may also be possible to use task-based exposure methods to represent the exposures of other workers in different areas whose exposures result from having performed the same tasks.

Persons performing engineering surveys typically use more sophisticated acoustical equipment in addition to sound level meters. These may include octave-band analyzers and sound level recorders that furnish information on the frequency and intensity of the noise being emitted by machinery or other sound sources in various modes of operation. These measurements are used to assess options for applying engineering controls.

Engineering and Administrative Controls. **Engineering and administrative controls** are essential to achieving an effective hearing loss prevention program. These controls represent the first two echelons in the hierarchy of controls: (1) remove the hazard and (2) remove the worker. The use of these controls should reduce hazardous exposure to the point where the risk to hearing is eliminated or at least more manageable. Engineering controls are technologically feasible for most noise sources, but their economic feasibility must be determined on a case-by-case basis. In some instances, the application of a relatively simple noise-control solution reduces the hazard to the extent that the other elements of the program, such as audiometric testing and the use of hearing protection devices, are no longer necessary. In other cases, the noise-reduction process may be more complex and must be accomplished in stages over a period of time. Even so, with each reduction of a few decibels, the hazard to hearing is reduced, communication is improved, and noise-related annoyance is reduced as well.

Companies should always specify low noise levels when purchasing new equipment. Many types of previously noisy equipment are now available in noise-controlled versions. Consequently, a "buy quiet" purchasing policy should not require new engineering solutions in many cases.

For hearing loss prevention purposes, engineering controls are defined as any modification or replacement of equipment or related physical change at the noise source or along the transmission path (with the exception of hearing protectors) that reduces the noise level at the employee's ear. Typical engineering controls involve the following:

- Reducing noise at the source (e.g., installing a muffler)
- Interrupting the noise path (e.g., erecting acoustical enclosures and barriers)
- Reducing reverberation (e.g., installing sound-absorbing material)
- Reducing structure-borne vibration (e.g., installing vibration mounts and providing proper lubrication)

Assessing the applicability of engineering controls is a sophisticated process. First, the noise problem must be thoroughly defined. This necessitates measuring the noise levels and developing complete information on employee noise exposure and the need for noise reduction. Next, an assessment of the effect of these controls on overall noise levels should be made. Once identified and analyzed, the preceding controls can be considered. Choices are influenced—to some extent—by the cost of purchasing, operating, servicing, and maintaining the control. For this reason, engineering, safety, and supervisory personnel, as well as employees who operate, service, and maintain equipment, must be involved in the noise-control plan. Employees who work with the equipment on a daily basis should be asked to provide valuable guidance on such important matters as the positioning of monitoring indicators and panels, lubrication and servicing points, and control switches and the proper location of access doors for operation and maintenance. An acoustical consultant may be hired to assist in the design, implementation, installation, and evaluation of these controls.

Administrative controls—defined as changes in the work schedule or operations that reduce noise exposure—may also be used effectively. Examples include operating a noisy machine on the second or third shift, when fewer people are exposed, or shifting an employee to a less noisy job before a hazardous daily noise dose has been reached. Generally, administrative controls have limited use in industry, because employee contracts seldom permit shifting from one job to another. Moreover, the practice of rotating employees between quiet and noisy jobs, although it may reduce the risk of substantial hearing loss in a few workers, may actually increase the risk of small hearing losses in many workers. A more practical administrative control is to provide for quiet areas where employees can gain relief from workplace noise. Areas used for work breaks and lunchrooms should be located away from noise.

Personal Hearing Protection Devices. A **personal hearing protection device** (or "hearing protector") is anything that can be worn to reduce the level of sound entering the ear. Earplugs (Figure 13–6), earmuffs

FIGURE 13–6 Earplug hearing protection devices (on gloves).
Source: Luap Vision/Fotolia

FIGURE 13–7 Earmuff hearing protection devices.
Source: Siraphol/Fotolia

(Figure 13–7), and ear canal caps are the three principal types of devices. Each employee reacts individually to the use of these devices, and a successful hearing loss prevention program should be able to respond to the needs of each employee. Ensuring that these devices protect hearing effectively requires the coordinated effort of management, the hearing loss prevention program operators, and the affected employees.

Education and Motivation. Training is a critical element of a good hearing loss prevention program. To obtain sincere and energetic support from management and active participation by employees, it is necessary to educate and motivate both groups. A hearing loss prevention program that overlooks the importance of **education and motivation** is likely to fail because employees do not understand why it is in their best interest to cooperate, and management fails to make the necessary commitment. Employees and managers who appreciate the precious sense of hearing and understand the reasons for and the mechanics of the hearing loss prevention program are more likely to participate for their mutual benefit, rather than viewing the program as an imposition.

Record Keeping. Records quite often get the least attention of any of the program's components. However, audiometric comparisons, reports of hearing protection device use, and the analysis of hazardous exposure measurements all involve **record keeping**. Unfortunately, records are often kept poorly because there is no organized system in place, and in many cases, those responsible for maintaining the records do not understand their value. People tend to assume that if they merely place records in a file or enter them into a computer, adequate record-keeping procedures are being followed.

Many companies have found that their record-keeping system is inadequate only when they discover that they need accurate information. This sometimes occurs during the processing of compensation claims.

Problems can be avoided by implementing an effective record-keeping system that includes the following components:

1. Management encourages that the system be kept active and accessible.
2. Hearing loss prevention program implementers make sure that all of the information entered is accurate and complete.
3. Employees validate the information.

Program Evaluation. Hearing loss prevention programs should be evaluated periodically to ensure their effectiveness.[2] Such **program evaluations** should have at least the following components: (1) training and education, (2) supervisor involvement, (3) noise measurement, (4) engineering and administrative controls, (5) monitoring and record keeping, (6) referrals, (7) hearing protection devices, (8) administration, and (9) audiometric testing. The following are checklists for some of these components.

Training and Education. Failures or deficiencies in hearing loss prevention programs can often be traced to inadequacies in the training and education of noise-exposed workers and of those who conduct elements of the program.

- Has training been conducted at least once a year?
- Was the training provided by a qualified instructor?
- Was the success of each training program evaluated?
- Is the content revised periodically?
- Are managers and supervisors directly involved?
- Are posters, regulations, handouts, and employee newsletters used as supplements?
- Are personal counseling sessions conducted for employees who have problems with hearing protection devices or experience hearing threshold shifts?

Supervisor Involvement. Data indicate that employees who refuse to wear hearing protection devices or who fail to show up for hearing tests frequently work for supervisors who are not totally committed to the hearing conservation program.

- Have supervisors been provided with the knowledge required to supervise the use and care of hearing protectors by subordinates?
- Do supervisors wear hearing protectors in appropriate areas?
- Have supervisors been counseled regarding what to do when employees resist wearing protectors or fail to show up for hearing tests?
- Are disciplinary actions enforced when employees repeatedly refuse to wear hearing protectors?

Noise Measurements. **Noise measurements**, to be useful, should be related to noise-exposure risks or the prioritization of noise-control efforts, rather than merely filed away. In addition, the results should be communicated to the appropriate personnel, especially when follow-up actions are required.

- Were the essential or critical noise studies performed?
- Was the purpose of each noise study clearly stated? Have workers exposed to hazardous noise been notified of their exposures and apprised of auditory risks?
- Are the results routinely transmitted to supervisors and other key individuals?
- Are results entered into health and medical records of noise-exposed workers?
- Are results entered into shop folders?
- If noise maps exist, are they used by the proper staff?
- Are noise measurement results considered when contemplating procurement of new equipment, modification of the facility, and relocation of workers?
- Have there been changes in areas, equipment, or processes that have altered noise exposure? Have follow-up noise measurements been conducted?
- Are appropriate steps taken to include (or exclude) workers whose exposures have changed significantly in the hearing loss prevention programs?

Engineering and Administrative Controls. Controlling noise by engineering and administrative methods is often the most effective means of reducing or eliminating the hazard. In some cases, engineering controls remove requirements for other components of the program, such as audiometric testing and the use of hearing protectors.

- Have noise-control needs been prioritized?
- Has the cost effectiveness of various options been addressed?
- Are workers and supervisors apprised of plans for noise-control measures?
- Are they consulted on various approaches?
- Will in-house resources or outside consultants perform the work?
- Have workers and supervisors been counseled on the operation and maintenance of noise-control devices?
- Are noise-control projects monitored to ensure timely completion?
- Has the full potential for administrative controls been evaluated? Are noisy processes conducted during shifts with fewer employees? Do workers have noise-protected lunch or break areas?

Monitoring Audiometry and Record Keeping. The skills of audiometric technicians, the status of the audiometer, and the quality of audiometric test records are crucial to the success of hearing loss prevention programs. Useful information may be ascertained from the audiometric records and from those who administer the tests as part of the process of monitoring audiometry and record keeping.

- Has the audiometric technician been adequately trained, certified, and recertified as necessary?
- Do on-the-job observations of the technicians indicate that they perform a thorough and valid audiometric test, instruct and consult the worker effectively, and keep appropriate records?
- Are records complete?
- Are follow-up actions documented?
- Are hearing threshold levels reasonably consistent from test to test? If not, are the reasons for inconsistencies investigated promptly?
- Are the annual test results compared with baseline to identify the presence of an OSHA standard threshold shift?
- Is the annual incidence of standard threshold shift greater than a few percentage points? If so, are problem areas pinpointed and remedial steps taken?
- Are audiometric deterioration trends being identified, both in individuals and in groups of employees?
- Do records show that appropriate audiometer calibration procedures have been followed?
- Is there documentation showing that the background sound levels in the audiometer room were low enough to permit valid testing?
- Are the results of audiometric tests being communicated to supervisors, managers, and workers?
- Has corrective action been taken if the rate of no-shows for audiometric test appointments was more than about 5 percent?
- Are workers who have incurred significant threshold shift notified in writing within 21 days?

Referrals. **Referrals** to outside sources for consultation or treatment are sometimes in order, but they can be an expensive element of the hearing loss prevention program and should not be undertaken unnecessarily.

- Are referral procedures clearly specified?
- Have letters of agreement between the company and consulting physicians or audiologists been executed?
- Have mechanisms been established to ensure that workers needing evaluation or treatment actually receive the service (i.e., transportation, scheduling, reminders)?
- Are records properly transmitted to the physician or audiologist and back to the company?
- If medical treatment is recommended, does the worker understand the condition requiring treatment, the recommendation, and methods of obtaining such treatment?
- Are workers being referred unnecessarily?

Hearing Protection Devices. When noise-control measures are not feasible, or until they are installed, hearing protection devices are the only way to prevent hazardous levels of noise from damaging the inner ear. Making sure that these devices are worn properly requires continuous attention on the part of supervisors and program implementers as well as noise-exposed employees.

- Have hearing protectors been made available to all employees whose daily average noise exposures are 85 dBA or above?
- Are workers given the opportunity to select from a variety of appropriate protectors?
- Are workers fitted carefully, with special attention to comfort?
- Are workers thoroughly trained, not only initially, but at least once a year?
- Are the protectors checked regularly for wear or defects and replaced immediately if necessary?
- If workers use disposable hearing protectors, are replacements readily available?
- Do workers understand the appropriate hygiene requirements?
- Have any workers developed ear infections or irritations associated with the use of hearing protectors? Are there any workers who are unable to wear these devices because of medical conditions? Have these conditions been treated promptly and successfully?
- Have alternative types of hearing protectors been considered when problems with current devices are experienced?
- Do workers who incur noise-induced hearing loss receive intensive counseling?
- Are those who fit and supervise the wearing of hearing protectors competent to deal with the many problems that can occur?
- Do workers complain that protectors interfere with their ability to do their jobs? Do they interfere with verbal instructions or warning signals? Are these complaints followed promptly with counseling, noise control, or other measures?
- Are workers encouraged to take their hearing protectors home if they engage in noisy recreational activities?

- Are more effective protectors considered as they become available?
- Is the effectiveness of the hearing protector program evaluated regularly?
- Have at-the-ear protection levels been evaluated to ensure that either overprotection or underprotection has been adequately balanced, according to the anticipated ambient noise levels?
- Is each hearing protector user required to demonstrate that he or she understands how to use and care for the protector? Are the results documented?

Administration. Keeping organized and current on administrative matters helps the program run smoothly.

- Have there been any changes in federal or state regulations? Have hearing loss prevention program policies been modified to reflect these changes?
- Are copies of company policies and guidelines regarding the hearing loss prevention program available in the offices that support the various program elements? Are those who implement the program elements aware of these policies? Do they comply with the policies?
- Are necessary materials and supplies being ordered with a minimum of delay?
- Are procurement officers overriding the hearing loss prevention program implementer's requests for specific hearing protectors or other hearing loss prevention equipment? If so, have corrective steps been taken?
- Is the performance of key personnel evaluated periodically? If such performance is found to be less than acceptable, are steps taken to correct the situation?
- Has the failure to hear warning shouts or alarms been tied to any accidents or injuries? If so, have remedial steps been taken?

IONIZING RADIATION

The widow of a construction worker who helped build the British Nuclear Fuels (BNF) Sellafield plant was awarded $286,500 when it was determined that his death from chronic myeloid leukemia was the result of overexposure to radiation. The Sellafield plant was constructed for the purpose of separating uranium from used fuel rods. After working at the plant for approximately nine months, the victim had received a total cumulative dose of almost 52 millisieverts of radiation, which exceeded the established limit for an entire 12-month period. BNF compensated the victim's wife and the families of 20 additional workers who died from causes related to radiation.

Terms and Concepts

An *ion* is an electrically charged atom (or group of atoms) that becomes charged when a neutral atom (or group of atoms) loses or gains one or more electrons as a result of a chemical reaction. If an electron is lost during this process, a positively charged ion is produced; if an electron is gained, a negatively charged ion is produced. To *ionize* is to become electrically charged or to change into ions. Therefore, **ionizing radiation** is radiation that becomes electrically charged or changed into ions. Types of ionizing radiation, as shown in Figure 13–8, include alpha particles, beta particles, neutrons, X-rays, gamma rays, high-speed electrons, and high-speed protons.

To understand the hazards associated with radiation, construction professionals need to understand the basic terms and concepts that follow:

- *Radiation* consists of energetic nuclear particles and includes alpha rays, beta rays, gamma rays, X-rays, neutrons, and high-speed electrons and protons.
- *Radioactive material* is material that emits corpuscular or electromagnetic emanations as the result of spontaneous nuclear disintegration.
- A *restricted area* is any area to which access is restricted in an attempt to protect employees from exposure to radiation or radioactive materials.
- An *unrestricted area* is any area to which access is not controlled because there is no radioactivity hazard present.
- A ***dose*** is the amount of ionizing radiation absorbed per unit of mass by part of the body or the whole body.
- ***Rad*** is a measure of the dose of ionizing radiation absorbed by body tissues stated in terms of the amount of energy absorbed per unit of mass of tissue. One rad equals the absorption of 100 ergs per gram of tissue.
- ***Rem*** is a measure of the dose of ionizing radiation to body tissue stated in terms of its estimated biological effect relative to a dose of one roentgen of X-rays.
- *Air dose* is the level of the dose that an instrument measures in the air at or near the surface of the body where the highest dosage occurs.
- *Personal monitoring devices* are devices worn or carried by an individual to measure radiation doses received. Widely used devices include film badges, pocket chambers, pocket dosimeters, and film rings.
- A ***radiation area*** is any accessible area in which radiation hazards exist that could deliver doses as follows: (1) Within one hour, a major portion of the body could receive more than 5 millirems or (2) within five consecutive days, a major portion of the body could receive more than 100 millirems.

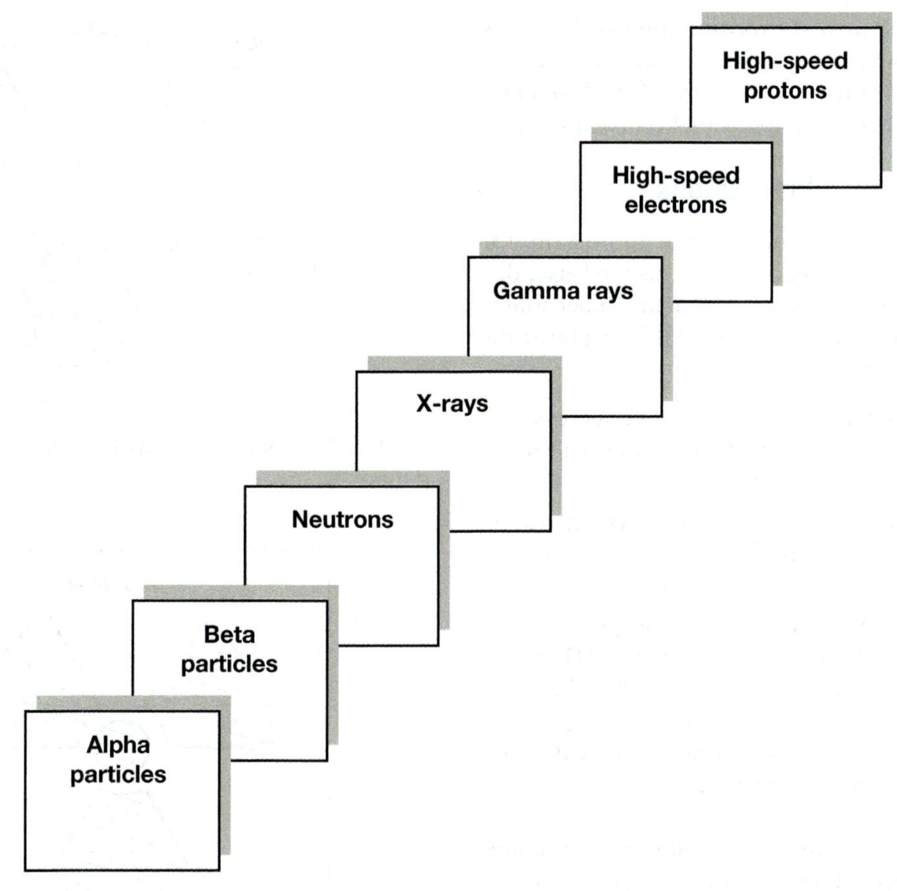

FIGURE 13–8 Types of ionizing radiation.

- A *high-radiation area* is any accessible area in which radiation hazards exist that could deliver a dose in excess of 100 millirems within one hour.

Exposure of Workers to Radiation

The exposure of employees to radiation must be carefully controlled and accurately monitored. Figure 13–9 shows the maximum doses for individuals in one calendar quarter. Employers are responsible for ensuring that these dosages are not exceeded.

There are exceptions to the amounts shown in Figure 13–9. According to OSHA, an employer may permit an individual in a restricted area to receive doses to the whole body greater than those shown in Figure 13–9 as long as the following conditions are met:

- During any calendar quarter, the dose to the whole body does not exceed 3 rems.
- The dose to the whole body—when added to the accumulated occupational dose to the whole body—shall not exceed $5(N - 18)$ rems, where N is the employee's age in years at the last birthday. For example, for age 30, the calculation would be as follows:

$$5 \times (30 - 18) = 5 \times 12 = 60 \text{ rems}$$

Body Region	Maximum Rems Per Calendar Quarter
Whole body	1.25
Head and trunk	1.25
Blood-forming organs	1.25
Lens of eyes	1.25
Gonads	1.25
Hands and forearms	18.75
Feet and ankles	18.75
Skin of whole body	7.50

FIGURE 13–9 Ionizing radiation exposure limits of humans.

- The employer maintains up-to-date past and current exposure records, which show that the addition of such a dose does not cause the employee to exceed the specified doses.[3]

Employers must ensure even more careful controls with individuals under 18 years of age. Such individuals may receive only doses that do not exceed 10 percent of those specified in Figure 13–9 in any calendar quarter.

OSHA is not the only agency that regulates radiation exposure. The Nuclear Regulatory Commission (NRC) is also a leading agency in this area. NRC regulations specify that the total internal and external dose for employees may not exceed 5 rems per year. This same revision established a total exposure limit of 0.6 rems over the entire course of a pregnancy. According to the NRC, the average radiation exposure of nuclear plant workers is less than 400 millirems (0.4 rems) annually.[4]

Personal Monitoring Precautions

Personal monitoring precautions are important for employees of companies that produce, use, release, dispose of, or store radioactive materials or any other source of ionizing radiation. Accordingly, OSHA requires the following precautions:

- Employers must conduct comprehensive surveys to identify and evaluate radiation hazards present in the workplace from any and all sources.
- Employers must provide appropriate personal monitoring devices, such as film badges, pocket chambers, pocket dosimeters, and film rings.
- Employers must require the use of appropriate personal monitoring devices by the following: (1) any employee who enters a restricted area where he or she is likely to receive a dose greater than 25 percent of the total limit of exposure specified for a calendar quarter, (2) any employee 18 years of age or less who enters a restricted area where he or she is likely to receive a dose greater than 5 percent of the total limit of exposure specified for a calendar quarter, and (3) any employee who enters a high-radiation area.[5]

Caution Signs and Labels

Caution signs and labels have always been an important part of safety and health programs. This is particularly true in companies where radiation hazards exist. The universal color scheme for caution signs and labels warning of radiation hazards is purple or magenta superimposed on a yellow background.

Both OSHA and the NRC require caution signs in radiation areas, high-radiation areas, airborne radiation areas, areas containing radioactive materials, and containers in which radioactive materials are stored or transported.[6]

Figure 13–10 shows the universal symbol for radiation. Along with the appropriate warning words, this symbol should be used on signs and labels. Figure 13–11 shows a warning sign and label that may be used in various radioactive settings. On containers, labels should also include the following information: (1) quantity of radioactive material, (2) type of radioactive material, and (3) date on which the contents were measured.

Evacuation Warning Signal

Companies that produce, use, store, or transport radioactive materials are required to have a signal-generating system that can warn of the need for evacuation. OSHA describes the **evacuation warning signal** system as follows:

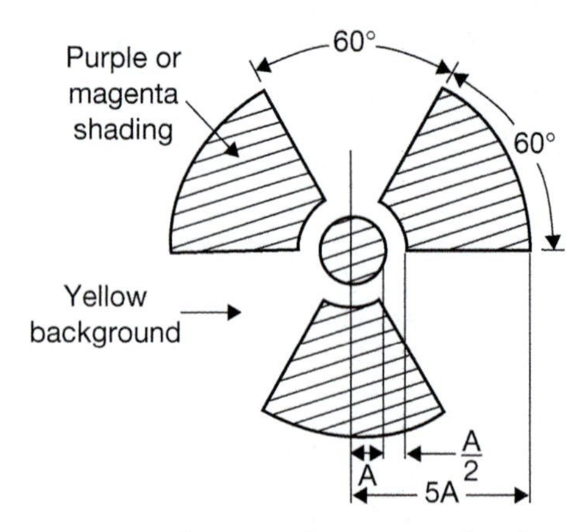

FIGURE 13–10 Specifications for universal radiation symbol.

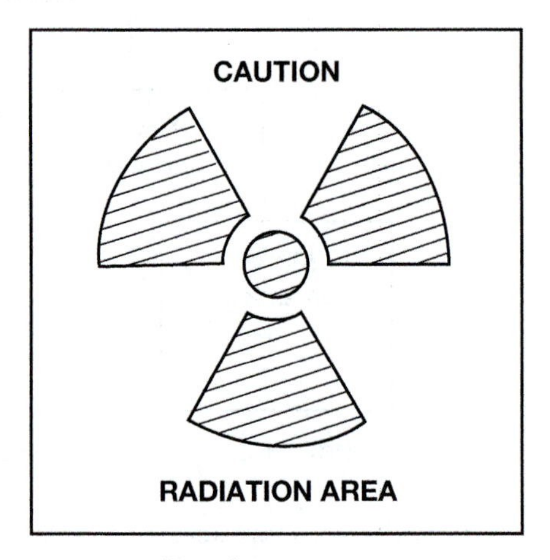

FIGURE 13–11 Sample radiation warning sign.

> The signal shall be a mid-frequency complex sound wave amplitude modulated at a subsonic frequency. The complex sound wave in free space shall have a fundamental frequency (f_1) between 450 and 500 hertz (Hz) modulated at a subsonic rate between 4 and 5 Hz. The signal generator shall not be less than 75 decibels at every location where an individual may be present whose immediate, rapid, and complete evacuation is essential.[7]

In addition to this basic requirement, OSHA also stipulates the following:

- A sufficient number of signal generators must be installed to cover all personnel who may need to be evacuated.
- The signal shall be unique, unduplicated, and instantly recognizable in the plant where it is located.
- The signal must be long enough in duration to ensure that all potentially affected employees are able to hear it.
- The signal generator must respond automatically without the need for human activation, and it must be fitted with backup power.[8]

Informing Personnel

It is crucial that construction companies using, storing, handling, or transporting radioactive materials keep workers informed about radiation hazards and the appropriate precautions for minimizing them. Consequently, OSHA has established specific requirements along these lines. They are summarized as follows:

- All workers must be informed of existing radiation hazards and where they exist, the extent of the hazards, and how to protect themselves from the hazards (precautions and PPE).
- All workers must be advised of any reports of radiation exposure requested by other employees.
- All employees must have ready access to any related company operating procedures.[9]

These requirements apply to all companies that do not have superseding requirements (i.e., companies regulated by the Atomic Energy Commission and companies in states with their own approved state-level OSHA plans).

Instruction and information are important in all safety and health programs. They are especially important in settings in which radiation hazards exist. Workers in these settings must be knowledgeable about radiation hazards and how to minimize them. Periodically updating instruction for experienced workers is as important as initial instruction for new workers and should not be overlooked. Often, it is the comfortable, experienced worker who overlooks a precaution and thereby causes an accident.

Storage and Disposal

Radioactive materials that are stored in restricted areas must be appropriately labeled, as described earlier in this chapter. Radioactive materials that are stored in unrestricted areas "shall be secured against unauthorized removal from the place of storage."[10] This requirement precludes the handling and transport, intentional or inadvertent, of radioactive materials by persons who are not qualified to move them safely.

A danger inherent in storing radioactive materials in unrestricted areas is that someone, such as a maintenance worker, may unwittingly attempt to move the container and damage it in the process. This could release doses that exceed prescribed acceptable limits.

The disposal of radioactive material is also a regulated activity. There are only three acceptable ways to dispose of radioactive waste: (1) transfer to an authorized recipient, (2) transfer in a manner approved by the Atomic Energy Commission, or (3) transfer in a manner approved by any state that has an agreement with the Atomic Energy Commission pursuant to Section 27(b) 42 U.S.C. 2021(b) of the Atomic Energy Act.[11] States having such agreements are listed in Figure 13–12.

The following states have agreed to dispose of radioactive waste pursuant to 27(b) 42 U.S.C. 2021(b) of the Atomic Energy Act.

Alabama	Mississippi
Arizona	New Hampshire
Arkansas	New York
Colorado	North Carolina
Florida	North Dakota
Georgia	Oregon
Idaho	South Dakota
Kansas	Tennessee
Kentucky	Texas
Louisiana	Washington
Maryland	

FIGURE 13–12 States that have an agreement with the Atomic Energy Commission.

Notification of Incidents

A radiation-related incident must be reported if it meets a specific set of requirements. This is known as notification of incidents. An *incident* is defined by OSHA as follows:

> Exposure of the whole body of any individual to 25 rems or more of radiation; exposure of the skin of the whole body of any individual to 150 rems or more of radiation; or exposure of the feet, ankles, hands, or forearms of any individual to 375 rems or more of radiation. The release of radioactive material in concentrations which, if averaged over a period of 24 hours, would exceed 5,000 times the limit specified.[12]

If an incident meeting one of these criteria occurs, the employer must notify the proper authorities immediately. Companies regulated by the Atomic Energy Commission are to notify the commission. Companies in states that have agreements with the Atomic Energy Commission (Figure 13–12) are to notify the state designee. All other companies are to notify the U.S. assistant secretary of labor. Telephone or telegraph notifications are sufficient to satisfy the immediacy requirement.

The notification requirements are eased to 24 hours in cases in which whole body exposure is between 5 and 24 rems; exposure of the skin of the whole body is between 39 and 149 rems; or exposure of the feet, ankles, hands, or forearms is between 75 and 374 rems.[13]

Reporting and Recording Overexposure Incidents

In addition to the immediate and 24-hour notification requirements explained in the previous section, employers are required to follow up with a written report within

30 days.[14] Written reports are required when a worker is exposed as set forth in the previous section or when radioactive materials are on hand in concentrations of greater than specified limits. Each report should contain the following material, as applicable:

- Extent of exposure of employees to radiation or radioactive materials.
- Levels of radiation and concentration of radiation involved.
- Cause of the exposure.
- Levels of concentration.
- Corrective action taken.

Whenever a report is filed concerning the overexposure of a worker, the report should also be given to that employee. The following note should be placed prominently on the report or in a cover letter: "You should preserve this report for future reference."

Records of the doses of radiation received by all monitored workers must be maintained and kept up-to-date. Records should contain cumulative doses for each monitored employee. Figure 13–13 is an example of a cumulative exposure report form of the kind that can be used to satisfy reporting requirements. Notice that the maximum quarterly dose per body or body region is indicated. Such records must be shared with monitored employees at least annually. A better approach is to advise monitored workers of their cumulative doses as soon as the new cumulative amount is recorded for that period.

Cumulative radiation records must be made available to former workers on request. Upon receiving a request from a former worker, employers must provide the information requested within 30 days. According to OSHA,

> Such report shall be furnished within 30 days from the time the request is made and shall cover each calendar quarter of the individual's employment involving exposure to radiation or such lesser period as may be requested by the employee. The report shall also include the results of any calculations and analysis of radioactive material deposited in the body of the employee. The report shall be in writing and contain the following statement: "You should preserve this report for future reference."[15]

NONIONIZING RADIATION

Nonionizing radiation is that radiation on the electromagnetic spectrum that has a frequency (hertz, cycles per second) of 10^{15} or less and a wavelength in meters of 3H × 10^{7} or less.[16] This encompasses visible, ultraviolet, infrared, microwave, radio, and AC power frequencies. Radiation at these frequency levels does not have sufficient energy to shatter atoms and ionize them. However, such radiation can cause blisters and blindness. In addition, there is mounting evidence of a link between nonionizing radiation and cancer. The warning symbol for radio frequency radiation is shown in Figure 13–14.

The greatest concerns about nonionizing radiation relate to the following sources: visible radiation, radio frequency and microwave radiation, ultraviolet radiation, infrared radiation, extremely low frequency radiation, and lasers. The main concerns in each of these areas are explained in the following paragraphs.

1. **Visible radiation** comes from light sources that create distortion. This can be a hazard to employees whose jobs require color perception. For example, 8 percent of the male population is red–green color blind and cannot properly perceive red warning signs.
2. The most common source of **ultraviolet radiation** is the sun. Potential problems from ultraviolet radiation include sunburn, skin cancer, and cataracts. Precautionary measures include special

Employee Name: ____________

Dates Covered: ____________

	Rems				Total
	Quarter 1	Quarter 2	Quarter 3	Quarter 4	Annual Rems
Whole body: head and trunk, blood-forming organs, lens of eyes, or gonads (1.25 rems/quarter max.)					
Hands and forearms, feet and ankles (18 rems/quarter max.)					
Skin of whole body (7.5 rems/quarter max.)					

FIGURE 13–13 Cumulative radiation exposure record.

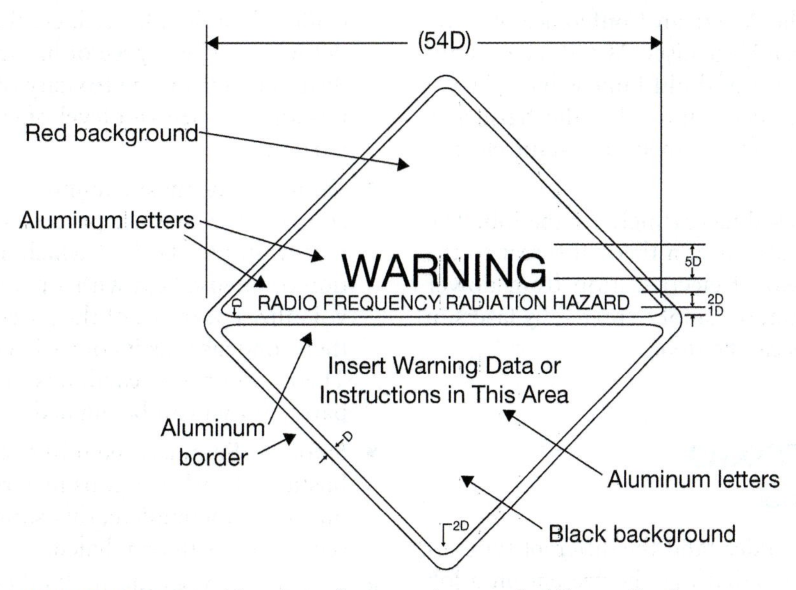

FIGURE 13–14 Specifications for warning symbol for radio frequency radiation.

sunglasses treated to block out ultraviolet rays and protective clothing. Other sources of ultraviolet radiation include lasers, welding arcs, and ultraviolet lamps.

3. **Infrared radiation** creates heat. Consequently, the problems associated with this kind of nonionizing radiation involve heat stress and dry skin and eyes. Primary sources of infrared radiation are high-temperature processes, such as the production of glass and steel.
4. **Radio frequency (RF) radiation** and **microwave (MW) radiation** are electromagnetic radiation in the frequency range of 3 kHz to 300 GHz. Usually, MW radiation is considered a subset of RF radiation, although an alternative convention treats RF and MW radiation as two spectral regions. Microwaves occupy the spectral region between 300 GHz and 300 MHz, whereas RF or radio waves range between 300 MHz and 3 kHz. RF and MW radiation are nonionizing in that there is insufficient energy (less than 10 eV) to ionize biologically important atoms. The primary health effects of RF and MW energy are considered to be thermal. The absorption of RF or MW energy varies with frequency. Microwave frequencies produce a skin effect—the person can literally sense the skin starting to feel warm. RF radiation may penetrate the body and be absorbed in deep body organs without the skin effect that can warn an individual of danger. Research has revealed other nonthermal effects. All of the standards of Western countries have, so far, based their exposure limits solely on preventing thermal problems. In the meantime, research continues. RF and MW radiation are used in aeronautical radios, citizen's band (CB) radios, cellular phones, processing and cooking foods, heat sealers, vinyl welders, high-frequency welders, induction heaters, flow solder machines, communications transmitters, radar transmitters, ion implant equipment, microwave drying equipment, sputtering equipment, glue curing, and power amplifiers used in metrology (calibration).
5. **Extremely low frequency (ELF) radiation** includes alternating current (AC) fields and nonionizing radiation from 1 to 300 Hz. Since ELF frequencies are low (wavelengths are on the order of 1,000 km), static electromagnetic fields are created. ELF fields are considered as separate, independent, nonradiating electric and magnetic fields. Electric and magnetic fields (EMFs) at 60 Hz are produced by power lines, electrical wiring, and electrical equipment. Electrical field strength is measured in volts per meter (V/m). Magnetic fields are created from the flow of current through wires or electrical devices and increase in strength as the current increases. Magnetic fields are measured in units of gauss (G) or tesla (T). Electrical equipment usually must be turned on for a magnetic field to be produced. Electrical fields are present even when equipment is turned off, as long as it is plugged in. Current research has focused on potential health effects of magnetic fields. Some inconclusive epidemiological studies have suggested that an increased risk of cancer is associated with high levels of magnetic field exposure. No similar associations have been reported for electrical fields. The levels of exposure to EMFs depend on the strength of the source, the

distance from the source, and the time spent in the magnetic field. The American Conference of Governmental Industrial Hygienists (ACGIH) has established occupational threshold limit values (TLVs) for static magnetic fields and sub-radio frequency (30 kHz and below) for magnetic and static electric fields.

6. **Lasers** are being used increasingly on the job. The hazards of lasers consist of a thermal threat to the eyes and the threat of electrocution from power sources. In addition, the smoke created by lasers in some applications can be toxic.

AIRBORNE TOXIC SUBSTANCES

Safety personnel must understand the different types of airborne toxic substances that may be present on a job site. Each type of toxic substance has a specific definition that must be understood in order to develop effective safety and health measures to protect against it. The most common types of airborne toxic substances are dusts, fumes, smoke, aerosols, mists, gases, and vapors (Figure 13–15).

- *Dusts.* **Dusts** are various types of solid particles that are produced when a given type of organic or inorganic material is scraped, sawed, ground, drilled, handled, heated, crushed, or otherwise deformed. The degree of hazard represented by dust depends on the toxicity of the parent material and the size and level of concentration of the particles.
- *Fumes.* The most common causes of **fumes** on construction sites are processes such as welding and torch cutting—both of which involve the interaction of intense heat with a parent material. The heat volatilizes portions of the parent material, which then condenses as it comes in contact with cool air. The result of this reaction is the formation of tiny particles that can be inhaled.
- *Smoke.* **Smoke** is the result of the incomplete combustion of carbonaceous materials. Because combustion is incomplete, tiny soot or carbon particles remain and can be inhaled.
- *Aerosols.* **Aerosols** are liquid or solid particles that are so small, they can remain suspended in air long enough to be transported over a distance. They can be inhaled.
- *Mists.* **Mists** are tiny liquid droplets suspended in air. Mists are formed in two ways: (1) when vapors return to a liquid state through condensation, and (2) when the application of sudden force or pressure turns a liquid into particles.

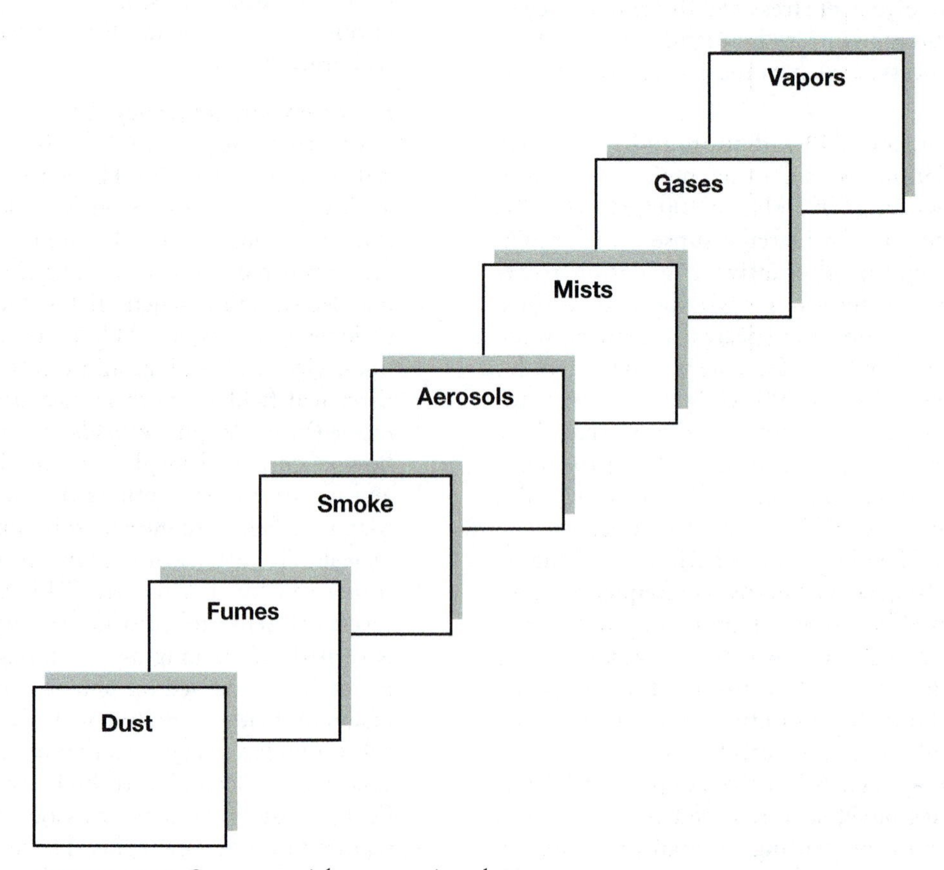

FIGURE 13–15 Common airborne toxic substances.

- *Gases.* Unlike other airborne contaminants that take the form of either tiny particles or droplets, **gases** are formless—actually formless fluids. Gases become particularly hazardous when they fill a confined, unventilated space. The most common sources of gases in a construction setting are welding and the exhaust from internal combustion engines.
- *Vapors.* Certain materials that are solid or liquid at room temperature and at normal levels of pressure will turn to **vapors** when heated or exposed to abnormal pressure. Evaporation is the most common process by which a liquid is transformed into a vapor.

Effects of Airborne Toxic Substances

Airborne toxic substances are also classified according to the type of effect they have on the body. The primary classifications are shown in Figure 13–16 and explained in the paragraphs that follow. With all airborne contaminants, concentration and duration of exposure are critical concerns.

Irritants. **Irritants** are substances that cause irritation to the skin, eyes, and the inner lining of the nose, mouth, throat, and upper respiratory tract. However, they produce no irreversible damage.

Asphyxiants. **Asphyxiants** are substances that can disrupt breathing so severely that suffocation results. Asphyxiants may be simple or chemical in nature. A simple asphyxiant is an inert gas that dilutes oxygen in the air to the point that the body cannot take in enough air to satisfy its needs for oxygen. Common simple asphyxiants include carbon dioxide, ethane, helium, hydrogen, methane, and nitrogen. Chemical asphyxiants, by chemical action, interfere with the passage of oxygen into the blood or the movement of oxygen from the lungs to body tissues. Either way, the end result is suffocation due to insufficient or no oxygenation. Common chemical asphyxiants include carbon monoxide, hydrogen cyanide, and hydrogen sulfide.

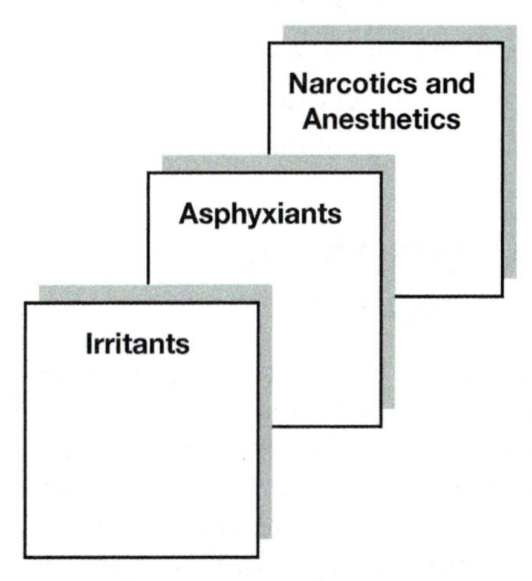

FIGURE 13–16 Airborne toxic substances.

Narcotics and Anesthetics. **Narcotics and anesthetics** are similar in that carefully controlled dosages can inhibit the normal operation of the central nervous system without causing serious or irreversible effects. This makes them particularly valuable in a medical setting. Dentists and physicians use narcotics and anesthetics to control pain before and after surgery. However, if the concentration of the dose is too high, narcotics and anesthetics can cause unconsciousness and even death. When this happens, death is the result of asphyxiation. Widely used narcotics and anesthetics include acetone, methyl-ethyl-ketone, acetylene hydrocarbons, ether, and chloroform.

HAZARD COMMUNICATIONS

Workers should be warned of chemical hazards by labels on containers or by **safety data sheets (SDSs)**, which summarize all pertinent information about a specific chemical. The **hazard communication** standard of the Occupational Safety and Health Act (OSH Act) requires that chemical suppliers provide users with an SDS for each chemical covered by the standard.

An SDS should contain the following information as appropriate: manufacturer's name, address, and telephone number; a list of hazardous ingredients; physical and chemical characteristics; fire and explosion hazard information; reactivity information; health hazard information; safety precautions for handling; and recommended control procedures.

An SDS must contain specific information in eight categories:

- *Section I: General information.* This section contains directory information about the manufacturer of the substance, including the following: manufacturer's name and address, telephone number of an emergency contact person, a nonemergency telephone number for information, and a dated signature of the person who developed or revised the SDS.
- *Section II: Hazardous ingredients.* This section should contain the common name, chemical name, and Chemical Abstracts Service (CAS) number for the substance. Chemical names are the scientific designations given in accordance with the nomenclature system of the International Union of Pure and Applied Chemistry. The CAS number is the unique number for a given chemical that is assigned by the Chemical Abstracts Service.
- *Section III: Physical and chemical characteristics.* Data related to the vaporization characteristics of the substance are contained in this section.

SAFETY DATA SHEET

SDS No. Q-1502

Section I. Product Identification

Product
XY-319ST (Developer)

Section Ia. Supplier's Name and Address

James Chemical Company
1170 Industrial Parkway
Chicago, Illinois
1-800-687-9712

Section II. Ingredients

Ingredients	CAS No.	Proportion	OSHA PEL	ACGIH TLV	Other Limits
Ferrite		>94%	Not listed	Not listed	None
Zinc oxide	1314-13-2				
Iron oxide	1309-37-1				
Copper oxide	1317-38-0				
Styrene-acrylate copolymer	25767-47-9	<5%	Not listed	Not listed	None
Carbon black	133-86-4	<0.4%	3.5 mg/m^3	3.5 mg/m^3	None

Section IIa. Hazardous Identification (Emergency Overview)

Developer is a black powder containing small amounts of toner and possessing no immediate hazard. There are no anticipated carcinogenic efforts from exposure, based on animal tests performed using toner.

FIGURE 13–17 A partial sample of an SDS.

- *Section IV: Fire and explosive hazard data.* Data related to the fire and explosive hazards of the substance are contained in this section. Special firefighting procedures are also included in this section.
- *Section V: Reactivity data.* Information concerning the stability of the substance as well as the potential for hazardous decomposition or polymerization of the substance is contained in this section.
- *Section VI: Health hazards.* This section contains a list of the symptoms that may be suffered as a result of overexposure to the substance. Emergency first aid procedures are also explained in this section.
- *Section VII: Safe handling and use.* This section explains special handling, storage, spill cleanup, and disposal methods and precautions related to the substance.
- *Section VIII: Control measures.* The types of ventilation, PPE, and special hygienic practices recommended for the substance are explained in this section.

Figure 13–17 is an example of an SDS.

PROCESS SAFETY MANAGEMENT

29 CFR 1926.64 provides guidelines for establishing a comprehensive process safety management program. The purpose of such a program is to prevent the unwanted release of highly hazardous chemicals that could lead to catastrophic results. The recommended elements of such a program include the following:

- Employee involvement
- Process safety information
- Process hazard analysis
- Operating procedures and practices
- Employee training
- Contractor screening
- Pre-start review
- Mechanical integrity program
- Non routine work authorizations
- Process change management
- Investigation of incidents
- Compliance audits of the program

HAZARDOUS WASTE OPERATIONS

The standard for **hazardous waste operations** (**HAZWOPER**) in construction is 29 CFR 1926.65. It gives organizations two options for responding to a chemical spill. The first is to evacuate all employees in the event of a spill and to call in professional emergency response personnel. Employers who use this option must have an emergency action plan in place. The second option is to respond internally. Employers using this option must have an emergency response plan in place.

1. *Emergency action plans.* An **emergency action plan (EAP)** should have at least the following elements: alarm system, evacuation plan, mechanism or procedure for emergency shutdown of the equipment, and procedure for notifying emergency response personnel.
2. *Emergency response plan.* Companies that opt to respond internally to chemical spills must have an **emergency response plan** that includes the provision of comprehensive training for employees. It is important to note that OSHA forbids the involvement of untrained employees in responding to a spill. The following topics are typical of those covered in up-to-date HAZWOPER courses:
   - Summary of key federal laws
   - Overview of impacting regulations
   - Classification and categorization of hazardous waste
     Definition of hazardous waste
     Characteristics
     Lists of hazardous wastes
   - Hazardous waste operations
     Definitions
     Levels of response
   - Penalties for noncompliance
     Civil penalty policy
   - Responses to spills
     Groundwater contamination
     Sudden releases
     Cleanup levels
     Risk assessment
     Remedial action
   - Emergency response
     Work plan
     Site evaluation and control
     Site-specific safety and health plan
     Information and training program
     Personal protective equipment
     Monitoring
     Medical surveillance
     Decontamination procedures
     Emergency response
     Other provisions
   - Contingency plans
     Alarm systems
     Action plan
   - Personal protective equipment (PPE)
     Developing a PPE program
     Respiratory equipment
     Protective clothing
     Donning PPE
     Doffing PPE
   - Safety data sheets (SDSs)
     Introduction
     Preparing SDSs
     SDS information
     Hazardous ingredients
     Physical and chemical characteristics
     Fire and explosion hazard data
     Reactivity data
     Health hazard data
     Precautions for safe handling and use
     Control measures
   - Site control
     Site maps
     Site preparation
     Work zones
     Buddy system
     Site security
     Communications
     Safe work practices
   - Hazardous waste containers
     Emergency control
     Equipment
     Tools
     Safety
   - Decontamination
     Types
     Decontamination plan
     Prevention of contamination
     Planning
     Emergencies
     Physical injury
     Heat stress
     Chemical exposure

Medical treatment area
Decontamination of equipment
Decontamination procedures
Sanitation of PPE
Disposal of contaminated materials

LEAD IN CONSTRUCTION

Lead is a toxic heavy metal that can be taken into the body by ingestion, if it somehow gets into a worker's mouth, or by inhalation, if introduced in the form of dust, mist, or fumes. If doses exceed a prescribed level, lead becomes toxic. Consequently, OSHA added a section to 29 CFR 1926.62 that explains the requirements for protecting construction workers who might be exposed to lead. For the purposes of this standard, lead includes metallic lead, all inorganic lead compounds, and organic lead soaps. The standard applies to all construction work—building, alterations, and repair—in which a worker might be exposed to lead. Figure 13–18 contains a list of the types of activities covered by this standard.

Overexposure to lead for an extended period of time can cause serious damage to several internal systems of the body, including the central nervous system and the urinary, reproductive, and blood-forming systems. Figure 13–19 contains a list of symptoms associated with lead exposure.

Provisions of the Standard

OSHA's standard for lead in construction sets forth requirements for permissible exposure limits, action levels, exposure monitoring and medical surveillance, and information and training. These requirements are summarized in this section.

- *Permissible exposure limits.* The **permissible exposure limits** require that no employee may be exposed to lead at airborne concentrations of more than 50 mg/m^3, averaged over an eight-hour period.

- Demolition or salvage of structures where lead or materials containing lead are present
- Removal or encapsulation of materials containing lead
- New construction, alteration, repair, or renovation of structures, substrates, or portions containing lead, or materials containing lead
- Installation of products containing lead
- Lead contamination from emergency cleanup
- Transportation, disposal, storage, or containment of lead or materials containing lead on the site or location at which construction activities are performed
- Maintenance operations associated with construction activities described above

FIGURE 13–18 Types of construction activities covered in 29 CFR 1926.62.

Loss of appetite
Constipation
Excessive tiredness
Headache
Fine tremors
Anxiety
Colic with severe abdominal pain
Muscle or joint pain or soreness
Metallic taste in mouth
Weakness
Nervous irritability
Hyperactivity
Numbness
Dizziness
Pallor
Insomnia

FIGURE 13–19 Symptoms of overexposure to lead.

- *Action level.* The **action level** is the level of exposure to lead at which a worker must begin taking the precautions prescribed in the standard. For construction, this level is 30 mg/m^3 calculated as an eight-hour TWA.
- *Exposure monitoring and medical surveillance.* Where a worker's initial exposure level exceeds the action level, employers must monitor by sampling the worker's regular, daily exposure to lead. If the initial determination proves that the worker's exposure is at or above the action level, but below the permissible exposure level (PEL), monitoring must be performed at least every six months. If the worker's exposure is above the PEL, monitoring must be performed at least quarterly and must continue until two consecutive measurements—taken at least seven days apart—are at or below the PEL. Medical surveillance must be made available to workers at no cost to them. Medical examinations must be made available to workers at no cost if they were exposed at or above the action level for more than 30 days in a year.
- *Information and training.* Employers are required to inform workers about lead hazards at job sites in accordance with 29 CFR 1926.59. This standard requires warning signs, labels, SDSs, and training. Figure 13–20 is an example of a warning sign that must be used at a construction site.

WARNING
LEAD WORK AREA
POISON
NO SMOKING OR EATING

FIGURE 13–20 Sample of lead exposure warning sign for a construction site.

SUBPART E: PERSONAL PROTECTIVE AND LIFESAVING EQUIPMENT

Subpart E covers all of the requirements employers must meet regarding the provision of PPE and lifesaving equipment for workers. Before selecting any specific type of PPE, employers are required to conduct a hazard survey to determine the types of PPE that might be needed to deal with hazards that cannot be completely eliminated by other methods (engineering or administrative controls). Employers are required to provide the necessary PPE to workers and to train them in the proper use of it. Subpart E contains the following sections:

- 1926.95 Criteria for personal protective equipment
- 1926.96 Occupational foot protection
- 1926.97 (Reserved)
- 1926.98 (Reserved)
- 1926.99 (Reserved)
- 1926.100 Head protection
- 1926.101 Hearing protection
- 1926.102 Eye and face protection
- 1926.103 Respiratory protection
- 1926.104 Safety belts, lifelines, and lanyards
- 1926.105 Safety nets
- 1926.106 Working over or near water
- 1926.107 Definitions applicable to this subpart

HEAD PROTECTION

A carpenter working on a catwalk drops a hammer. The falling hammer accelerates over the 20-foot drop and strikes a worker below. Had the victim not been wearing **head protection** in the form of a hard hat, he might have sustained serious injuries from the impact, Figure 13–21.

Objects that fall, are slung from a machine, or otherwise become projectiles pose a serious hazard to the heads of workers. Consequently, protecting workers from projectiles requires the use of appropriate PPE and strict adherence to safety rules by all employees. Approximately 120,000 people sustain head injuries on the job each year. Falling objects are involved in many of these accidents. These injuries occurred in spite of the fact that many of the victims were wearing hard hats. Such statistics have been the driving force behind the development of tougher, more durable hard hats.

FIGURE 13–21 Hard hat use is critical in construction.
Source: Ilkercelik/Fotolia

Originally introduced in 1919, the hard hats first used for head protection in a construction setting were inspired by the helmets worn by soldiers in World War I. Such early versions were made of varnished resin-impregnated canvas. As material technology evolved, hard hats were made of vulcanized fiber, then aluminum, and then fiberglass. Today's hard hats are typically made from the thermoplastic material polyethylene, using the injection-molding process. Basic hard hat design has not changed radically since before World War II. Hard hats are designed to provide limited protection from impact, primarily to the top of the head, and thereby reduce the amount of impact transmitted to the head, neck, and spine.

The American National Standards Institute (ANSI) standard for hard hats is Z89.1. OSHA subsequently adopted this standard as its hard hat standard. Hard hats can help reduce the risk associated with falling or projected objects, but only if they are worn. The use of hard hats in certain settings in which falling objects are likely has been mandated by federal law since 1971.

Protection from Hard Hats

OSHA requires that hard hats be worn when workers are in areas where there is the potential for injury to the head from falling objects and exposure to electrical conductors that might come in contact with a worker's head. Hard hats are classified according to the type of

SAFETY FACTS & FINES

Personal protective equipment can be especially important when working with electricity. A company in Hamburg, New York, was fined $57,125 when an employee was severely burned by an arc blast produced when the power tool he was using contacted a live electrical part. OSHA cited the company for the following violations: (1) failure to require employees to wear nonconductive head protection when appropriate, (2) failure to deenergize parts or equipment before working near them, and (3) failure to provide sufficient work space when working near electrical equipment.

impact they are designed to withstand and their ability to reduce the danger of exposure to electrical current. To comply with the requirements of ANSI Z89.1, which OSHA has adopted, hard hats must meet or exceed the following requirements:

- *Type I hard hats.* Reduce the force of a blow to the top of the head only. This type of hard hat is designed to protect workers from falling objects.
- *Type II hard hats.* Reduce the force of a blow to the top of the head and the force of a lateral blow or off-center blow. This type of hard hat is designed to protect workers from blows to the top and side of the head.
- *Class E (electrical) hard hats.* Reduce exposure to high-voltage conductors and provide dielectric protection up to 20,000 volts (phase to ground). Note that this protection pertains to the head only.
- *Class G (general) hard hats.* Reduce exposure to low-voltage conductors and offer dielectric protection up to 2,200 volts (phase to ground). Note that this protection pertains to the head only.
- *Class C (conductive) hard hats.* Not designed to protect workers from electricity. They protect from impact and can in some instances provide better breathability than other hard hats.

EYE AND FACE PROTECTION

Eye and face protection are critical in construction work, Figure 13–22. The recognized standard for eye and face protection is ANSI Z87.1. The latest edition of this standard is ANSI Z87.1 - 2010. This edition replaces the earlier version that was updated in 2003. OSHA typically adopts this ANSI standard for eye and face protection. Key elements of the standard include the following:

- *Choosing protection based on the hazard.* Employers are required to assess the workplace to determine what types of eye and face hazards are present (e.g., impact, optical radiation, chemical splash, dust, etc.) and choose PPE that will effectively protect against those hazards. The standard contains a pull-out selection chart showing the recommended protection devices for specific hazards.
- *Criteria for compliance.* Eye and face PPE is classified as being either "Non-Impact Rated" or "Impact Rated." Testing must use the European small and medium head-form sizes. PPE that claims to protect against splash/droplet, dust, or fine dust must meet specific performance and marking requirements specified in the standard. Impact rated devices must provide lateral coverage. Device must pass a test in which a hot steel rod makes contact with the PPE without causing the device to ignite. Devices that are impact rated must carry a "+" as part of the manufacturers marking and an optical radiation or shade marking as specified in the standard. For example, a welding lens with a 3.0 shade welding filter would be marked "3." In addition to the manufacturer's markings, devices must carry the designation "Z87" if not impact rated and "Z87+" if impact rated. Figures 13–23, 13–24, and 13–25 are examples of the types of devices available for eye and face protection.

FIGURE 13–22 Eye protection is critical in construction.
Source: BillionPhotos.com/Fotolia

Assessing the Workplace for Eye Hazards

The type of eye protection needed in a given setting depends on the type of work done in that setting and the corresponding hazards. Before establishing a vision

FIGURE 13–23 Safety glasses that wrap around for lateral protection.
Source: Coprid/Fotolia

FIGURE 13–24 Safety glasses.
Source: Naypong/Fotolia

FIGURE 13–25 Safety goggles with hard hat.
Source: Zorandim75/Fotolia

protection program, it is necessary to assess the workplace. OSHA recommends using the following questions in making a workplace assessment:

- Do employees perform tasks or work near others who perform tasks that may produce airborne dust or flying particles?
- Do employees handle or work near others who handle hazardous liquid chemicals or blood?
- Do employees work in situations that may expose their eyes to chemical or physical irritants?
- Do employees work in situations that may expose their eyes to intense light or lasers?

On the basis of the answers to these questions, a vision protection program can be developed to protect workers. That program should meet certain requirements as recommended by OSHA. These requirements are summarized in the next section.

Requirements for Choosing Vision Protection Devices

There are many different types of eye protection devices available that vary in terms of function, style, fit, lens, and other options. Figure 13–26 matches typical eye and face hazards with recommended personal protective devices. OSHA recommends applying the following criteria when selecting vision protection devices:

Selecting the Right Eye and Face Protection Devices

Flying particles, chips, sand: Face shield
Hot sparks: Face shield
Heat: Reflective face shield
Molten metal: Face shield
Chemical splash: Face shield
Ultraviolet light and infrared heat: Welding helmet or welding shield with shaded lens

FIGURE 13–26 Selecting the right personal protective equipment.

- Select only those that meet the standards set forth in ANSI Z87.1-1989.
- Select devices that protect against the specific hazards identified in the assessment.
- Select devices that are as comfortable as possible to wear.
- Select devices that do not restrict vision in any way.
- Select devices that are durable, easy to clean, and easy to disinfect.
- Select devices that do not interfere with the functioning of other personal protective equipment.

Training

Once the job site has been assessed and eye protection devices have been selected, it is important to provide employees with training in the proper use of the devices. This accomplishes the following: First, it ensures that the eye protection devices are used properly. Second, it shows employees that they have a critical role to play in the protection of their eyes. OSHA recommends training that covers the following topics:

- Why it is important to use the eye protection devices?
- How the devices protect the eyes?
- Limitations of the devices
- When the devices should be used?
- How the devices should be used?
- How straps are adjusted for both effectiveness and comfort?
- How the employee can identify signs of wear that may lessen the effectiveness of the devices?
- How and when the devices should be cleaned and disinfected?

First Aid for Eye Injuries

Even with proper eye protection, there is still a risk that an employee may sustain an injury. Even the best vision protection program is not perfect. When this happens, the following guidelines for first aid apply:

- Be gentle with the employee. Do not add to the injury with rough treatment.
- Do not attempt to remove objects embedded in the eyeball.
- Rinse the eyes with a copious amount of water for 15 to 30 minutes to remove the chemicals.
- Cover both eyes after the rinsing has been completed. Call for professional help.
- Never press on an injured eye or put any pressure on it (as when covering the eyes).
- Do not allow the employee to rub his or her eyes.

HEARING PROTECTION DEVICES

Where noise hazards exist, workers should be required to use appropriate **hearing protection devices (HPDs)**. Four classifications of HPDs are widely used: (1) enclosures, (2) earplugs, (3) superaural caps, and (4) earmuffs.

Enclosures are devices that completely encompass the employee's head—much like the helmets worn by jet pilots. *Earplugs* (also known as *aural plugs*) are devices that fit into the ear canal. Custom-molded earplugs are designed and molded for the individual employee. Premolded earplugs are generic in nature, are usually made of a soft rubber or plastic substance, and can be reused. *Formable earplugs* can be used by anyone. They are designed to be formed individually to a person's ears, used once, and then discarded. *Superaural caps* fit over the external edge of the ear canal and are held in place by a headband. *Earmuffs*, also known as *circumaurals*, cover the entire ear with a cushioned cup that is attached to a headband. Earplugs and earmuffs are able to reduce perceived noise by 20–30 dB. By combining earplugs and earmuffs, noise can be reduced by an additional 3–5 dB.

The effectiveness of HPDs can be enhanced through the use of technologies that reduce noise levels. These *active noise reduction (ANR)* technologies reduce noise by manipulating sound and signal waves. Such waves are manipulated by creating an electronic mirror image of sound waves that tends to cancel out the unwanted noise in the same way that negative numbers cancel out positive numbers in a mathematical equation. Using ANR in conjunction with enclosure devices or earmuffs can be an especially effective strategy.

Traditional or passive HPDs can distort or muffle sounds at certain frequencies, particularly high-pitched sounds. *Flat-attenuation HPDs* solve this problem by using electronic devices to block all sound frequencies equally; this eliminates, or at least reduces, the distortion and muffling problems. Flat-attenuation HPDs are especially helpful for employees in settings in which high-pitched sound is present and should be heard and for employees who have already begun to lose their ability to hear such sounds. The ability to hear high-pitched sounds is significant because warning signals and the voices of alarmed co-workers can be high pitched.

A benefit of ANR technologies is *optimization*. The amount of noise protection can be adjusted so that workers can hear as much as they should, but not too much. Too much noise can cause workers to suffer a hearing loss. Too little noise can mean that they may not hear warning signals.

TORSO PROTECTION

Often the types of hazards present on construction sites pose risks to the worker's torso. Hot metal, hot liquids, or chemicals may splash on a worker's torso. Flying objects may hit a person in the back or ribs. Radiation may be absorbed by a worker's torso. Acid may splash on a worker's chest or back. The torso is vulnerable to these hazards; consequently, protective clothing is available. **Torso protection** clothing includes vests, aprons, jackets, coveralls, and full-body suits.

In selecting protective clothing, it is important to know the types of hazards present. One material might work well for a given hazard but be ineffective for another. The following selection criteria match hazards and materials:

- *Fire resistance.* Wool and treated cotton
- *Cuts and bruises.* Cotton duck (a closely woven fabric)
- *Dry heat and flame.* Leather
- *Acids and chemicals.* Rubber, neoprene, and plastics
- *Dust.* Disposable material, such as treated paper

HAND AND ARM PROTECTION

In the United States, there are more than 500,000 hand injuries every year. Hand injuries are serious and costly for both employers and workers. The selection of **hand and arm protection** (gloves) should be based on a comprehensive assessment of the tasks performed for a given job, the hazards present, and the duration of exposure to the hazards.

With the assessment completed, employers should review specification information from manufacturers of safety gloves and select the gloves that are best suited for the individual situation. Selecting just the right gloves for the job has historically been one of employers' greatest difficulties. For example, take the issue of fit. A poorly fitted set of gloves cannot offer the degree of protection that a responsible employer or worker wants. Yet, because manufacturers have not developed a consistent set of measurements for sizing gloves, the only way to determine whether a pair fits properly is for the employee to try them on.

Fit is just one of the problems when selecting gloves. Other critical features include the protection capability, comfort, and tactile sensitivity of the gloves. Often, greater comfort and tactile sensitivity can mean less protection; correspondingly, greater protection can mean less comfort and tactile sensitivity.

In an attempt to simplify the process of selecting the right gloves, the ANSI and the Industrial Safety Equipment Association (ISEA) developed a joint hand-protection standard ANSI/ISEA 105-1999. This standard simplifies glove selection by (1) defining characteristics of protection in a variety of critical areas, including cut, puncture, and abrasion resistance; protection from cold and heat; chemical resistance (including both permeation and degradation); viral penetration resistance; dexterity and liquid tightness; and flame and heat resistance and (2) standardizing the tests used to measure all of these various characteristics.

Common Glove Materials

Depending on the potential individual hazards in a given situation, the right gloves for the application may be made of a variety of materials (Figure 13–27). The most widely used materials in making gloves are as follows:

- *Leather.* Offers comfort, excellent abrasion resistance, and minimum cut resistance.
- *Cotton.* Offers comfort, minimal abrasion resistance, and minimum cut resistance.
- *Aramids.* Offers comfort, good abrasion resistance, excellent cut resistance, and excellent heat resistance.
- *Polyethylene.* Offers comfort, excellent abrasion resistance, and minimal cut resistance. Gloves made of this material should not be subjected to high temperatures.
- *Stainless steel cord (wrapped in synthetic fiber).* Offers comfort, good abrasion resistance, and optimal cut resistance.
- *Chain link or metal mesh.* Offers very little comfort, but maximum abrasion and cut resistance.
- *Butyl rubber.* Offers little comfort, but has excellent resistance to heat, ozone, tearing, and certain chemicals, including alcohols, aldehydes, ketones, esters, nitriles, gases, amides, acids, and nitroglyceride compounds.
- *Viton rubber.* Offers little comfort, but performs well against chemicals that butyl rubber fails to protect against, including aliphatic, halogenated, and aromatic compounds. Like butyl gloves, viton gloves also perform well in handling alcohols, gases, and acids.

FIGURE 13–27 Protective gloves.
Source: Artem Shadrin/Fotolia

FOOT AND LEG PROTECTION

Foot and toe injuries account for almost 20 percent of all disabling workplace injuries in the United States. There are more than 180,000 foot and toe injuries in the workplace each year. Consequently, foot and leg protection are important. The major kinds of injuries to the foot and toes are from the following:

- Impact from sharp or heavy objects that fall on feet or legs (this type accounts for 60 percent of all injuries)
- Compression when feet are rolled over by or pressed between heavy objects
- Punctures through the sole of the foot
- Conductivity of electricity or heat through foot
- Electrocution of foot from contact with an energized, conducting material
- Slips on unstable walking surfaces
- Hot liquid or metal splashed into shoes or boots
- Temperature extremes

The key to protecting workers' feet and toes is to match the protective measure with the hazard. This involves the following steps: (1) identify the various types of hazards present in the workplace, (2) identify the types of footwear available to counter the hazards, and (3) require that proper footwear be worn. Shoes selected should meet all applicable ANSI standards and have a corresponding ANSI rating.

Modern safety boots that provide comprehensive foot and toe protection are available. The best safety boots provide all of the following types of protection:

- *Steel toe* for impact protection
- *Rubber or vinyl* for chemical protection
- *Puncture-resistant soles* for protection against sharp objects
- *Slip-resistant soles* for protection against slippery surfaces
- *Electricity-resistant material* for protection from electric shock

Employers are not required to provide footwear for employees, but they should provide training on foot protection. The training must cover at least the following topics:

- Conditions when footwear should be worn
- Type of footwear needed in a given situation
- Limitations of protective footwear
- Proper use of protective footwear

In addition to falling, puncture, and chemical hazards, there are also hazards associated with prolonged standing. Prolonged standing or walking is common in construction. It can cause lower back pain, sore feet, swelling in the legs, general muscular fatigue, and other health problems. The following precautions can minimize standing hazards:

- *Antifatigue mats.* **Antifatigue mats** provide cushioning between the feet and hard working surfaces, such as concrete floors (Figure 13–28). This cushioning effect can reduce muscle fatigue and lower back pain. However, too much cushioning can be just as bad as too little. Consequently, it is important to test mats on a trial basis before buying a large quantity. Mats that become slippery when wet should be avoided. In areas where chemicals are used, be sure to select mats that will hold up to saturation by chemicals.
- *Shoe inserts.* When antifatigue mats are not feasible because employees must move from area to area and, correspondingly, from surface to surface, **shoe inserts** can help. Such inserts are worn inside the shoe and provide the same type of cushioning that mats provide. Shoe inserts can help reduce lower back, foot, and leg pain. It is important to ensure proper fit. If inserts make an employee's shoes too tight, they do more harm than good. In such cases, employees may need to wear a slightly larger shoe.
- *Foot rails.* **Foot rails** added to machines such as saws can help relieve the hazards of prolonged standing. Foot rails allow employees to elevate one foot at a time by 4 or 5 inches. The elevated foot rounds out the lower back, thereby relieving some of the pressure on the spinal column. Placement of a rail is important. It should not be placed in a position that inhibits movement or becomes a tripping hazard.

FIGURE 13–28 Anti-fatigue mats.
Source: Kalpis/Fotolia

FALL PROTECTION, LIFELINES, AND SAFETY NETS

The OSH Act mentions fall protection in several places. In addition, the ANSI publishes a Fall Protection Standard (ANSI Z359: *Safety Requirements for Personal Fall Arrest Systems, Subsystems, and Components*). The most comprehensive fall protection standard is OSHA's fall protection standard for the construction industry (Subpart M of 29 CFR 1926), which is covered in detail in Chapter 15, Subpart M.

OSHA's current fall protection standard sets the general *trigger height* at 6 feet. This means that any construction employee working higher than 6 feet off the ground must use a fall protection device, such as a safety harness and **lifeline** (Figure 13–29). Figure 13–30

FIGURE 13–29 Personal fall arrest harness.
Source: Nikolayshubin/Fotolia

Language of Fall Protection

Fall protection has a language of its own. In order to understand OSHA regulations and other fall protection guidelines, it is necessary to know the following terms:

- *Anchorage.* A secure point of attachment for lifelines, lanyards, or deceleration devices.
- *Body belt.* A strap with means both for sharing it about the waist and for attaching it to a lanyard, lifeline, or deceleration device.
- *Body harness.* Straps that may be secured about the person in a manner that distributes the fall arrest forces over at least the thighs, pelvis, waist, chest, and shoulders, with a means for attaching the harness to other components of a personal fall arrest system.
- *Connector.* A device used to couple (connect) parts of a personal fall arrest system or positioning device system together.
- *Hole.* A void or gap 2 inches (5.1 cm) or more in the least dimension in a floor, roof, or other walking or working surface.
- *Lanyard.* A flexible line of rope, wire rope, or strapping that generally has a connector at each end for connecting the body belt or body harness to a deceleration device, lifeline, or anchorage.
- *Lifeline.* A component consisting of a flexible line for connection to an anchorage at one end to hang vertically (vertical lifeline) or for connection to anchorages at both ends to stretch horizontally (horizontal lifeline). Serves as a means for connecting other components of a personal fall arrest system to the anchorage.
- *Low-slope roof.* A roof having a vertical slope of 4 inches or less per 12 inches.
- *Opening.* A gap or void 30 inches (76 cm) or more high and 18 inches (46 cm) or more wide, in a wall or partition, through which employees can fall to a lower level.
- *Personal fall arrest system.* A system including but not limited to an anchorage, connectors, and a body belt or body harness used to arrest an employee in a fall from a working level. The use of a body belt for fall arrest is prohibited.
- *Platform.* A working space for persons, elevated above the surrounding floor or ground.
- *Positioning device system.* A body belt or body harness system rigged to allow an employee to be supported on an elevated vertical surface, such as a wall, and work with both hands free while leaning backwards.
- *Rope grab.* A deceleration device that travels on a lifeline and automatically, by friction, engages the lifeline and locks to arrest a fall.
- *Self-retracting lifeline or lanyard.* A deceleration device containing a drum-wound line that can be slowly extracted from, or retracted onto, the drum under minimal tension during normal employee movement and which, after onset of a fall, automatically locks the drum and arrests the fall.

FIGURE 13–30 Construction professionals should be familiar with these fall protection terms.

contains the language of fall protection for construction professionals.

Regulations from OSHA 1926 that apply specifically to fall protection in scaffolding work are as follows:

- 1926.451(g)(2) reads, "The employer shall have a competent person determine the feasibility and safety of providing fall protection for employees erecting or dismantling supported scaffolds. Employers are required to provide fall protection for employees erecting or dismantling supported scaffolds where the installation and use of such protection is feasible and does not create a greater hazard."
- 1926.502(d)(15) reads, "Anchorages used for attachment of personal fall arrest equipment shall be independent of any anchorage being used to support or suspend platforms and capable of supporting at least 5,000 pounds (22.2 kN) per employee attached, or shall be designed, installed, and used as follows:
  - (i) as part of a complete personal fall arrest system that maintains a safety factor of at least two; and
  - (ii) under the supervision of a qualified person."
- 1926.502(d)(16) reads, "Personal fall arrest systems, when stopping a fall, shall:
  - (i) limit maximum arresting force on an employee to 900 pounds (4 kN) when used with a body belt;
  - (ii) limit maximum arresting force on an employee to 1,800 pounds (8 kN) when used with a body harness;
  - (iii) be rigged such that an employee can neither free-fall more than 6 feet (1.8 m) nor contact any lower level;
  - (iv) bring an employee to a complete stop and limit maximum deceleration distance an employee travels to 3.5 feet (1.07 m); and
  - (v) have sufficient strength to withstand twice the potential impact energy of an employee free-falling a distance of 6 feet (1.8 m), or the freefall distance permitted by the system, whichever is less."

Ladder Safety in Construction

Jobs that involve the use of ladders introduce their own set of safety problems—one of which is an increased potential for falls. Ladders should be inspected before every use, and employees who use them should follow a set of standard rules.

Inspecting Ladders. Taking a few moments to look over a ladder carefully before using it can prevent a fall. The following guidelines are recommended:

- See if the ladder has the manufacturer's instruction label on it.
- Determine whether the ladder is strong enough.
- Read the label specifications about weight capacity and applications.
- Look for the following conditions: cracks on side rails; loose rungs, rails, or braces; or damaged connections between rungs and rails.
- Check for heat damage and corrosion.

- Check wooden ladders for moisture that may cause them to conduct electricity.
- Check metal ladders for burrs and sharp edges.
- Check fiberglass ladders for signs of *blooming* (deterioration of exposed fiberglass).

Dos and Don'ts of Ladder Use. Many accidents involving ladders result from improper use. Following a simple set of rules for the proper use of ladders can reduce the risk of falls and other ladder-related accidents. Use the following guidelines:

Do

- Check for slipperiness on shoes and ladder rungs.
- Secure the ladder firmly at the top and bottom.
- Set the ladder's base on a firm, level surface.
- Apply the 4:1 ratio (base should be 1 foot away from the wall for every 4 feet between the base and the support point).
- Face the ladder when climbing up or down.
- Barricade the base of the ladder when working near an entrance.

Don't

- Lean a ladder against a fragile, slippery, or unstable surface.
- Lean too far to either side while working (stop and move the ladder).
- Rig a makeshift ladder; use the real thing.
- Allow your waist to go any higher than the last rung when reaching upward on a ladder.
- Carry tools in your hand while climbing a ladder.
- Place a ladder on a box, table, or bench to make it reach higher.

GUIDELINES FOR RESPIRATORS

OSHA's respiration standard (29 CFR 1926.103) for construction refers companies to OSHA's general industry rule for **respirators** (29 CFR 1910.134). This standard requires companies to use NIOSH-approved respirators. The respirator is one of the most important types of PPE available to individuals who work in hazardous environments (Figure 13–31). Because the performance of a respirator can mean the difference between life and death, NIOSH publishes strict guidelines regulating the manufacture of respirators. The standard with which manufacturers must comply is 42 CFR Part 84. In addition, construction professionals must ensure that employees are provided with respirators that meet all of the specifications set forth in 42 CFR Part 84.

FIGURE 13–31 Air filtering respirator.
Source: Gennadiy Poznyakov/Fotolia

There are two types of respirators: **air filtering** and **air supplying**. Air filtering respirators filter toxic particulates out of the air. To comply with 42 CFR Part 84, an air-filtering respirator must protect its wearer from the most penetrating aerosol size of particle, which is 0.3 microns of aerodynamic mass in median diameter. The particulate filters used in respirators are divided into three classes—each class has three levels of efficiency: 95 percent, 99 percent, and 99.97 percent.

- **Class N respirators** are used only in environments that contain no oil-based particulates because they are not oil resistant. They may be used in atmospheres that contain solid or nonoil contaminants.

SAFETY FACTS & FINES

When employees are required to work in dangerous conditions, it is important that they—and their employer—know what the dangers are. As a result of negligence in this area, a company in Newark, New Jersey, was fined $183,900. OSHA's attention was drawn to the company by two incidents, mere days apart, in which employees were harmed while working in confined spaces. In the first incident, there was one death and two injuries. In the second, two employees lost consciousness. After an investigation, the firm was fined for failure to provide supervisors with adequate confined space training, failure to prevent unauthorized entry into such spaces, and failure to identify hazards before employees entered confined spaces.

- **Class R respirators** may be used in atmospheres that contain any contaminant. However, the filters in Class R respirators must be changed after each shift if oil-based contaminants are present because they are oil resistant but not oil proof.
- **Class P respirators** may be used in any atmosphere containing any particulate contaminant. They are oil proof.

If there is any question about the viability of an air filtering respirator in a given setting, employees should use air-supplying respirators. This type of respirator works in much the same way as an air tank for a scuba diver. Air from the atmosphere is completely blocked out, and fresh air is provided via a self-contained breathing apparatus.

In addition to the NIOSH regulations regarding respirators (42 CFR 84), there are regulations published by OSHA in 29 CFR 1910.134. Key provisions of the OSHA regulations are as follows:

- Respirators, when they are required, must be provided by the employer.
- Medical evaluations must be provided for respirator users.
- Fit testing must be conducted according to standards.
- Respirators must be used in reasonably foreseeable emergency situations.
- Respirators must be properly cleaned and maintained.
- Adequate air quality, quantity, and flow for atmosphere-supplied respirators must be ensured.
- Training and evaluation programs must be provided to ensure effectiveness.

Air Safety Program Elements

Companies with job sites in which fumes, dust, gases, vapors, or other potentially harmful particulates are present should have an air-safety program as part of their overall safety and health program. The program should have at least the following elements:

- Accurate hazard identification and analysis procedures to determine what types of particulates are present and in what concentration.
- Standard operating procedures (in writing) for all elements of the air-safety program.
- Respirators that are appropriate in terms of the types of hazards present and that are described in 42 CFR Part 84.
- Training, including fit testing, limitations, use, and maintenance of respirators.
- Standard procedures for routine operation and storage of respirators.

PROTECTING WORKERS FROM HEAT HAZARDS

The key question that must be answered by construction professionals concerning employees whose work may subject them to heat stress is as follows[17]:

> What are the conditions to which most adequately hydrated, unmedicated, healthy workers may be exposed without experiencing heat strain or any other adverse effects?

The American Council of Government Industrial Hygienists (ACGIH) publishes a comprehensive manual to help construction professionals answer this question for the specific situations and conditions they face. This manual, entitled *TLVs and BEIs: Threshold Limit Values for Chemical Substances and Physical Agents and Biological Exposure Indices*, provides reliable guidance and should be in every construction professional's library. In addition to using the information contained in this manual, construction companies should have a comprehensive heat stress management program in place and apply sound professional judgment.

Definitions

Heat stress is the net heat load to which a worker may be exposed from the combined contributions of metabolic cost of work; environmental factors (i.e., air temperature, humidity, air movement, and radiant heat exchange); and clothing requirements. A mild or moderate heat stress may cause discomfort and may adversely affect performance and safety, but it is not harmful to health. As the heat stress approaches human tolerance limits, the risk of heat-related disorders increases.

Heat strain is the overall physiological response resulting from heat stress. The physiological adjustments are dedicated to dissipating excess heat from the body. Acclimatization is a gradual physiological adaptation that improves an individual's ability to tolerate heat stress.

Recognizing Heat Strain

Construction professionals, supervisors, and workers should know how to recognize heat strain. The following factors are signs of excessive heat strain. Exposure to heat stress should be stopped immediately for any employee experiencing any of these symptoms (see Figure 13–32):

- Fatigue
- Nausea or vomiting
- Headache
- Light-headedness
- Clammy, moist skin
- Pale or flushed complexion
- Fainting when trying to stand
- Rapid pulse

FIGURE 13–32 Recognizable signs of heat exhaustion.

- A sustained rapid heart rate (180 beats per minute minus the employee's age in years). For example, a 40-year-old employee has a sustained heart rate of 150 beats per minute. This is a problem because the heart rate exceeds 140 beats per minute (180 − 40).
- Core body temperature of more than 38.5°C (101.3°F).
- Recovery heart rate one minute after a peak work effort of more than 110 beats per minute.
- Sudden and severe fatigue, nausea, dizziness, or light-headedness.

The symptoms of heat strain can be assessed on the spot. Other symptoms can only be monitored over time. Employees are at greater risk of excessive heat strain if they experience any of the following:

- Profuse perspiration that continues for hours
- Weight loss of more than 1.5 percent of body weight during one work shift
- Urinary sodium excretion of less than 50 moles in 24 hours

Clothing

Heat is best removed from the body when there is free movement of cool dry air over the skin's surface. This promotes the evaporation of perspiration from the skin, which is the body's principal cooling mechanism. Clothing impedes this process—some types more than others. Encapsulating suits and clothing that is impermeable or highly resistant to the flow of air and water vapor multiply the potential for heat strain.

When assessing heat stress hazards in the workplace, consider the added effect of clothing. For example, the **wet-bulb globe temperature (WBGT)** working conditions should be increased by 3.5°C for employees wearing cloth overalls. This factor increases to 5°C with double cloth overalls. In other words, consider the WBGT to be 3.5°C higher than it registers when an employee is wearing cloth overalls or their equivalent (5°C with double cloth overalls).

Because the WBGT is influenced by air temperature, radiant heat, and humidity, it can be helpful in establishing a threshold for making judgments about working conditions. WBGT values can be calculated using the following formula:

Exposed to Direct Sunlight

$$WBGT = 0.7\ T_{nwb} \text{ to } 0.2\ T_g + 0.1\ T_{db}$$

where T_{nwb} = natural wet bulb temperature
T_g = globe temperature
T_{db} = dry bulb (air) temperature

Not Exposed to Direct Sunlight

$$WBGT = 0.7\ T_{nwb} + 03\ T_g$$

These formulas for WBGT provide a beginning point for making judgments. The WBGT must be adjusted for clothing, work demands, and the worker's acclimatization state. The key is to ensure that employees never experience a core body temperature of 38°C or higher.

Heat Stress Management

Construction professionals should continually emphasize the importance of paying attention to recognizable symptoms of heat stress. In addition, they should ensure that a comprehensive heat stress management program is in place. Such a program should consist of both general and specific controls.

General Controls. The following general controls are recommended:

- Provide accurate verbal and written instructions, training programs, and other information about heat stress and strain.
- Encourage drinking small volumes (approximately 8 oz.) of cool water about every 20 minutes.
- Permit self-limitation of exposure. Encourage co-worker observation to detect signs and symptoms of heat strain in others.
- Counsel and monitor those employees who take medications that may compromise normal cardiovascular, blood pressure, body temperature regulation, renal, or sweat gland functions as well as those who abuse or who are recovering from the abuse of alcohol and other intoxicants.
- Encourage healthy lifestyles, ideal body weight, and electrolyte balance.
- Adjust expectations of those returning to work after absence from heat stress situations and encourage consumption of salty foods (with approval of the employee's physician if employee is on a salt-restricted diet).
- Consider medical screening to identify those susceptible to systemic heat injury.

Specific Controls. The following specific controls are recommended:

- Establish engineering controls that reduce the metabolic rate, provide general air movement, reduce process heat and water vapor release, and shield radiant heat sources, among others.
- Consider administrative controls that set acceptable exposure times, allow sufficient recovery, and limit physiological strain.
- Consider personal protection devices that have been demonstrated to be effective for the specific work practices and conditions at the location.

WORKING OVER OR NEAR WATER

OSHA's 29 CFR 1926.106 covers working over or near water. Since this type of situation is common with construction projects, construction professionals, and students should be familiar with the following requirements:

- Personnel working over or near water where the danger of drowning exists must be provided with either U.S. Coast Guard approved life jackets or buoyant work vests.
- Prior to or after each use the life jackets or buoyant work vests must be inspected for defects that could impede their proper performance. PPE found to have defects may not be used again.
- Ring buoys with a minimum of 90 feet of line must be readily available for emergency use. The distance between ring buoys may not exceed 200 feet.
- At least one life-saving skiff must be immediately available for emergency use.

Summary

Subpart A of 29 CFR 1926 explains the purpose and scope of the standards, variance from safety and health standards, and enforcement of the standards. Key issues related to Subpart A are as follows: (1) Does the company have contracts that are subject to the Contract Work Hours and Safety Standards Act (Section 107)? (2) Do the regulations in 29 CFR 1926 apply to the company? (3) Does the company need a variance from a given safety and health standard? (4) Has an OSHA inspector ever visited one of the company's job sites? (5) Has the company ever undergone administrative adjudication for violation of an OSHA regulation?

The goal of Subpart B is to ensure that contractors provide workers with a safe and healthy work environment. Key issues related to Subpart B are as follows: (1) Does the company pursue contracts with federal agencies? (2) Does the company serve as a prime contractor and subcontract portions of the work? (3) Do any statutes under Reorganization Plan 14 apply to the company? (4) What types of duties does the company delegate to subcontractors? (5) How does the company ensure that subcontractors are meeting their delegated obligations?

Subpart C covers a variety of general safety and health provisions, most of which are dealt with in more depth later in the standard. Key issues related to Subpart C are as follows: (1) Does the company have an accident-prevention program in place? (2) Does the company conduct safety and health training programs for its workers? (3) Does the company maintain comprehensive and accurate records of accidents, injuries, exposures, and corresponding medical treatments? (4) Is first aid and medical treatment readily available at all job sites? (5) Is adequate fire protection equipment available at all job sites? (6) Is proper housekeeping ensured at all job sites? (7) Is adequate illumination provided where needed? (8) Are sanitation needs adequately provided for at all job sites? (9) Is the necessary personal protective equipment (PPE) provided and properly used? (10) Are all exits properly marked and accessible? (11) Does the company maintain a comprehensive emergency action plan (EAP)?

Subpart D covers numerous health and environmental concerns. Key issues related to Subpart D are as follows: (1) Does the company provide onsite first aid at all job sites? (2) Are medical services readily available at all job sites? (3) Are sanitation facilities available in the right numbers at all job sites? (4) Are workers protected from hazardous noise levels? (5) Are workers protected from radiation hazards? (6) Are workers protected from gases, vapors, fumes, dusts, and mists? (7) Is proper illumination provided at all necessary locations? (8) Is proper ventilation provided at all necessary locations? (9) Are safety data sheets (SDSs) made available where needed? (10) Are hazardous waste operations properly controlled at all job sites? (11) Does the company have an emergency response plan for handling and storage of hazardous waste?

Subpart E covers all of the requirements employers must meet regarding the provision of PPE. Key issues related to Subpart E are as follows: (1) Does the company survey its job sites to determine if PPE is necessary? (2) Does the company make the necessary PPE available to workers? (3) Does the company require the use of PPE as appropriate? (4) Does the company train workers in the proper use of PPE? (5) Do supervisors ensure that PPE is used properly when needed? (6) Are falling or flying hazards present at job sites? (7) Are electrical hazards present at job sites? (8) Are noise or eye hazards present at construction sites? (9) Is work performed at heights or over water? (10) Are airborne contaminants present?

Key Terms and Concepts

Action level
Aerosols
Airborne contaminants
Antifatigue mats
Asphyxiant
Attenuation
Caution signs and labels
Class N respirators
Class P respirators
Class R respirators
Decibel (dB)
Dose
Dosimeter
Dusts
Education and motivation
Extremely low frequency (ELF) radiation
Emergency action plan (EAP)
Emergency response plan
Engineering and administrative controls
Evacuation warning signal
Eye and face protection
First aid training programs
Foot and leg protection
Foot rails
Fumes
Gases
Hand and arm protection
Hazard communication
Hazardous noise
Hazardous waste operations (HAZWOPER)

Head protection
Hearing protection devices (HPDs)
Hearing threshold levels (HTLs)
Heat strain
Heat stress
Impulsive noise
Infrared radiation
Ionizing radiation
Irritants
Lasers
Lateral protection
Lead
Lifeline
Safety data sheets (SDSs)
Medical services and first aid
Microwave (MW) radiation
Mists
Monitoring audiometry
Monitoring hearing hazards
Narcotics and anesthetics
Noise measurements
Nonionizing radiation
Notification of incidents
Permissible exposure limits
Personal hearing protection device
Personal protective equipment (PPE)
Program evaluations
Rad
Radiation area
Radio frequency (RF) radiation
Record keeping
Referrals
Rem
Respirator, air filtering
Respirator, air supplying
Shoe inserts
Significant threshold shift (STS)
Smoke
Time-weighted average (TWA)
Torso protection
Ultraviolet radiation
Vapors
Visible radiation
Wet-bulb globe temperature (WBGT)

Review Questions

1. What is the major thrust of 29 CFR 1926 Subpart A?
2. Give three issues for construction companies that grow out of Subpart A.
3. What is the primary focus of 29 CFR 1926 Subpart B?
4. Give three issues for construction companies that grow out of Subpart C.
5. Describe the contents of a basic first aid training program that might be provided for construction workers.
6. In addition to providing workers with the necessary first aid training, what else should construction companies do to ensure that proper medical services are available when needed?
7. Define the following terms from the language of hearing loss:
   - Continuous noise
   - Exchange rate
   - Hearing threshold level
   - Noise-induced hearing loss
   - Time-weighted average
8. Name four factors that affect the level of risk associated with noise hazards.
9. List three engineering controls that might be used to reduce the risk of hearing loss on a job site.
10. Describe the basic components of program evaluation for a hearing loss program.
11. Name four types of ionizing radiation.
12. Describe the requirements for an evacuation warning signal for companies that transport, store, or use radioactive materials.
13. What are OSHA's requirements for notification of a radiation-related incident?
14. What is nonionizing radiation? Name four types.
15. List and define four different types of airborne contaminants.
16. Explain three classifications of airborne toxic substances.
17. What is a safety data sheet (SDS)? What does one contain?
18. In broad terms, what are the requirements of HAZWOPER?
19. What is the permissible exposure limit for airborne concentrations of lead?
20. Explain why hard hats with lateral protection are better than the traditional hard hat.
21. What questions should construction companies ask when they assess job sites for eye hazards?
22. When the eye hazard comes from molten metal, what type of protection device should be used?
23. Describe the following types of hearing protection devices:
    - Enclosures
    - Earplugs
    - Superaural caps
24. What types of clothing are available for torso protection?
25. What material might be used for making work gloves to provide maximum protection against abrasions and cuts? What material might be used for making work gloves to provide maximum protection against chemicals that butyl rubber does not protect against?
26. Explain the steps to be used in matching protective measures with hazards to the feet of workers.
27. Explain three precautions that can be taken to alleviate the hazards associated with prolonged standing.
28. Describe in broad terms the requirements of OSHA's fall protection standard (29 CFR 1926 Subpart M).
29. Describe the two types of respirators.
30. List the elements of an air safety program.
31. Summarize OSHA's requirements for working over or near water.

Critical Thinking and Discussion Activities

1. Supervisor A: "I don't think we should have anyone trained in first aid on any job site. All this first aid training does is create a bunch of amateur doctors who cause more harm than good. Besides, what if one of our first aid-qualified workers gets infected with HIV while trying to help an injured worker?" Supervisor B: "I don't agree. If we wait until the EMTs arrive every time we have a worker injured, someone someday is going to die who could have been saved." Who is right in this debate? What is your opinion?
2. A manager at XYZ Construction Company is having an argument with his high school–aged son over noise at rock concerts. His son wants to go to a rock concert this weekend, but his dad won't allow it unless the young man agrees to wear some type of ear protection. The son says, "Ear plugs at a rock concert? Right. I'll be laughed out of the auditorium." The father responds, "Four or more hours of amplified music in such close quarters can result in hearing damage. No ear plugs, no concert." Who is right in this argument? What is your opinion?
3. "I'm throwing this cellular telephone and our microwave oven away this minute," says the construction safety student to her college roommate. "Why? We'll starve without our microwave, and you practically live on that cell phone," responded her roommate. "Because I don't want to die of cancer!" Is the construction safety student overreacting? What is your opinion?
4. CEO of a residential roofing business: "This fall protection standard is ridiculous. I can't afford to buy lifelines for all of my employees. Besides, they work faster without lifelines." OSHA inspector: "The lifelines won't cost you nearly as much as an injured employee who falls off a roof and breaks his neck." What is your opinion of OSHA's fall protection standard? Should residential contractors be exempted?
5. Carpenter: "Work gloves are too cumbersome. They just get in the way and I'm not going to wear them." Supervisor: "One of these days you are going to wish you had worn gloves. The inconvenience is worth the protection." Are work gloves justified, or should companies let the workers decide whether or not to wear them? What is your opinion?
6. "I think we should change the filters in the respirators before letting the next shift get started," said supervisor A. "No way. We are behind on this job already. Changing filters will take a half-hour off our work time. Besides, we don't know that any oil-based contaminants are present on this site," said supervisor B. "That's the point. We don't know that they aren't present either," said supervisor A. Should the supervisors take the time to change the filters in the respirators, or is one of them just overreacting? What is your opinion?

Application Activities

1. Contact a construction company in your community and ask how they handle medical services and first aid. Is their program acceptable? How could it be improved?
2. Conduct the research necessary to answer the following question: What is the minimum level of noise and the least amount of exposure time to it that results in probable hearing loss?
3. Conduct the research necessary to answer this question: Are construction workers at risk of cancer if they work on a job site that is under the main transmission lines for the local electrical power company? They work in these conditions for at least eight hours per day for six months.
4. Contact a construction company in your community and ask if they make SDSs available to workers on any of their job sites. Get a copy of an SDS and examine it.
5. Observe the workers at several different construction sites. Are any wearing hard hats with lateral protection? Are any working without hard hats?
6. Observe the workers at a multistory job site. Are any of them using lifelines?

Endnotes

1. Retrieved January 12, 2016, from http://www.cdc.gov/niosh/96-110a.html
2. NIOSH, Appendix B, 1–6.
3. 29 CFR 1926.53.
4. Nuclear Regulatory Commission (NRC), 10 CFR Part 20.
5. Ibid.
6. Ibid.
7. 29 CFR 1910.1096(f)(i)(ii).
8. Ibid.
9. 29 CFR 1910.1096(i).
10. 29 CFR 1910.1096(j).
11. 27(b) 42 U.S.C., 2021(b).
12. 29 CFR 1910.1096(l)(ii).
13. 29 CFR 1910.1096(l)(2).
14. 29 CFR 1910.1096(m)(2).
15. 29 CFR 1910.1096(o)(i).
16. Retrieved January 12, 2016, from http://www.acgih.org
17. Ibid.

CHAPTER FOURTEEN

SUBPARTS F THROUGH J AND RELATED SAFETY PRACTICES

LEARNING OBJECTIVES

- Explain the most important aspects of Subpart F.
- Explain the most important aspects of Subpart G.
- Explain the most important aspects of Subpart H.
- Explain the most important aspects of Subpart I.
- Explain the most important aspects of Subpart J.

Subparts F through J contain many critical requirements of the Occupational Safety and Health Administration's (OSHA) construction safety standard (29 CFR 1926). Included in this chapter are requirements and corresponding practices related to the following topics: fire protection and prevention; signs, signals, and barricades; material handling, storage, use, and disposal; power and hand tools; and welding and cutting.

SUBPART F: FIRE PROTECTION AND PREVENTION AND RELATED SAFETY PRACTICES

Subpart F covers the requirements for **fire protection**, fire prevention, handling and use of flammable and combustible liquids, handling and use of liquid petroleum, and use of **temporary heating devices**. Subpart F contains the following sections:

- 1926.150 Fire protection
- 1926.151 Fire prevention
- 1926.152 Flammable and combustible liquids
- 1926.153 Liquefied petroleum gas (LP gas)
- 1926.154 Temporary heating devices
- 1926.155 Definitions applicable to this subpart
- 1926.156 Fixed extinguishing systems, general
- 1926.157 Fixed extinguishing systems, gaseous agent
- 1926.158 Fire detection systems
- 1926.159 Employee alarm systems

FIRE PROTECTION AND PREVENTION

Fire hazards are conditions that favor fire development or growth. Three elements are required to start and sustain a fire: (1) oxygen, (2) fuel, and (3) heat. Because oxygen is naturally present in most earth environments, fire hazards usually involve the mishandling of fuel or heat.

Fire, or **combustion**, is a chemical reaction between oxygen and a combustible fuel. Combustion is the process by which fire converts fuel and oxygen into energy, usually in the form of heat. By-products of combustion include light and smoke. For the reaction to start, a source of ignition, such as a spark or an open flame or a sufficiently high temperature, is needed. Given a sufficiently high temperature, almost every substance will burn. The **ignition temperature**, or **combustion point**, is the temperature at which a given fuel can burst into flame.

Fire is a chain reaction. For combustion to continue, there must be a constant source of fuel, oxygen, and heat. Look at Figure 14–1. The flaming mode is represented by the tetrahedron on the left (heat, oxidizing agent, and reducing agent) that results from a chemical chain reaction, and the smoldering mode is represented by the triangle on the right. **Exothermic chemical**

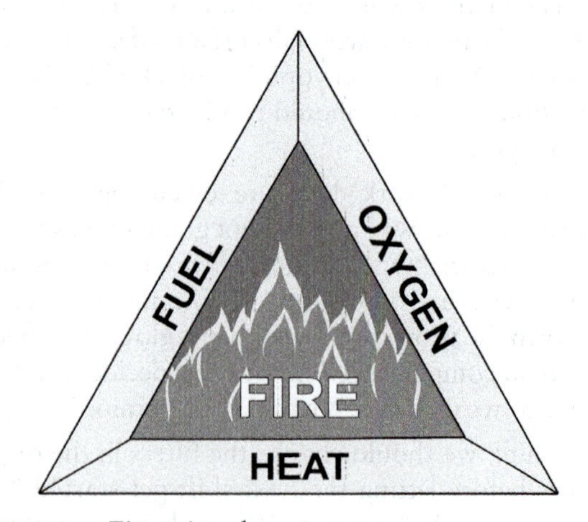

FIGURE 14–1 Fire triangle.
Source: LCosmo/Fotolia

reactions create heat. Combustion and fire are exothermic reactions and can often generate large quantities of heat. **Endothermic reactions** consume more heat than they generate. An ongoing fire usually provides its own source of heat. *Cooling* is one of the principal ways to control a fire or put it out.

Sources of Fire Hazards

Almost everything in a construction environment can burn. Metal furniture, machines, plaster, and concrete block walls are usually painted, and most paints and lacquers easily catch fire. Oxygen is almost always present. Therefore, the most effective method of fire suppression is passive—the absence of sufficient heat. Within our environment, various conditions elevate the risk of fire and so are termed *fire hazards*.

For identification, fires are classified according to their properties, which relate to the nature of the fuel. The properties of the fuel directly correspond to the best means of combating a fire (Figure 14–2). Without a source of fuel, there is no fire hazard. However, almost everything in our environment can be a fuel. Fuels occur as solids, liquids, vapors, and gases (Figure 14–3).

Flammable liquids have a flash point below 100°F. **Combustible liquids** have a flash point at or higher than 100°F. Flammable and combustible liquids can be broken down into subcategories as follows:

- *Flammable liquids.* Class 1-A (Flash point below 73°F, boiling point below 100°F), Class 1-B: (Flash point 73°F, boiling point at or above 100°F), and Class 1-C (Flash point at or above 73°F, but boiling point below 100°F).
- *Combustible liquids.* Class II (Flash point at or above 100°F, but below 140°F), Class III-A (Flash point at or above 140°F, but below 200°F), and Class III-B (Flash point at or above 200°F).

Class A	Solid materials such as wool, plastic, textiles, and their products: paper, housing, clothing.
Class B	Flammable liquids and gases.
Class C	Electrical (live electricity situations, not including fires in other materials started by electricity).
Class D	Combustible, easily oxidized metals, such as aluminum, magnesium, titanium, and zirconium.
Special categories	Extremely active oxidizers or mixtures; flammables containing oxygen, nitric acid, hydrogen peroxide; solid missile propellants.

FIGURE 14–2 Classes of fire.

FIGURE 14–3 Most construction material can be fuel for a fire.

Source: Andrew Seymour/Fotolia

The National Fire Protection Association (NFPA) has devised the NFPA 704 system for quick identification of hazards presented when substances burn (Figure 14–4). The NFPA's red, blue, yellow, and white diamond is used on product labels, shipping cartons, and buildings. Ratings within each category are 0–4, in which 0 represents no hazard and 4 the most severe hazard level. The colors refer to a specific category of hazard:

- Red = Flammability
- Blue = Health
- Yellow = Reactivity
- White = Special information

Although we do not think of electricity as burning, both natural and generated electricity play a large role in causing fires. Lightning strikes cause many fires every year. In the presence of a flammable gas or liquid mixture, one spark can produce a fire.

Electrical lines and equipment can cause fires in several ways: by a short circuit that provides an ignition spark, by arcs, or by resistances generating a heat buildup. Electrical switches and relays commonly arc as contact is made or broken.

Space heaters frequently have hot sides, tops, backs, and bottoms in addition to the heat-generating face. Hotplates, coffeepots, and coffeemakers often create heated surfaces. Many types of electric lighting generate heat, which is transferred to the lamp housing.

Engines produce heat, especially in their exhaust systems. Compressors produce heat through friction, which is transferred to their housings. Boilers produce hot surfaces, as do steam lines and equipment that use steam as power. Radiators, pipes, flues, and chimneys all have hot surfaces. Wood stock that has been cut by a blade heats up as the blade gets hot. Surfaces exposed to direct sunlight become hot and transmit their heat by

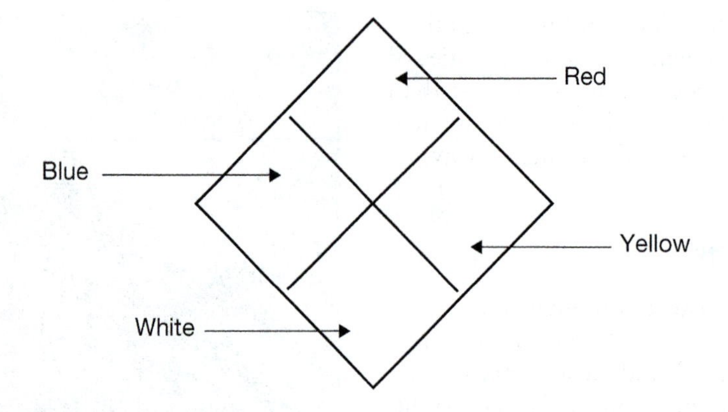

Flammability has a red background and is the top quarter of the diamond.

0 No hazard. Materials are stable during a fire and do not react with water.
1 Slight hazard. Flash point is well above normal ambient temperature.
2 Moderate hazard. Flash point is slightly above normal ambient temperature.
3 Extreme fire hazard. Gases or liquids can ignite at normal temperature.
4 Extremely flammable gases or liquids with very low flash points.

Health has a blue background and is the left quarter of the diamond.

0 No threat to health.
1 Slight health hazards. Respirator is recommended.
2 Moderate health hazard. Respirator and eye protection are required.
3 Extremely dangerous to health. Protective clothing and equipment are required.
4 Imminent danger to health. Breathing or skin absorption may cause death. A fully encapsulating suit is required.

Reactive has a yellow background and is the right quarter of the diamond.

0 No hazard. Material is stable in a fire and does not react with water.
1 Slight hazard. Materials can become unstable at higher temperatures or react with water to produce a slight amount of heat.
2 Moderate or greater hazard. Materials may undergo violent chemical reaction but will not explode. Materials react violently with water directly or form explosive mixtures with water.
3 Extreme hazard. Materials may explode if given an ignition source or have violent reactions with water.
4 Constant extreme hazard. Materials may polymerize, decompose, explode, or undergo other hazardous reactions on their own. Area should be evacuated in event of a fire.

Special information has a white background and is the bottom quarter of the diamond.

This area is used to note any special hazards presented by the material.

FIGURE 14–4 Identification of fire hazards.

conduction to their other side. Heated surfaces, such as those described here, are potential sources of fire.

Fire Safety Programs

Companies that are interested in protecting their employees from fire hazards should remember the Boy Scouts' motto: "Be prepared." The best way to be prepared is to establish a comprehensive fire safety program that encompasses all of the functional activities required for being prepared. A comprehensive fire safety program should have at least the following components: assessment, planning, awareness/prevention, and response.

An effective way to develop, implement, and maintain a fire safety program is to establish a cross-functional fire safety committee. "Cross-functional" means that it should have members from all of the companies' various functional units and all subcontractors. It should also have at least one executive-level manager to ensure and demonstrate that level of support. This approach has several advantages, including the following: (1) It focuses the eyes and ears of a broad cross section of the workforce on fire safety, (2) it ensures a broad base of diverse input, and (3) it ensures executive-level commitment.

Assessment. **Assessment** of job sites for fire hazards should be continuous and ongoing. Although the company's "competent person" has primary responsibility for this, committee members also should be involved and should involve the subcontractors or trades that they represent. Members of the safety committee should be trained in the fundamentals of fire hazard assessment. They should then pass this knowledge on to workers in their units and teams. In this way, all employees are involved in continually looking for fire hazards and in communicating their concerns to the safety committee.

Planning. OSHA requires that an organization's emergency fire safety plan have at least the following components:

- Emergency escape procedures and routes
- Critical "shutdown" procedures
- Employee headcount procedures
- Rescue and medical procedures
- Procedures for reporting fires and emergencies
- Important contact personnel for additional information

Once the plan is in place, it should be reviewed at least annually and updated as necessary.

Awareness and Prevention. After the fire safety committee has completed the emergency plan and upper management has approved it, workers must become acquainted with it. All workers should receive awareness training so that they understand their role in carrying out the emergency plan. The fire safety committee should evaluate the training program periodically, using guidelines such as the following:

- Do all workers know the role they play in implementing the emergency plan?
- How are disabled employees provided for?
- Do all workers understand the escape plans? Evacuation procedures?
- Do all new and temporary workers receive training?
- Are all workers informed when the plan is revised?
- Is a comprehensive drill undertaken at least once each year?
- Are all workers familiar with the sound of the alarm system?
- Is the alarm system checked periodically?
- Are sufficient fire detection devices in place? Are they tested periodically?
- Do all workers know the most likely causes of fires?

Response. Emergencies can occur in even the safest companies. Therefore, it is important that workers understand the emergency plan and periodically practice **response**. Just knowing what the plan says is not sufficient. People do not always think clearly in emergency situations. They will, however, do what they have learned to do through practice. Consequently, one of the fire safety committee's most important responsibilities is to arrange periodic drills so that workers can automatically respond properly.

Fire Protection Strategies

Fire is a real and present danger at any construction site. Consequently, construction companies are required by 29 CFR 1926.150 to have a fire protection program and to provide the necessary firefighting equipment at their job sites. Firefighting equipment must be properly maintained, easily accessible, and regularly inspected. Portable fire extinguishers must be maintained in accordance with NFPA document No. 10A: "Maintenance and Use of Portable Fire Extinguishers." In addition, a water supply of sufficient volume, pressure, and duration must also be available. The source of the water supply may be permanent or temporary.[1]

Fire Extinguishing Systems

Fire extinguishing systems include standpipe and hose systems, which provide the pressurized water for firefighting. Hoses for these systems usually vary from 1 inch to 2.5 inches in diameter.

Portable fire extinguishers are classified by the types of fire they can most effectively reduce. Figure 14–5 describes the four major fire extinguisher classifications. Blocking or shielding the spread of fire can include covering the fire with an inert foam, inert powder, nonflammable gas, or water with a thickening agent added. The fire may suffocate under such a covering. Flooding a liquid fuel with nonflammable liquid can dilute it and reduce its hazard potential.

Fire Brigades. Fire brigades consist of employees who have been trained to respond in the event of a fire. Requirements for fire brigades are as follows:

1. *Scope.* If an employer forms a fire brigade, the requirements in this section apply.
2. *Fire planning.* Have prefire **planning** conducted by the local fire department and the workplace fire brigade so that they may become familiar with the workplace and process its hazards. Involvement

Fire Class	Extinguisher Contents	Mechanism	Disadvantages
A	Foam, water, dry chemical	Cooling, smothering, dilution, breaks the fire, reaction chain	Freezing, if not kept heated
B	Dry chemical, bromotrifluoromethane, and other halogenated compounds, foam, CO_2, dry chemical	Chain-breaking smothering, cooling, shielding	Halogenated compounds are toxic
C	Bromotrifluoromethane, CO_2, dry chemical	Chain-breaking smothering, cooling, shielding	Halogenated compounds are toxic; chemical shielding fires may ignite after CO_2 dissipates
D	Specialized powders, such as graphite, sand, limestone, soda ash, sodium chloride	Cooling, smothering	Expensive cover of powder may be broken with resultant reignition

FIGURE 14–5 Fire extinguisher characteristics.

with the local fire department or fire-prevention bureau is encouraged to facilitate coordination and cooperation between members of the fire brigade and those who may be called upon for assistance during a fire emergency.

3. *Organizational statement.* The organizational statement should contain the following information: a description of the duties that the fire brigade members are expected to perform, the line authority of each fire brigade officer, the number of the fire brigade officers and number of training instructors, and a list and description of the types of awards or recognition that brigade members may be eligible to receive.
4. *Physical capability.* The physical capability requirement applies only to those fire brigade members who perform interior structural firefighting. "Physically capable" can be defined as being able to perform the duties specified. Physical capability can also be determined by physical performance tests or by a physical examination when the examining physician is aware of the duties that the fire brigade member is expected to perform. Employees who cannot meet the physical capability requirement may still be members of the fire brigade, as long as such employees do not perform interior structural firefighting. It is suggested that fire brigade members who are unable to perform interior structural firefighting be assigned less stressful and physically demanding fire brigade duties (e.g., certain types of training, record keeping, fire-prevention inspection and maintenance, and fire pump operations).
5. *Training and education.* Training and education must be commensurate with those functions that the fire brigade is expected to perform (i.e., those functions specified in the organizational statement). Such a performance requirement provides the necessary flexibility to design a training program that meets the needs of individual fire brigades. At a minimum, hands-on training should be conducted annually for all fire brigade members. However, for those fire brigade members who are expected to perform interior structural firefighting, some type of training or education session must be provided at least annually.
6. *Firefighting equipment.* It is important to remove from service and replace any firefighting equipment that is damaged or unserviceable. This prevents fire brigade members from using unsafe equipment by mistake. Firefighting equipment, except portable

SAFETY FACTS & FINES

Failure to promptly and properly remove construction debris from job sites can put workers at risk. A construction company in Atlanta, Georgia, was fined $108,000 for failing to properly remove debris from a job site and for allowing that debris to block entrances and exits. Specifically, the company was cited for the following: (1) poor removal of debris, (2) obstructions to entrances and exits, and (3) lack of guardrails. OSHA inspectors claimed that these violations could have led to serious injuries to workers, or even death.

The following fire protection strategies can be used on any job site:

- Assess the job site to identify all present and potential fire hazards.
- Make sure all workers have been trained to recognize, eliminate, and reduce fire hazards.
- Maintain convenient access to firefighting equipment at all job sites.
- Ensure that firefighting equipment is periodically inspected and properly maintained.
- Ensure that an adequate water source is available and that portable fire extinguishers are provided.
- Ensure that water sources and extinguishers are protected from freezing in cold environments.
- Make sure that fire extinguishers are recharged after every use.
- Ensure that fire extinguishers are not removed or tampered with, unless they are being inspected or used.

fire extinguishers and respirators, must be inspected at least annually. Portable fire extinguishers and respirators must be inspected at least monthly (see Figure 14–6).

7. *Protective clothing.* It is not the intention of the standards to require employers to provide a full ensemble of protective clothing for every fire brigade member without consideration given to the types of hazardous environments to which the fire brigade member may be exposed. It is the intention of the standards to require adequate protection for those fire brigade members who may be exposed to fires in an advanced stage or to smoke, toxic gases, and high temperatures. Therefore, the protective clothing requirements apply only to those fire brigade members who perform interior structural firefighting operations. In addition, the protective clothing requirements do not apply to the protective clothing worn during outside firefighting operations (brush and forest fires, crash crew operations) or other special firefighting activities. The protective clothing to be worn during these types of firefighting operations must protect against the hazards that may be encountered by the fire brigade members.
8. *Respiratory protective devices.* Respiratory protection must be worn by fire brigade members while working inside buildings or confined spaces where toxic products of combustion or an insufficient oxygen supply is likely to be present; respirators are also to be worn during emergency situations involving toxic substances. When fire brigade members respond to emergency situations, they may be exposed to unknown contaminants in unknown concentrations. Therefore, it is imperative that fire brigade members wear proper respiratory protective devices during these situations. In addition, there are many instances in which toxic products of combustion are still present during mop-up and overhaul operations. Therefore, fire brigade members should continue to wear respirators during these operations.

FIGURE 14–6 Portable fire extinguishers should be readily accessible and periodically inspected.

Source: Oleksandr Delyk/Fotolia

Fire-Prevention Strategies

Even when a company has taken all of the necessary precautions in terms of fire protection, the potential for fires still exists. Consequently, **fire-prevention strategies** must be implemented at all construction sites. These strategies fall into four broad categories: housekeeping, tools and equipment, flammable and combustible liquids, and electrical hazards.

Housekeeping. Housekeeping involves keeping the job site neat, orderly, and properly "picked up." This means putting things back where they belong rather than leaving them scattered around the job site. It also means that workers should clean up after themselves every day. Housekeeping strategies include the following:

1. Keep construction sites clear of scraps, refuse, spilled material, and process waste.
2. Remove scrap and debris from the job site at least daily.
3. Remove empty packing crates, plastic containers, paper wrappers, and similar items from the job site immediately (place them in a metal trash disposal container or dumpster).
4. Keep a sufficient number of metal dumpsters available to accommodate all of the trash that is generated each day (Figure 14–7).

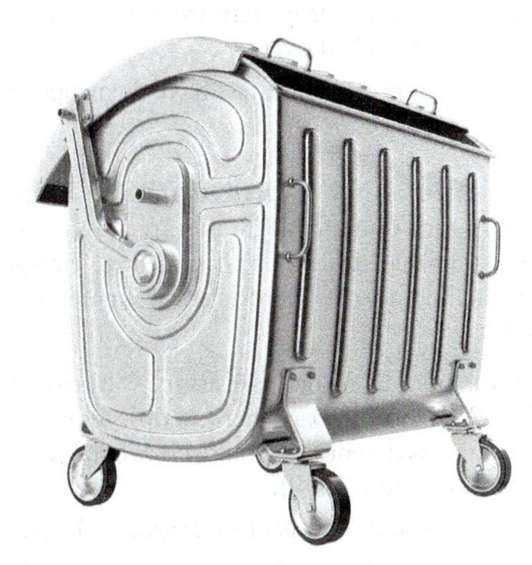

FIGURE 14–7 Combustible waste should be placed in metal bins or dumpsters.

Source: GeorgSV/Fotolia

FIGURE 14–8 Trash should be collected at least daily from job sites and placed in a metal dumpster.
Source: Studio Porto Sabbia/Fotolia

5. Place oily rags and other combustible waste materials in metal bins or dumpsters (Figure 14–8).
6. Make sure that stacked building materials do not block ingress and egress for workers or firefighters.
7. Make sure that firefighters have sufficient access to storage areas on the construction site.

Tools and Equipment. Power tools, portable generators, and portable heaters are all hazardous devices because they generate heat. Fire-prevention strategies related to tools and equipment include the following:

1. Use only nonsparking tools in areas where flammable materials are used or stored.
2. Ensure that tools, machines, and equipment are properly maintained (cleaned and lubricated regularly) and frequently inspected.
3. Keep portable heaters away from flammable or combustible materials.
4. Ensure that portable heaters are secured so that they cannot be knocked over.
5. Ensure that portable heaters have the proper guards on them.

Flammable and Combustible Liquids. Flammable and combustible liquids are among the most common fire hazards on construction sites. Fire-prevention options for storing and using these materials are as follows:

1. Store materials in flame-resistant cabinets that are isolated from places where people work. Proper drainage and venting should be provided for such cabinets (Figure 14–9).
2. Store materials in tanks below ground level.
3. Store materials on the first floor of multistory buildings.
4. Substitute less flammable supplies when possible.

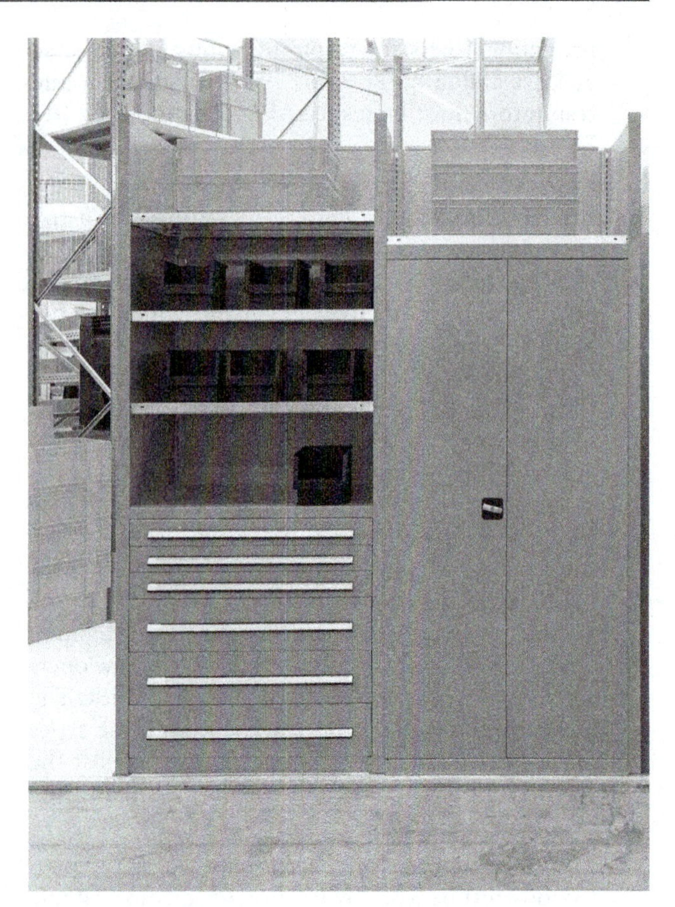

FIGURE 14–9 Flammable materials should be stored in fireproof facilities away from heat sources.
Source: Markobe/Fotolia

5. Prohibit smoking near any possible fuels.
6. Store fuels away from areas where electrical sparks from equipment, wiring, or lightning may occur.
7. Keep fuels separate from areas where there are open flames. These may include welding torches, heating elements, or furnaces.
8. Isolate fuels from tools or equipment that may produce mechanical or static sparks.
9. Clean up spills of flammable liquids as soon as they occur. Properly dispose of the materials used in the cleanup.
10. Keep work areas free from extra supplies of flammable materials (e.g., paper rags, boxes). Have only what is needed on hand, with the remaining inventory properly stored.
11. Run electrical cords along walls, rather than across aisles or in other trafficked areas. Cords that are walked on can become frayed and dangerous.
12. Turn off the power and completely deenergize equipment before conducting maintenance procedures.

Electrical Hazards. **Electrical hazards** are common on construction sites. Power tools and electrically powered machines and equipment are fundamental to modern construction techniques. Because of the potential for electric sparks and because of the heat generated by power tools and equipment, electricity is a major hazard at most job sites. Fire-prevention strategies related to electricity include the following:

1. Repair defects in electrically powered tools, machines, and equipment immediately.
2. Lock out and tag out electrically powered machines and equipment that have malfunctioned until they have been repaired.
3. Do not overload circuits.
4. Avoid the overuse of extension cords.
5. Run electrical cords along walls and ceilings so that they are not stepped on. Constant walking on electrical cords can cause them to fray.
6. Provide appropriate guards for all portable illumination devices. Locate them high and out of the way so that they are not tripped over or run into by workers.
7. Make sure that electrical equipment is turned off when not in use.

SUBPART G: SIGNS, SIGNALS, AND BARRICADES AND RELATED SAFETY PRACTICES

Subpart G covers the various requirements for using signs, signals, and barricades to warn of hazardous areas or conditions. Signs, signals, and barricades are used to warn both workers and pedestrians that hazards are present. Subpart G contains the following sections:

- 1926.200 Accident-prevention signs and tags
- 1926.201 Signaling
- 1926.202 Barricades
- 1926.203 Definitions applicable to this subpart

SIGNS AND SIGNALS

Numerous types of **signs and signals** can be used to help prevent accidents and incidents. The most widely used are as follows:

- *Danger signs.* **Danger signs** are used when the hazard is imminent and just entering the area in question is likely to result in an injury or a toxic exposure. The predominant color on a danger sign is red (Figure 14–10).
- *Caution signs.* **Caution signs** are used when an area is potentially, but not imminently, dangerous. The predominant color on a caution sign is yellow (Figure 14–11).
- *Exit signs.* **Exit signs** should be prominently placed and should have red letters (at least 6 inches high) on a white background.
- *Safety instruction signs.* Instruction signs should have a green upper panel with white lettering for the principal message being conveyed. Other parts of the message should consist of black letters on a white background.

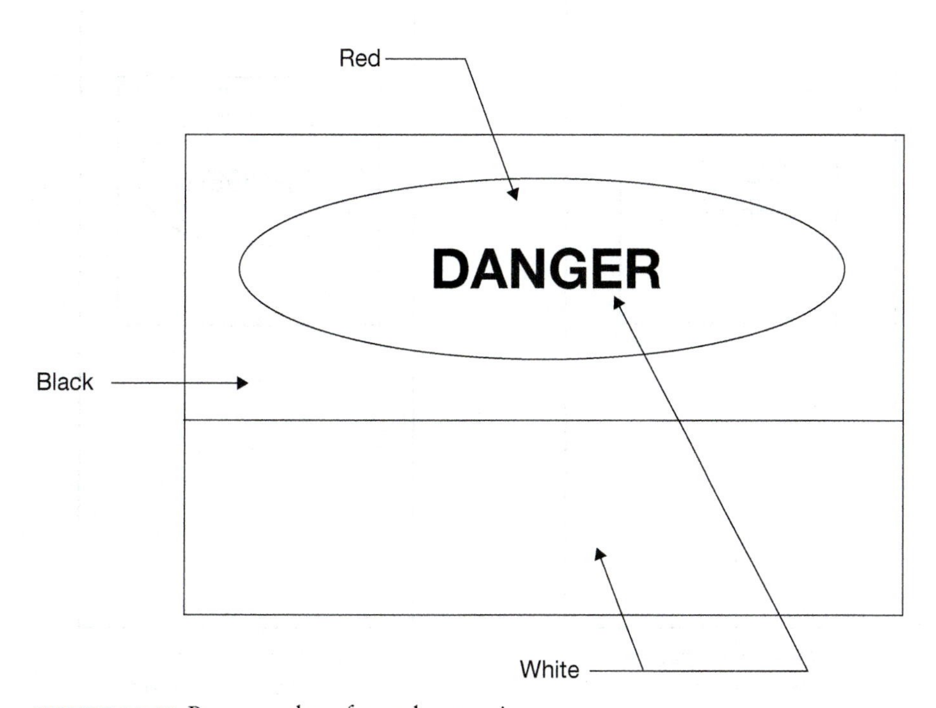

FIGURE 14–10 Proper colors for a danger sign.

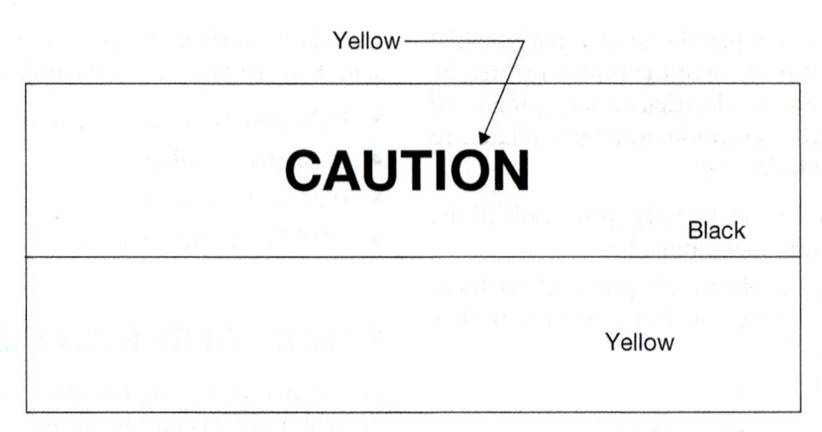

FIGURE 14–11 Proper colors for a caution sign.

- *Directional signs.* **Directional signs** should have the principal message in white on a black background (white directional symbol). Additional messages on the sign should consist of black letters on a white background.
- *Traffic signs.* **Traffic signs** must conform to ANSI D6.1: *Manual on Uniform Traffic Control Devices for Streets and Highways*. Traffic signs should be posted or placed at the very beginning of the hazardous area in question (Figure 14–12).
- *Accident-prevention signs.* **Accident-prevention tags** are used for temporarily "tagging out" tools, machines, and equipment that are defective (Figure 14–13).
- *Out-of-order tags.* **Out-of-order tags** are used to temporarily "tag out" tools, machines, and equipment that are in need of repair or maintenance (Figure 14–14).

FIGURE 14–12 Traffic signs at construction sites must conform to ANSI D6.1.
Source: Djvstock/Fotolia

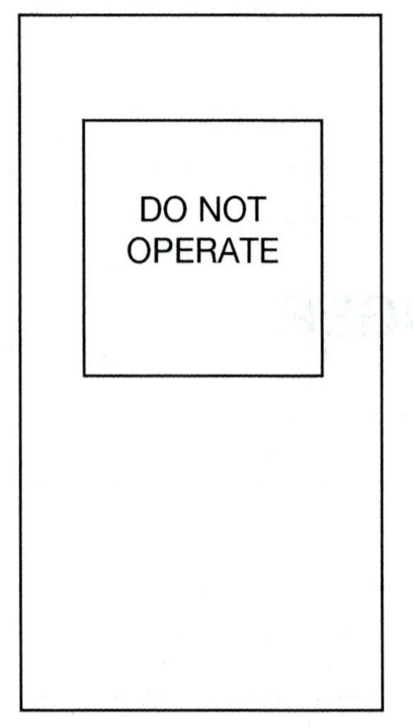

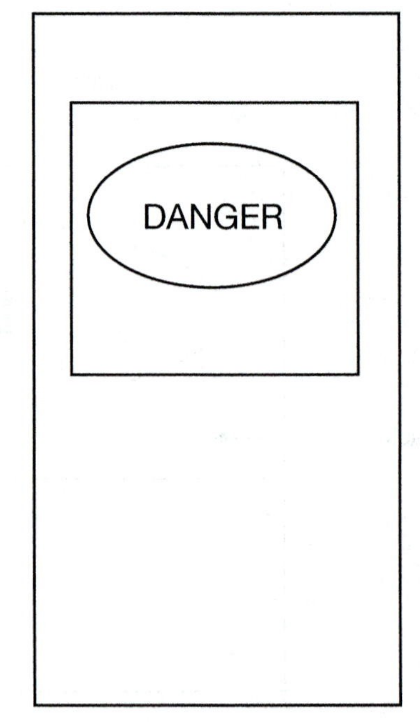

FIGURE 14–13 Specifications for accident-prevention tags.

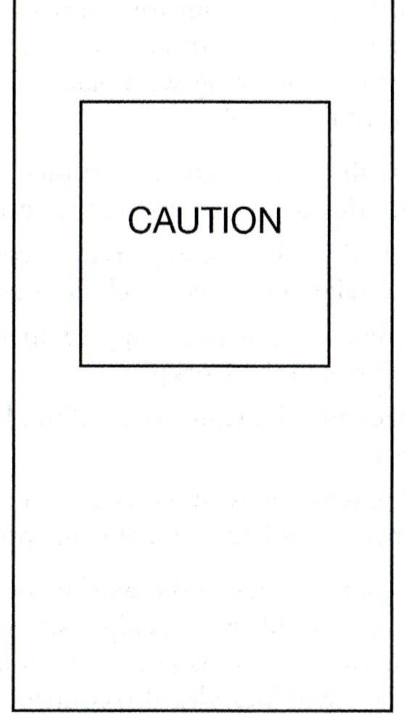

FIGURE 14–14 Specifications for accident-prevention signs.

BARRICADES AND BARRIERS

Barricades and barriers are not used just to warn people to stay out of an area, but to actually prevent them from entering. Barricades used on construction sites must conform to ANSI D6.1: *Manual on Uniform Traffic Control Devices for Streets and Highways*. Barricades may be made of a variety of materials, but the most commonly used—jersey barriers—are made of reinforced concrete. Barriers may be less substantial than barricades. They warn people of hazardous areas or conditions. Barriers can consist of warning tape, plastic fencing, wire ropes, or rubber cones, Figure 14–15.

FIGURE 14–15 Sample barrier.
Source: Roman_23203/Fotolia

SUBPART H: MATERIAL HANDLING, STORAGE, USE, AND DISPOSAL AND RELATED SAFETY PRACTICES

Subpart H covers the requirements for handling, storage, use, and disposal of materials. Of particular importance in this subpart are the requirements for rigging, stacking, racking, and securing materials to keep them from slipping or falling. Also covered are the requirements for the safe and healthy disposal of waste materials, such as wood scraps, nails, and other by-products of construction processes. Subpart H contains the following sections:

- 1926.250 General requirements for storage
- 1926.251 Rigging equipment for material handling
- 1926.252 Disposal of waste materials

HANDLING AND STORING MATERIALS

Handling material is one of the most dangerous tasks construction personnel perform. Material handling incidents are a factor in almost 40 percent of the lost-time accidents in the construction industry. Such accidents often occur as the result of poor handling techniques, insufficient planning, poor rigging, and improper storage techniques. Consequently, safe **handling and storing materials** strategies are important. This section explains various techniques that can be used to ensure safe handling and storage of materials at construction sites.

Handling Techniques and Concerns

Construction workers need to know that handling materials is a hazardous activity. They also need to know the types of injuries most commonly associated with material handling. These injuries and their most common causes are shown in Figure 14–16. The most serious of the injuries associated with handling materials is back injury. Back injuries that result from improper lifting are among the most common in the workplace. In fact, back injuries account for approximately $12 billion in workers' compensation costs annually. The following statistics concerning workplace back injuries show the scope of the problem.

- Of all workers' compensation claims, lower back injuries account for 20–25 percent.
- Of all workers' compensation costs, 33–40 percent are related to lower back injuries.
- Every year, there are approximately 46,000 back injuries in the workplace.
- Back injuries cause 100 million lost workdays each year.
- Approximately 80 percent of the population experience lower back pain at some point in their lives.[2]

Back injuries in the workplace are typically caused by improper lifting, reaching, sitting, and bending. Lifting hazards (such as poor posture), ergonomic factors, and personal lifestyles also contribute to back problems. Consequently, a company's overall safety and health program should have a back safety and proper lifting component.

Common Material Handling Injuries and Their Causes

Injury Cause	Cause
• Back strains, sprains	• Improper lifting techniques and carrying loads that are too large and bulky
• Fractures, bruises	• Flying or falling material and catching hands in pinch points
• Cuts, abrasions	• Improperly secured, stored, or stacked materials

FIGURE 14–16 Material handling injuries that are common in construction.

Back Safety and Lifting Educational Program. Prevention is critical in back safety. Consequently, construction professionals need to know how to establish back safety programs that overcome the hazards of lifting and other activities. The following strategies are essential:

- *Display poster illustrations.* Posters that illustrate the proper body mechanics for lifting, reaching, sitting, and bending techniques can be displayed strategically throughout job sites.
- *Preemployment screening.* Preemployment screening can identify people who already have back problems when they apply. More than 40 percent of back injuries occur in the first year of employment, and the majority of these injuries are lifting related.
- *Regular safety inspections.* Periodic inspections of job sites can identify potential problem areas so that corrective action can be taken immediately. Occasionally bringing a worker's compensation consultant in to assist with an inspection can help identify hazards that company personnel may miss.
- *Education and training.* Education and training designed to help employees understand how to lift, bend, reach, stand, walk, and sit safely can be the most effective preventive measure undertaken. Companies that provide back safety training report a significant decrease in back injuries.
- *Use external services.* A variety of external health-care agencies can help companies extend their programs. Local health-care-providing agencies and organizations, the services they can provide, and a

contact person in each agency should be identified. Maintaining a positive relationship with these external service people can increase the services available to employers.

- *Map out the prevention program.* The first five steps should be written down and incorporated in the company's overall safety and health program. The written plan should be reviewed periodically and updated as needed.

Proper Lifting Techniques. Back injuries are unfortunately common in the construction industry. What is even more unfortunate is that the majority of them can be prevented by observing some simple rules of thumb relating to proper lifting. The cut-off point for what workers should be allowed to lift without assistance is 51 pounds. However, even at that weight or below, objects should meet the following criteria:

- The object should be within 7 inches of the front of the worker's body.
- The object should be waist high and directly in front of the worker's body.
- Lifting should require no twisting of the worker's body.
- The object should have a handle(s) on it.
- Contents of the object cannot shift once it is lifted.

When trying to lift objects that do not meet the criteria listed above, workers should be required to: (1) Get help from another worker(s), (2) decrease the weight of the object if that is possible, or (3) use some type of mechanical device for lifting such as a hand truck. Even when these rules of thumb are followed to the letter, it is important to ensure that workers have been properly trained concerning safe lifting. One of the most effective ways to prevent back injuries is to teach workers **proper lifting techniques**. The following lifting safety instructions should be taught as part of a company's safety program.

Plan Ahead

- Determine if you can lift the load. Is it too heavy or too awkward?
- Decide if you need assistance.
- Check your route to see whether it has obstructions and slippery surfaces.

Lift with Your Legs, Not Your Back

- Bend at your knees, keeping your back straight.
- Position your feet close to the object.
- Center your body over the load.
- Lift straight up smoothly; don't jerk.
- Keep your torso straight; don't twist while lifting or after the load is lifted.
- Set the load down slowly and smoothly with a straight back and bent knees; don't let go until the object is on the floor.

Push, Don't Pull

- Pushing puts less strain on your back; don't pull objects.
- Use rollers under the object whenever possible.

Storage Techniques and Concerns

Improperly storing construction materials can and often does cause accidents. Consequently, **storage techniques and concerns** are important. The following techniques can be used for reducing the hazards associated with string and stacking materials:

1. When storing materials in tiers, they must be secured to prevent slipping, sliding, or collapsing of the stack. There are several techniques for doing this, including stacking, racking, blocking, and interlocking.
2. Do not store incompatible materials together (e.g., bags of cement stacked with lumber or bricks).
3. Bagged materials should be cross-keyed at least every 10 bags high and stacked by stepping back the layers.
4. Brick stacks are not to exceed 7 feet in height. Loose brick stacks should be tapered 2 inches in every foot of height above the 4-foot level.
5. Masonry blocks stacked higher than 6 feet should be tapered back one-half block per tier above the 6-foot level, Figure 14–17.
6. Lumber should be stacked on a level surface with solidly supported sills so that it is stable and self-supporting.

FIGURE 14–17 Stacking of masonry blocks.
Source: Mariusz Niedzwiedzki/Fotolia

7. Lumber that is handled mechanically may be stacked to a limit of 20 feet in height. Lumber that is to be handled manually is limited to a stacking height of 16 feet. Nails must be removed from used lumber before it is stacked.
8. Structural steel, poles, bar stack, PVC pipe, and other cylindrically shaped materials should be stacked and blocked to prevent tilting or spreading (Figure 14–17).
9. Materials stored inside of a building under construction should be kept at least 6 feet away from hoist ways or floor openings. These materials should not be stored within 10 feet of a wall that does not exceed the height of the stack of stored material.
10. Maximum safe load limits of floors, expressed in pounds per square foot, should be posted conspicuously within buildings and other types of structures (unless the floor is on grade).
11. Materials are not to be stored on scaffolds or walkways except in the amounts needed immediately for operations that are under way. Such amounts should never exceed the maximum safe load for the scaffold or walkway.
12. Aisles should be kept clear and properly maintained to allow for the safe movement of people or equipment that is moving material.
13. When differences exist in elevation of working areas or roads, means are to be provided (e.g., ramps, blocking, grading) to accommodate movement between the levels.
14. Workers must use personal fall arrest systems (Subpart M) whenever they are required to work on stored material in silos, hoppers, tanks, or other storage areas in which height is a factor.
15. Areas in which housekeeping equipment and supplies are stored should be kept clear of accumulated materials that might create hazards from tripping, fire, explosions, or pests.
16. Vegetation should be controlled around stored materials to prevent the introduction of hazards.
17. When portable and powered dock boards are used, the load should never exceed their maximum safe capacity. Portable dock boards must be secured in position either by anchors or by devices designed to prevent slippage. Portable dock boards must be equipped with handholds. Protective techniques must be applied to prevent movement of railroad cars while dock boards or bridge plates are in position.
18. Workers should be attentive to proper lifting techniques when handling materials manually (see previous section). Use handles or holders to reduce the pinching hazards. Gloves and other protective clothing should be worn when handling materials with sharp or rough edges (Figure 14–18).

FIGURE 14–18 Work gloves can prevent hand injuries when handling sharp or rough material.
Source: Sondem/Fotolia

RIGGING OF MATERIALS

Poorly or improperly rigged material can create one of the most dangerous situations workers can face on a construction site. The rigging of materials requires special skills not possessed by just any construction worker. To rig materials properly, an individual (competent person) must know two things: (1) proper rigging procedures and (2) how to select the right equipment for the job. In addition to having this knowledge, it is important that the competent person inspect rigging equipment before and during each use. Rigging equipment is to be loaded in accordance with Tables H-1 through H-20 in 29 CFR 1926.251 (Subpart H). Figure 14–19 shows a portion of Table H-19 from Subpart H, which contains information on safe working loads for shackles. When not in use, rigging equipment should be removed from the work area and properly stored to prevent the introduction of hazards.

Loads must be rigged so that they are level; environmental conditions such as wind can affect the status of the load. The competent person should inspect all slings, fasteners, and attachments before and during every use.

SAFE WORKING LOADS FOR SHACKLES

Material size (inches)	Pin diameter (inches)	Safe working load (tons)
1/2	5/8	1.4
5/8	3/4	2.2
3/4	7/8	3
7/8	1	4.3
1	1⅛	5.6
1⅛	1¼	6.7
1¼	1	8.2

FIGURE 14–19 Portion of Table H-19 of 29 CFR 19.26.251. Ton weight used is 2,000 lbs.
Source: From Occupational Safety & Health Administration. Published by U.S Department of Labor.

Types of Rigging Materials

The most widely used rigging materials are as follows: welded alloy steel chains, wire ropes, synthetic rope, web slings, and shackles and hooks.

Welded Alloy Steel Chains. **Welded alloy steel chains** must have permanently affixed, durable identification information, including size, grade, rated capacity, and sling manufacturer. Such chain slings must conform to the values shown in Table H-1. Whenever any point of any chain exceeds the wear limits specified in Table H-2, it must be removed from service.

Wire Rope. **Wire rope** and wire rope slings must conform to the values set forth in Tables H-3 through H-14. When the sizes, classifications, and grades of wire rope are not covered in one of these tables, the manufacturer's specifications may be used, provided that a safety factor of 5 is added in.

Web Slings. **Web slings** made of synthetic material must be permanently marked with the manufacturer's name, stock number, rate loads for various types of hitches, and type of material. Rated capacities must not be exceeded. The safety factor of web slings should be 5–1 in accordance with ANSI B30.9A. The slings must be load tested to at least twice the rated load. Nylon, polyester, and polypropylene web slings, as well as slings with aluminum fittings, should not be used where caustic materials in any form (e.g., fumes, vapors, sprays, mists, liquid acids) are present.

Shackles and Hooks. The manufacturer's specifications should be used to determine the safe loads for various types of **shackles and hooks**. If manufacturer's specifications are not available, the company should test hooks to twice the planned safe working load and keep accurate, dated records of the tests (Figure 14–20).

FIGURE 14–20 Hooks should be tested to twice the planned safe workload.
Source: Petovarga/Fotolia

DISPOSAL OF WASTE MATERIALS

Proper **disposal of waste materials** on construction sites is critical. Failure to properly dispose of waste materials is one of the most frequently cited violations for which OSHA fines or otherwise penalizes construction companies. The following requirements should be observed when disposing of waste materials on job sites:

1. Use a chute of wood or an equivalent material any time waste materials are dropped a distance of 20 feet or more to a location outside an exterior wall of a building or structure. A chute is a slide enclosed on all four sides through which material can be moved from a higher to a lower place.
2. When waste material is dropped through holes in the floor, the place where it lands should be enclosed by barricades not less than 42 inches high and not less than 6 feet back from the edge of the landing site. Signs warning of the hazards of falling material must be posted on all levels where the hazards are present. Waste materials should not be removed from the landing zone until they are no longer being dropped from above.
3. All scrap lumber, waste material, and rubbish should be removed from the immediate work area as work progresses (but not while it is still being thrown down from one higher elevation to a lower elevation). When waste materials are burned, the burning must comply with all applicable local fire regulations.
4. All solvent waste, oily rags, and flammable materials must be kept in fire-resistant covered containers until removed from the construction site.

SAFETY FACTS & FINES

Failure to keep building exits open and free of impediments can be costly. A company in Bronx, New York, was fined $111,150 for failing to keep fire exits unlocked and unobstructed. As a result of this failure, employees would have had no means of escape in the event of a fire or some other type of emergency requiring evacuation. OSHA inspected the facility after receiving an anonymous tip about locked fire exits.

SUBPART I: HAND AND POWER TOOLS AND RELATED SAFETY PRACTICES

Hand and power tools are the stock-in-trade of the construction industry. Subpart I covers the safe use of hand and power tools and applies to those owned by the company, by employees, and by subcontractors. The principal focus of this subpart is proper guards for tools. Subpart I contains the following sections:

- 1926.300 General requirements
- 1926.301 Hand tools
- 1926.302 Power-operated hand tools
- 1926.303 Abrasive wheels and tools
- 1926.304 Woodworking tools
- 1926.305 Jacks—lever and ratchet, screw, and hydraulic
- 1926.306 Air receivers
- 1926.307 Mechanical power transmission apparatus

PNEUMATIC TOOLS

Pneumatic tools are powered by compressed air. Pneumatic tools that are widely used in construction include the following: hammers, drills, sanders, and chippers (Figure 14–21). No person should be allowed to use a pneumatically powered tool without first receiving the necessary training. The following safety procedures apply when using such tools:

1. Pneumatic tools that operate at pressures in excess of 100 psi and that shoot fasteners (e.g., nails, staples) must be equipped with a special protective device to prevent the ejection of fasteners when the muzzle of the tool is anywhere except pressed against the work surface.

FIGURE 14–21 Example of a pneumatic tool that is widely used in construction.
Source: Ftfoxfoto/Fotolia

2. Workers should always wear the appropriate eye and face protection devices when using pneumatic tools.
3. Pneumatic tools should be pointed only at the work surface, never at another person.
4. Pneumatic tools that shoot fasteners must be equipped with a safety clip or retainer to prevent attachments from being unintentionally shot from the tool.
5. Pneumatic tools should be checked continually to ensure that they are securely fastened to the air hose.
6. All air hoses for pneumatic tools more than one-half inch in inside diameter must be equipped with a device to reduce pressure in the event of a hose failure.
7. When operating a pneumatic jackhammer, workers must wear the appropriate personal protective gear (eye, face, feet, and hearing protection).
8. If compressed air is used for cleaning purposes on anything other than concrete forms, mill scale, or similar situations, the compressed air pressure must be reduced to less than 30 psi. Also, the manufacturer's safe operating pressure for hoses, valves, and filters must not be exceeded.
9. Airless spray guns that atomize paints or other fluids at 1,000 psi or more must be equipped with an easily visible and accessible device to prevent an accidental release.

POWDER-ACTUATED GUNS

Only properly trained workers should be allowed to handle or use **powder-actuated guns**. Such a tool is exactly what its name implies—a gun—and should be treated as such. The following safety procedures apply when using powder-actuated guns:

1. All powder-actuated guns used in construction must meet the requirements of ANSI A10.3: "Safety Requirements for Explosive-Actuated Fastening Tools."
2. Powder-actuated guns are not to be used in an environment in which the atmosphere is explosive or flammable.
3. Before using a powder-actuated gun, it should be inspected thoroughly to make sure that it is safe and operable.
4. Powder-actuated guns should be pointed only at the work surface, never at another person or piece of property.
5. Powder-actuated guns should be loaded only when they are ready for immediate use. They should be unloaded when not in use.

WARNING!
Powder-Actuated Guns in
Use in This Area

FIGURE 14–22 Warning signs should be posted when powder-actuated guns are in use.

6. Appropriate personal protective equipment (PPE) should be used when operating powder-actuated guns, including eye and face protection.
7. When a powder-actuated tool misfires, it should be held in the operating position for a full 30 seconds. At this point, it should be fired again. If the tool still misfires, it should be held in the operating position for another 30 seconds and then unjammed according to the manufacturer's specifications.
8. Defective or inoperable tools should be put out of service and properly tagged out until repaired.
9. When powder-actuated tools are being used, warning signs should be posted conspicuously in the area (Figure 14–22).
10. Improper use of a powder-actuated gun is sufficient cause for the revocation of an operator's certification card.
11. Powder-actuated guns should be stored in a container by themselves, never mixed with other tools.

POWER TOOLS (GENERAL)

Safety precautions apply when using any kind of **general power tools**. The main precautions that should be observed when using power tools are as follows:

1. Workers should always wear the appropriate PPE to protect their eyes, face, hands, and feet.
2. Damaged or worn power tools should be removed from service and tagged out appropriately.
3. Power tools should not be used without the proper guards in place (Figure 14–23), and all power tools should be equipped with a deadman switch.
4. The following types of hand-held power tools should be equipped with a positive "on-off" control: platen sanders, small grinders (with wheels of 2-inch diameter or less), planers, laminate trimmers, nibblers, shears, scroll saws, and jigsaws (with blades that are one-quarter of an inch wide or less).
5. The following types of hand-held power tools should be equipped with a momentary contact "on-off" control: drills, tapers, fastener drivers, and grinders with wheels more than 2 inches in diameter.
6. All rotating or reciprocating tools must be guarded to prevent contact by a worker (Figure 14–23). Guards must meet the specifications set forth in ANSI B15.1.

FIGURE 14–23 Notice the guard on the hand-held power saw.
Source: Tyler Olson/Fotolia

7. Power tools without a positive means of accessory holding must be equipped with a constant pressure switch that shuts the tool down if the pressure is released.

POWER TOOLS (ELECTRICAL)

The various precautions listed in the previous section apply to hand-held **electrical power tools**. In addition, the following precautions apply:

1. All electrical power tools should be equipped with an approved double-grounding capability or be grounded in accordance with the requirements set forth in 29 CFR 1926 Subpart K.
2. Electrical power tools should be equipped with a three-wire cord with ground and be grounded and double insulated *or* be powered by a low-voltage isolation transformer.
3. A ground fault circuit interrupter should be used, or the tool must be double insulated.
4. The third prong on an electrical plug should not be removed under any circumstances.
5. Electrical power tools should be operated in accordance with their manufacturer's specifications.
6. Electrical power tools should be stored in a dry place when not in use.
7. Electrical power tools should be used only in dry conditions.
8. Proper illumination should be provided when using electrical power tools.

POWER TOOLS (HYDRAULIC)

In addition to the precautions explained earlier for power tools in general, the following precautions should be observed when using **hydraulic power tools**:

(1) The fluid used in the tools must be approved under Schedule 30 of the U.S. Bureau of Mines, (2) the fluid must be fire resistant, and it must retain its operating characteristics at the most extreme temperatures to which it will be exposed, and (3) manufacturer's specifications for safe operating pressures must not be exceeded (this applies to hoses, valves, pipes, fittings, and filters).

WOODWORKING TOOLS

A number of precautions apply specifically to woodworking tools. In general, all woodworking tools must meet applicable requirements of ANSI 01.1: "Safety Code for Woodworking Machinery" (Figure 14–24). In addition, the following precautions apply:

1. Fixed tools must be equipped with a disconnect switch that can be locked or tagged in the "off" position.
2. Operating speed must be permanently marked on circular saws with blades more than 20 inches in diameter.
3. Operating speed must be permanently marked on circular saws that operate at speeds exceeding 10,000 peripheral feet per minute.

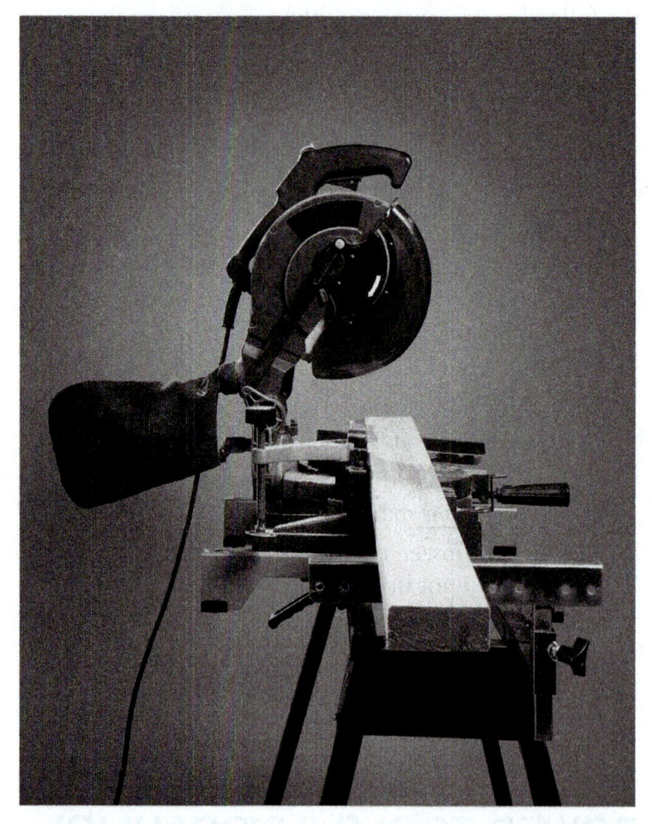

FIGURE 14–24 This woodworking tool must meet the requirements of ANSI 01.1: "Safety Code for Woodworking Machinery."
Source: Vibe Images/Fotolia

4. Saws should not be operated at speeds exceeding the maximum speed that is permanently marked on them. If a saw is retensioned for a different speed, the new speed should be permanently marked on it before use.
5. Automatic feeding devices should be used with fixed woodworking machines whenever possible. Feeder attachments should have feed rolls and other moving parts covered and properly guarded.
6. Portable circular saws should be equipped with guards above and below the base plate or shoe. The upper guard must cover the saw to the depth of the teeth, except for the minimum allowance needed for bevel cuts. The lower guard must cover the saw to the depth of the teeth except for the minimum allowance needed for contact with the work and retraction from the work.
7. On radial saws, the upper hood must completely enclose the upper portion of the blade. It must be constructed in such a way and of such a material that it protects workers from any type of flying debris produced by the tool. The saw must be equipped with a device to guard the lower portion of the blade that automatically adjusts to the thickness of the stock.
8. The table for circular crosscut saws and circular hand-fed ripsaws must be equipped with a hood that is constructed in such a way and of such materials that it will protect workers from flying debris produced by the tool.

SUBPART J: WELDING AND CUTTING AND RELATED SAFETY PRACTICES

Subpart J covers the requirements for welding, cutting, and brazing. The principal topics covered in this subpart are as follows: gas welding and cutting, arc welding, fire prevention, ventilation requirements, handling and use of compressed gas cylinders, and welding materials. Subpart J is of critical importance for construction professionals because welding, cutting, and brazing are such potentially dangerous tasks. More than 5 percent of all eye injuries in the construction industry are associated with welding, cutting, and brazing. Subpart J contains the following sections:

- 1926.350 Gas welding and cutting
- 1926.351 Arc welding and cutting
- 1926.352 Fire prevention
- 1926.353 Ventilation and protection
- 1926.354 Welding, cutting, and heating—preservative coatings

GENERAL PRECAUTIONS

A number of general precautions should be observed by construction personnel regardless of the type of welding or cutting operation in question. These precautions are as follows:

1. No construction personnel should be allowed to undertake welding, cutting, or brazing tasks without first being properly trained and earning the proper level of qualification.
2. Before beginning any welding or related task, construction personnel should inspect the work area to identify and eliminate fire hazards.
3. Before beginning a welding or related task, construction personnel should make sure that provisions have been made for ventilation.
4. Construction personnel should wear the proper eye, face, hand, torso, and foot protection and take every precaution to avoid contact with sparks, hot slag, and hot surfaces.
5. Compressed gas cylinders should be placed and secured in an upright position, handled with extreme care, and stored at least 20 feet away from flammable materials and heat sources (Figure 14–25).
6. When storing oxygen cylinders, keep them at least 20 feet away from combustible materials and compressed gas cylinders. When this is not possible, isolate the oxygen cylinders using a barrier at least 5 feet high made of noncombustible material that has at least a 30-minute fire-resistance rating (Figure 14–26).
7. Before beginning welding or cutting tasks, remove all combustible materials from the area. If this is not possible, cover the combustible items with a noncombustible material. Also, when noncombustible items cannot be removed from the area—even when they have been properly covered—an employee should be posted near the welder as a fire watch. This worker should be equipped with a fire extinguisher and should have been trained in its proper use.
8. Construction personnel other than the welder who are working in the vicinity should wear the proper eye protection.
9. When opening valves on tanks with regulators, construction personnel should stand away from regulators (to the rear or side, never in front). An internal failure could cause the adjustment screw to fly off under high pressure, creating an impact hazard.

SPECIFIC PRECAUTIONS: GAS WELDING AND CUTTING

In addition to the general precautions explained in the previous section, a number of more specific precautions apply when gas welding and cutting tasks are being done. These precautions include the following:

1. The valve cap should always be securely in place when transporting or handling compressed gas cylinders.
2. Cylinders should be secured to a pallet, cradle, or sling board for hoisting. Never hoist cylinders freely by using slings or magnets.
3. Ensure that cylinders do not receive a sharp impact by being dropped, bumped, or struck or by any

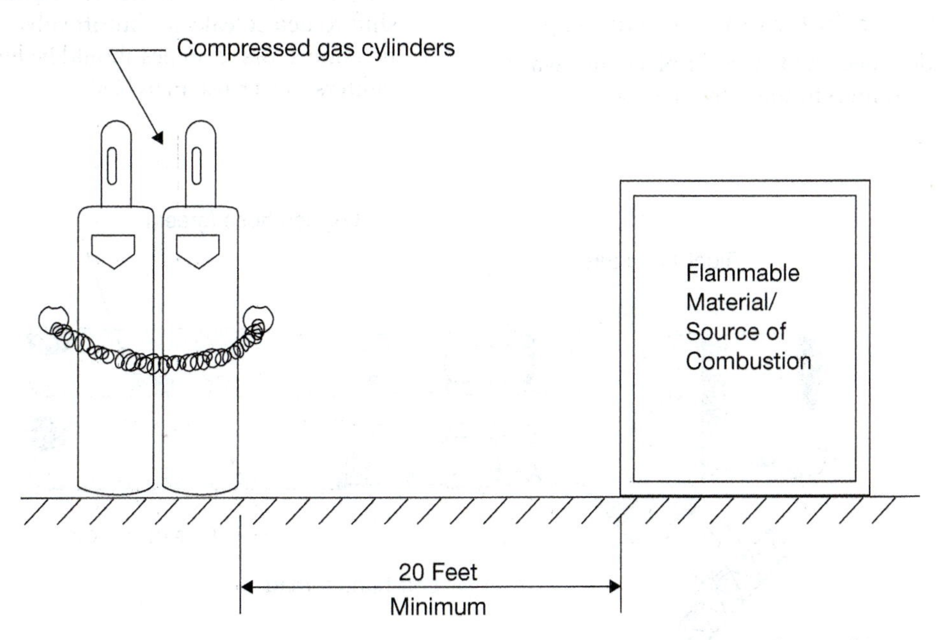

FIGURE 14–25 Compressed gas cylinders should be kept away from flammable materials.

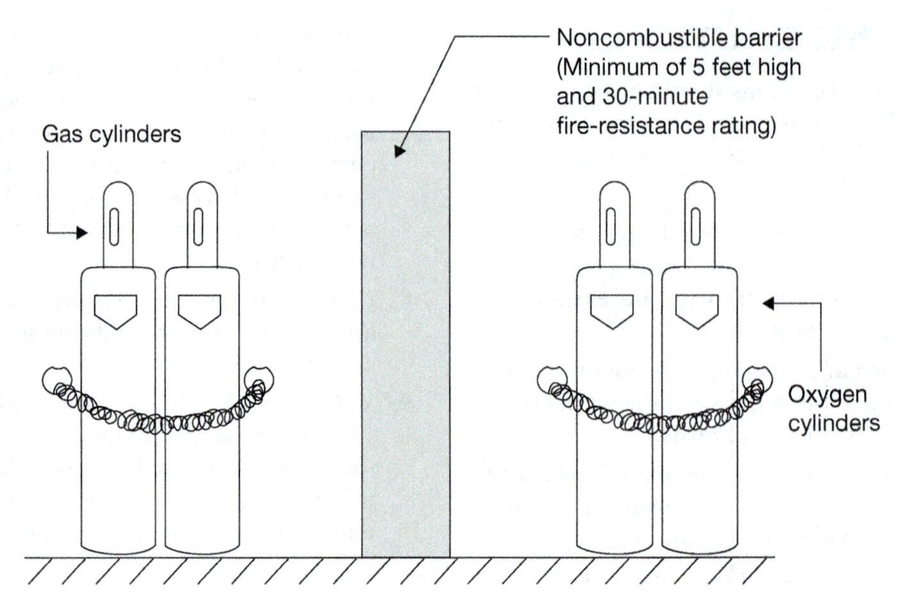

FIGURE 14–26 A noncombustible barrier can be used for isolating oxygen cylinders.

other means. If cylinders must be moved by hand, they should be tilted and rolled on their bottom edges.

4. When cylinders are to be transported by a powered vehicle, they should be secured in an upright position.
5. The fuel, gas, and oxygen hoses must be easily and clearly distinguishable from each other. They may be distinguished using different colors or different surface textures on the hoses. In addition, the oxygen and gas hoses must not be interchangeable (Figure 14–27).
6. Never lift cylinders by the valve protection caps.
7. When cylinders become frozen in place, use warm water (never boiling) to thaw them loose.
8. When a cylinder is empty, is being moved, or is not in use, the cylinder valve should be closed.
9. If oxygen and fuel hoses are taped together, not more than 4 inches of every foot of hose may be taped.
10. Gas and oxygen hoses should be inspected before every shift and removed from service if any defect is found. Hoses that show evidence of wear must be tested to twice the normal pressure and at least to 300 psi.
11. Torch tip openings should be kept unclogged and clean using appropriate devices designed for this purpose. Torches should be inspected before each shift to detect leaks in shutoff valves, couplings, and tip connectors. Torches should be lighted by friction lighters (never use matches).

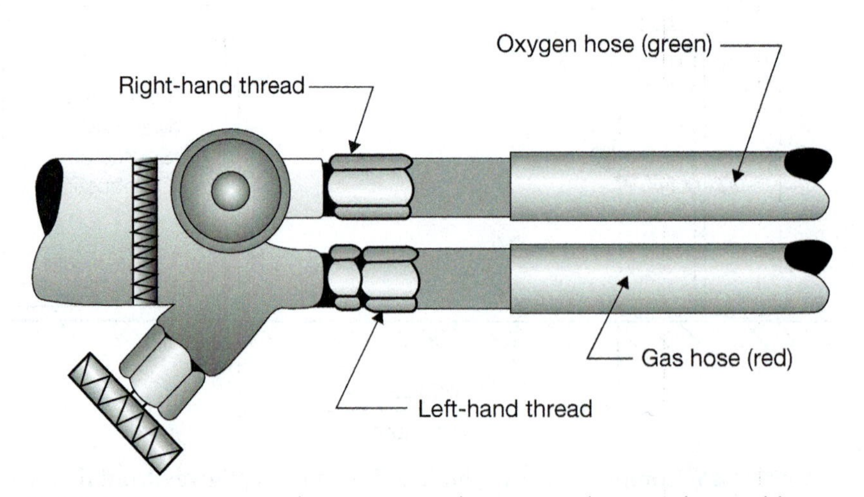

FIGURE 14–27 Hoses to the cutting torch must not be interchangeable.

SPECIFIC PRECAUTIONS: ARC WELDING AND CUTTING

Several different types of arc welding processes are used in construction; the most widely used are as follows:

- *Shielded Metal Arc Welding.* **Shielded metal arc welding,** more commonly referred to as "stick" welding, is the most widely used type of arc welding in construction. The process is based on creating an electric arc between a covered electrode and the work surface. A shielding atmosphere is created by the decomposition of the flux (the electrode covering). Filler material comes from the metal core of the electrode as it melts (Figure 14–28).
- *Gas Metal Arc Welding.* **Gas metal arc welding,** often referred to as "MIG" welding, is based on maintaining an electric arc between an electrode and the work surface and shielding the arc with a gas that may or may not be inert. The wire electrode is fed to the weld continuously, where it melts, creating the weld (Figure 14–29).
- *Gas Tungsten Arc Welding.* **Gas tungsten arc welding,** often referred to as "TIG" welding, is similar to "MIG" welding, except that the tungsten electrode is not consumed. Instead, a consumable welding rod is introduced into the process (Figure 14–30).

A number of precautions specific to these various types of arc welding should be observed by construction personnel. These precautions include the following:

1. Construction personnel should use only manual electrode holders that have been designed specifically for arc welding.
2. All parts that carry current should be properly insulated against the maximum voltage that may be encountered.
3. Cables used with arc welding equipment must be properly insulated, flexible, and equipped to handle the maximum amount of current required for the job in question.
4. All arc welding equipment and cables should be inspected carefully before each shift. Defective equipment or cables should be removed from service immediately.
5. Arc welding operations should be shielded appropriately to protect other personnel in the area from eye damage or flying debris.

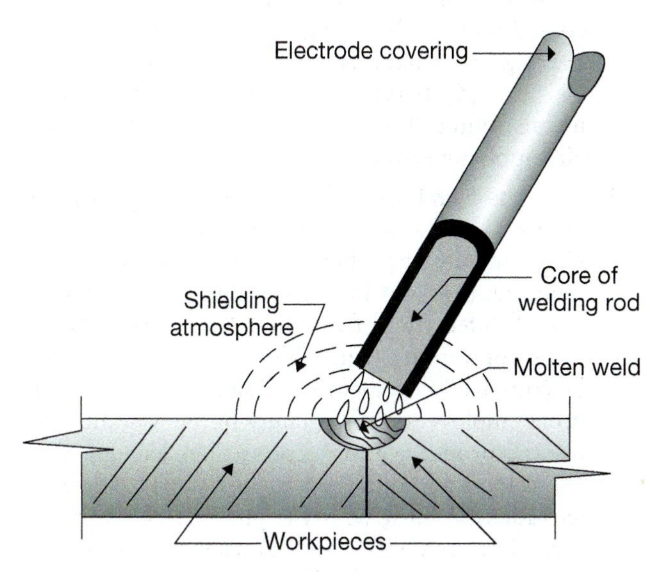

FIGURE 14–28 Shielded metal arc welding, or "stick" welding.

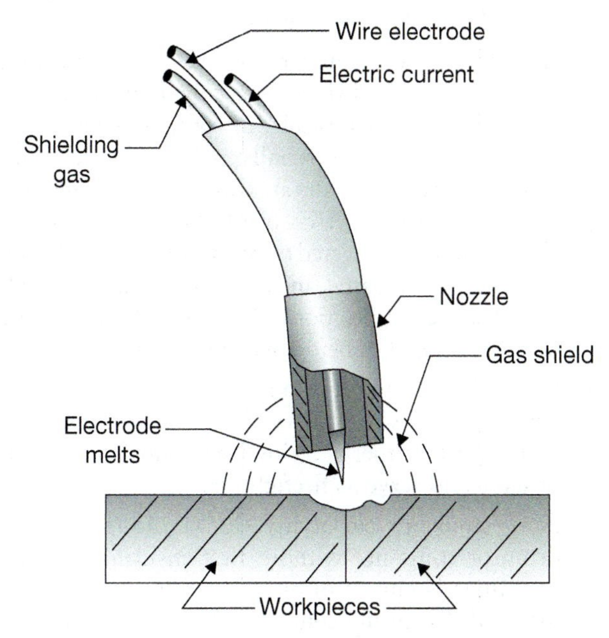

FIGURE 14–29 Gas metal arc welding, or "MIG" welding.

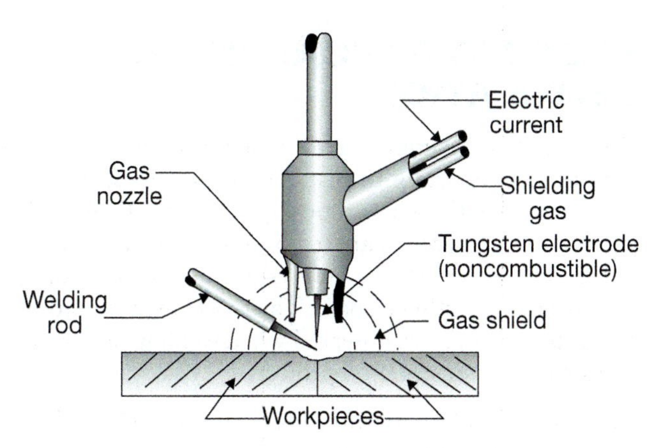

FIGURE 14–30 Tungsten metal arc welding, or "TIG" welding.

SPECIFIC PRECAUTIONS: FIRE PREVENTION

Fire is always a concern when welding or cutting is taking place on a construction site. Many of the materials used in construction are flammable or combustible. Consequently, workers must be especially attentive to **fire prevention** when welding or cutting tasks are to be done. The following precautions should be observed:

1. Make sure that an operable fire extinguisher is nearby and easily accessible.
2. Use only friction lighters to light welding and cutting torches. Never use a device that produces flame (e.g., matches or a cigarette lighter).
3. Keep any type of ignition source away from gas cylinders (including welding arcs).
4. Create a safe environment for welding by moving material to be welded to a specially designated location. If this is not possible, move flammable materials and objects—especially paints, thinner, and solvents—away from the area where the welding will take place. Any material or objects that cannot be moved should be covered or otherwise protected from the heat, sparks, flame, and arcs associated with welding. Heavy concentrations of dust should be cleaned up and disposed of before welding tasks begin.
5. When there is any question as to the safety of the environment where the welding or cutting takes place, post properly trained personnel nearby to observe and serve as firefighters.
6. Never apply a weld or any kind of heat to a metal drum or container without first ensuring that the container is empty of any potentially flammable material. Clean the container thoroughly if it contained flammable or toxic liquids. Also, vent the container and leave an opening to release any built-up pressure that might occur as heat is applied.

SPECIFIC PRECAUTIONS: VENTILATION

Ventilation is essential when welding or cutting. In any setting, it is important to ensure that the fumes, smoke, and other potentially hazardous by-products of the welding or cutting processes are removed by proper and adequate ventilation. This is especially true when welding or cutting in a confined space. The following ventilation precautions should be observed when welding or cutting takes place in construction projects:

1. Mechanical ventilation may be either general or local exhaust systems.
2. Adequate ventilation produces the number of air exchanges necessary to keep welding fumes and smoke within the acceptable breathing limits set forth in 29 CFR 1926 Subpart D: "Occupational Health and Environmental Controls."
3. When using local exhaust ventilation, the welder must be able to move the exhaust hood so that it is placed as close as possible to the welding surface. Local systems must hold fumes and smoke to levels within the safe breathing range set forth in 29 CFR 1926 Subpart D: "Occupational Health and Environmental Controls."
4. Air that is contaminated by welding or cutting processes must be removed to the open air or otherwise kept removed from the source of air intake.
5. All replacement air must be clean, respirable, and within the safe breathing range.
6. Pure oxygen must not be used for ventilation, cooling, or cleaning (including cleaning of clothing, the welding area, or the welded material).
7. When welding or cutting in a confined space, all of the precautions just described apply. However, when adequate ventilation cannot be achieved without blocking the means of egress, workers must be provided with air supplying respirators as described in 29 CFR 1926 Subpart E: "Personal Protective and Life Saving Equipment." In addition, an employee must be stationed outside the opening of the space close enough to maintain communication with the worker inside the confined space. This individual must be capable of initiating rescue procedures when necessary. When the opening to the confined space is small, the worker inside must wear a lifeline and a safety belt so that he or she can be removed from the space if necessary.
8. When welding or cutting the following materials in a confined space, welders must use air supplying respirators: (1) metals containing lead or metals coated with lead-bearing materials, (2) cadmium-bearing or coated metals, (3) mercury-coated metals, and (4) beryllium-containing metals (these metals require both local exhaust ventilation and air supplying respirators).
9. Inert gas metal arc welding produces higher levels of ultraviolet radiation than other types of welding. Consequently, the following special precautions are required: (1) Chlorinated solvents must be kept at least 200 feet away from the welding area, (2) surfaces conditioned with chlorinated solvents must be completely dry before welding, (3) welders and other employees in the vicinity of the welding must use the proper eye protection (filter lens goggles), (4) welders and other workers exposed to ultraviolet radiation must wear the proper PPE to protect their skin from burns, and (5) when welding or cutting stainless steel, local exhaust ventilation or air supplying respirators are required. Figure 14–31 contains a list of the most common potential health hazards associated with welding and cutting.

POTENTIAL HEALTH HAZARDS IN WELDING AND CUTTING
Chemical Hazards
Zinc
Cadmium
Beryllium
Iron oxide
Mercury
Lead
Fluorides
Chlorinated solvents
Phosgene
Carbon monoxide
Nitrogen oxides
Ozone
Physical Hazards
Ultraviolet radiation
Infrared radiation
Intense visible light

FIGURE 14–31 Welding introduces numerous potential hazards to the workplace.

Summary

Subpart F covers the requirements for fire protection, fire prevention, handling and use of flammable and combustible liquids, handling and use of liquid petroleum, and use of temporary heating devices. Comprehensive fire safety programs for construction companies have the following components: assessment, planning, **awareness and prevention**, and response. Fire-prevention strategies fall into the following categories of activity: housekeeping, tools and equipment, flammable and combustible liquids, and electrical hazards. Subpart G covers the various requirements for using signs, signals, and barricades to warn of hazardous areas or conditions. Signs, signals, and barricades are used to warn both workers and pedestrians. The most widely used signs and signals are danger, caution, exit, safety instruction, and directional and traffic signs. Accident-prevention and out-of-order tags are also widely used.

Subpart H covers the requirements for handling, storage, use, and disposal of materials. Of particular importance in this subpart are the requirements for rigging, stacking, racking, and securing of materials to keep them from slipping or falling. Also important are the requirements for the safe and healthy disposal of waste materials. Subpart I covers the safe use of hand and power tools and applies to those owned by the company, by employees, and by subcontractors. The major focus of this subpart is proper guards for tools. Specific precautions apply for pneumatic, electric, and hydraulic power tools as well as for woodworking tools.

Subpart J covers the requirements for welding and cutting. The principal topics covered in this subpart are as follows: gas welding and cutting, arc welding, fire prevention, ventilation requirements, handling and use of compressed gas cylinders, and welding materials. This subpart is of particular importance to construction professionals because more than 5 percent of the eye injuries in construction are associated with welding and cutting.

Key Terms and Concepts

Accident-prevention tags
Assessment
Awareness and prevention
Back safety and lifting educational program
Barricades and barriers
Caution signs
Combustible liquids
Combustion
Combustion point
Danger signs
Directional signs
Disposal of waste materials
Electrical hazards
Endothermic reactions
Exit signs
Exothermic chemical reactions
Fire extinguishing systems
Fire prevention
Fire-prevention strategies
Fire protection
Flammable liquids
Gas metal arc welding
Gas tungsten arc welding
Handling and storing materials
Ignition temperature
Out-of-order tags
Planning
Pneumatic tools
Powder-actuated guns
Power tools (electrical)
Power tools (general)
Power tools (hydraulic)
Proper lifting techniques
Response
Shackles and hooks
Shielded metal arc welding
Signs and signals
Storage techniques and concerns
Temporary heating devices
Traffic signs
Ventilation
Web slings
Welded alloy steel chains
Wire rope

Review Questions

1. Define the term *fire.*
2. What is an exothermic chemical reaction?
3. Describe the characteristics of Class C fires.
4. List the components of a comprehensive fire safety program.
5. What are the disadvantages of foam, water, and dry chemical fire extinguishers?
6. What are the requirements for fire brigades, if construction companies decide to have them?
7. List the four fire-prevention strategies under each of the following categories: housekeeping, tools and equipment, flammable and combustible liquids, and electrical hazards.
8. When should danger signs be used?
9. Describe the proper specifications for a safety instruction sign.
10. What are the elements of a back safety lifting program?
11. Describe the proper lifting techniques that should be used by construction personnel.
12. What should be done to ensure safety when masonry blocks are stacked higher than 6 feet?
13. Describe the four most commonly used rigging materials.
14. How should solvent waste, oily rags, and flammable materials be secured until removed from the construction site?
15. What are the requirements for pneumatic tools that operate at pressures in excess of 100 psi?
16. What are the requirements when using compressed air for cleaning anything other than concrete forms of mill scale?
17. How should a construction worker handle a misfire when using a powder-actuated gun?
18. What power tools are required to have a positive *on-off* control?
19. List the three specific requirements that apply to hydraulic tools.
20. Describe the type of guards that portable circular saws should have.
21. What should construction workers do if welding is required in an area where there are combustible materials?
22. Describe how the gas and oxygen hoses for gas welding and cutting torches can be made distinguishable from each other.
23. What is the proper procedure for dealing with a hose that shows evidence of wear?
24. What is gas tungsten arc welding, and how does it differ from gas metal arc welding?
25. How can a construction worker create a safe environment for welding on a construction site?

Critical Thinking and Discussion Activities

1. A construction company has just begun a new condominium, built on a vacant lot between two existing condominiums. The condominium association managers for the two existing buildings have asked for a meeting with you, the project superintendent. They are concerned about fire hazards. There was a fire during the construction of one of their condominiums. Discuss the various fire-prevention and fire protection strategies that you plan to use on this project, as if you are sharing them with these condominium association managers. Make sure to include prevention strategies related to welding, because this project requires welding.
2. The same two condominium association managers are concerned that curious owners in their buildings might wander onto the new job site and be injured. Discuss the various ways that signs, signals, and barricades will be used to prevent this.
3. While discussing fire safety and barricades with the condominium association managers, the issue of waste disposal comes up. The managers are worried that construction waste will find its way from the construction site onto the adjoining properties they manage. Discuss how construction waste will be contained and removed on this construction site.

Application Activities

1. Find a local construction company that will work with you on this activity. Obtain the company's permission to walk through a job site and conduct an assessment of fire hazards. Does the company have an emergency fire safety plan? If so, analyze the plan and critique it. Is anything missing? How would you improve it?
2. Locate a road construction project in or near your community. Park your car and observe the project during a workday. Assess the company's use of signs, signals, and barricades to warn of hazardous conditions. Could you improve the company's use of signs, signals, and barricades? If so, how?
3. Locate a construction company that will give you permission to observe as materials are rigged for handling and movement. Critique in writing,

without naming the company, the rigging procedures and equipment.

4. Observe the work at a construction site in your community to determine how the company disposes of construction waste materials. Critique the company's procedures in writing, without naming the company.
5. Observe the work at a construction site in your community to determine if hand and power tools are being used with the proper guards. Are construction workers wearing the appropriate personal protective equipment? Write up your assessment without using the company's name.
6. Locate a construction company that will give you permission to observe welding being done at one of its job sites. What type of welding is being done? Critique the safety precautions taken, and write up your assessment without using the company's name.

Endnotes

1. National Fire Protection Association, *Life Safety Code* (NFPA 101) (Quincy, MA: NFPA, 2001), 101–119.
2. National Safety Council, *Accident Facts* (Chicago: NSC, 2015), 39.

CHAPTER FIFTEEN

SUBPARTS K THROUGH O AND RELATED SAFETY PRACTICES

LEARNING OBJECTIVES

- Summarize the most important elements of Subpart K.
- Explain the most important requirements in Subpart L.
- Explain the most important requirements in Subpart M.
- Summarize the most important elements of Subpart N.
- Explain the most important elements of Subpart O.

Subparts K through O contain some of the most comprehensive requirements of 29 CFR 1926. All of the requirements are important, but the student of construction safety should pay special attention to the electrical requirements set forth in Subpart K, because electrocution is one of the leading causes of injury and death in the construction industry.

SUBPART K: ELECTRICAL REQUIREMENTS AND RELATED SAFETY PRACTICES

Subpart K covers the temporary and permanent installation and the ongoing operation of electrical power and electrically powered devices and equipment in construction projects. Subpart K does *not* apply to *permanently installed* electrical power put in place before the possible electrical confusion with current construction project. There are four major areas of emphasis in Subpart K: (1) safety requirements for installation, (2) safety-related work practices, (3) safety-related maintenance and environmental considerations, and (4) safety requirements for special equipment. Subpart K contains the following sections:

- General
  - 1926.400 Introduction
  - 1926.401 (Reserved)
  - (Note: *Reserved* means temporarily unused; held open for later use.)
- Installation Safety Requirements
  - 1926.402 Applicability
  - 1926.403 General requirements
  - 1926.404 Wiring design and protection
  - 1926.405 Wiring methods, components, and equipment for general use
  - 1926.406 Specific purpose of equipment and installations
  - 1926.407 Hazardous (classified) locations
  - 1926.408 Special systems
  - 1926.409 to 1926.415 (Reserved)
- Safety-Related Work Practices
  - 1926.416 General requirements
  - 1926.417 Lockout and tagging of circuits
  - 1926.418 to 1926.430 (Reserved)
- Safety-Related Maintenance and Environmental Considerations
  - 1926.431 Maintenance of equipment
  - 1926.432 Environmental deterioration of equipment
  - 1926.433–1926.440 (Reserved)
- Safety Requirements for Special Equipment
  - 1926.441 Battery locations and battery charging
  - 1926.442–1926. 448 (Reserved)
- Definitions
  - 1926.449 Definitions applicable to this subpart

OVERVIEW OF ELECTRICITY

Electricity is the flow of negatively charged particles, called **electrons**, through an electrically conductive material. Electrons orbit the nucleus of an atom, which is located approximately in the atom's center. The negative charge of the electrons is neutralized by particles called **neutrons**, which act as temporary energy repositories for the interactions between positively charged particles, called **protons**, and electrons.

Figure 15–1 shows the basic structure of an atom, with the positively charged nucleus in the center. The electrons are shown as energy bands of orbiting, negatively charged particles. Each ring of electrons contains a particular quantity of negative charges. The basic characteristics of a material are determined by the number of electron rings and the number of electrons in the outer

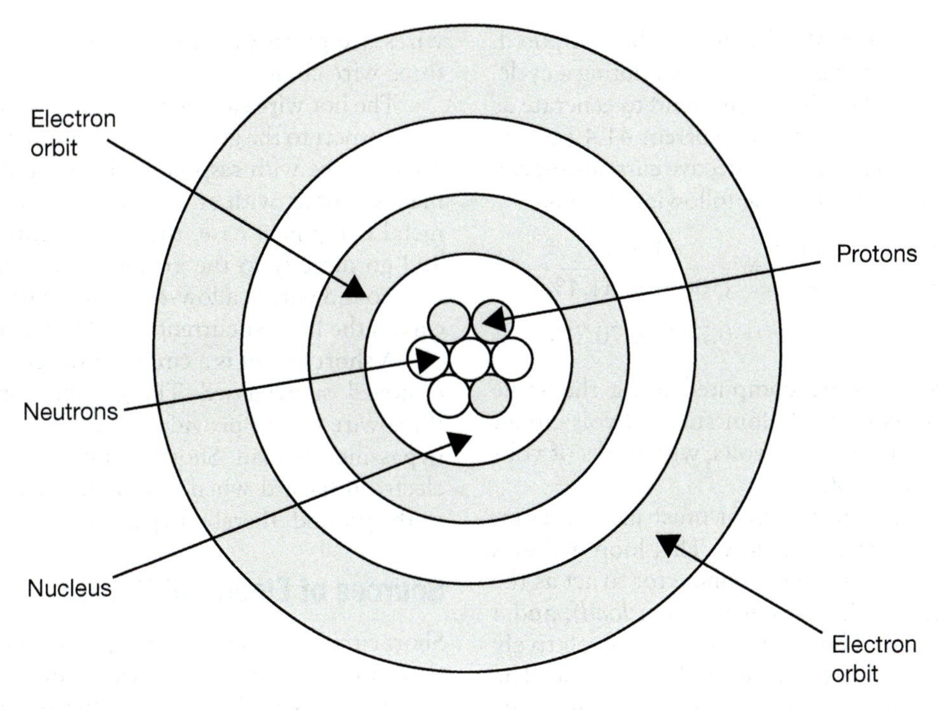

FIGURE 15–1 Structure of an atom.

rings of its atoms. A *positive charge* is present when an atom (or group of atoms) in a material has too many electrons in its outer shell. In all other cases, the atom or material carries a *negative charge.*

Electrons that are freed from an atom and are directed by external forces to travel in a specific direction produce *electrical current,* also called *electricity.* **Conductors** are substances that have many free electrons at room temperature and can pass electricity. *Insulators* do not have a large number of free electrons at room temperature and do not conduct electricity. Substances that are neither conductors nor insulators can be called **semiconductors.**

Electrical current passing through the human body causes a shock. The quantity and path of this current determines the level of damage to the body. The path of this flow of electrons is from a negative source to a positive point, since opposite charges attract one another.

When a surplus or deficiency of electrons exists on the surface of a material, **static electricity** is produced. This type of electricity is called *static* because there is no positive material nearby to attract the electrons and cause them to move. Friction is not required to produce static electricity, although it can increase the charge of existing static electricity. When two surfaces of opposite static electricity charges are brought into close range, a discharge, or spark, occurs. The spark from static electricity is often the first clue that such static exists. A common example is the spark that comes from rustling woolen blankets in dry, heated indoor air.

The *potential difference* between two points in a circuit is measured in volts. The higher the **voltage,** the more likely it is that electricity will flow between the negative and positive points.

Pure conductors offer little **resistance** to the flow of electrons. Insulators, on the other hand, have a very high resistance to electricity. Semiconductors have a medium-range resistance to electricity. The higher the resistance, the lower the flow of electrons. Resistance is measured in **ohms.**

Electrical current is produced by the flow of electrons. The unit of measurement for current is **amperes (**or *amps*). One amp is a current flow of 6.28×10^{18} electrons per second. Current is usually designated by the letter *I. Ohm's law* describes the relationship among volts, ohms, and amps. One ohm is the resistance of a conductor that has a current of 1 amp under the potential of 1 volt. Ohm's law is stated as follows:

$$V = IR$$

where

V = potential difference in volts
I = current flow in amps
R = resistance to current flow in ohms

Power is measured in wattage (or *watts*) and can be determined from Ohm's law:

$$W = VI \text{ or } W = I^2R$$

where

W = power in watts

The most common uses of electricity in construction are supplied by *alternating current* (AC). In the United States, standard AC circuits cycle 60 times per second. The number of cycles per second is known as *frequency* and is measured in *hertz.* Since voltage cycles AC current, an *effective current* for AC circuits can be computed, which is slightly less than the peak current during a cycle.

A *direct current* (DC) has been found to generate as much heat as AC that has a peak current 41.4 percent higher than the DC. The ratio of effective current-to-peak current can be determined by the following formula:

$$\frac{\text{(Effective current in amps)}}{\text{(Peak current in amps)}} = \frac{(100\%)}{(100\% + 41.4\%)}$$
$$= 0.707, \text{ or } 70.7\%$$

Effective voltages are computed using the same ratios as effective current. A domestic 110-volt circuit has an effective voltage of 110 volts, with peaks of voltage that are over 150 volts.

The path of electrical current must make a complete loop for the current to flow. This loop includes the source of electrical power, a conductor to act as the path, a device to use the current (called a *load*), and a path to the ground. The earth maintains a relatively stable electrical charge and is a good conductor. The earth is considered to have zero potential because of its massive size. Any electrical conductor pushed into the earth is said to have *zero potential*. The earth is used as a giant common conductor back to the source of power.

Electrical Hazards

Electrical hazards occur when a person makes contact with a conductor carrying a current and simultaneously contacts the ground or another object that includes a conductive path to the ground. The person's body completes the circuit loop by providing a load for the circuit and, thereby, enables the current to pass through him or her. People can be protected from this danger by insulating the conductors, insulating the people, or isolating the danger from the people.

The National Electrical Code (NEC) is published by the National Fire Protection Association (NFPA). This code specifies industrial and domestic electrical safety precautions. The NEC categorizes industrial locations and gases relative to their degree of fire hazard and describes in detail the safety requirements for industrial and home wiring. The NEC has been adopted by many jurisdictions as the local electrical code. The National Board of Fire Underwriters sponsors Underwriters Laboratories (UL). UL determines whether equipment and materials for electrical systems are safe in the various NEC location categories. UL provides labels for equipment that state what has been approved as safe within the tested constraints, which are the limitations for safe use.

Typical 110-volt circuit wiring has a *hot wire* carrying current, a *neutral wire*, and a *ground wire*. The neutral wire may be called a **grounded conductor**, with the ground wire being called a *grounding* conductor. Wires have color-coded insulation coating: neutral wires usually are white, hot wires are red or black, and ground wires are green or bare. Figure 15–2 shows a typical three-wire circuit.

The hot wire carries an effective voltage of 110 volts with respect to the ground. The neutral wire carries nearly zero voltage with respect to the ground. If the hot wire makes contact with an unintended conductor, such as a metal equipment case, the current can bypass the load and go directly to the ground. With the load skipped, the ground wire is a low-resistance path to the earth and carries the highest current possible for that circuit.

A **short circuit** is a circuit in which the load has been removed or bypassed. The ground wire in a standard three-wire circuit provides a direct path to the ground, bypassing the load. Short circuits are another source of electrical hazard when a human body is the conductor to the ground, thereby bypassing the load.

Sources of Electrical Hazards

Short circuits are one of many potential electrical hazards that can cause electrical shock. Another hazard is water, which considerably decreases the resistance of materials, including the human body. The resistance of wet skin can be as low as 450 ohms, whereas dry skin may have an average resistance of 600,000 ohms.[1] According to Ohm's law, the higher the resistance, the lower the current flow. When the current flow is reduced, the probability of electrical shock is also reduced. The major causes of electrical shock are as follows:

- Contact with bare wire carrying current; the wire's insulation may have deteriorated, or it may normally be bare
- Working with electrical equipment that lacks the UL label for safety inspection
- Electrical equipment that has not been properly grounded; failure of the equipment can lead to short circuits
- Working with electrical equipment on damp floors or other sources of wetness
- Static electricity discharge
- Using metal ladders to work on electrical equipment; these ladders can provide a direct line from the power source to the ground, causing a shock
- Working on electrical equipment without ensuring that the power has been shut off
- Lightning strikes

Electrical Hazards to Humans

The greatest danger to humans suffering electrical shock results from current flow. The voltage determines whether a particular person's natural resistance to current flow will be overcome. Some levels of current **freeze** a person to the conductor, so that the person cannot voluntarily release his or her grasp. **Let-go current** is the

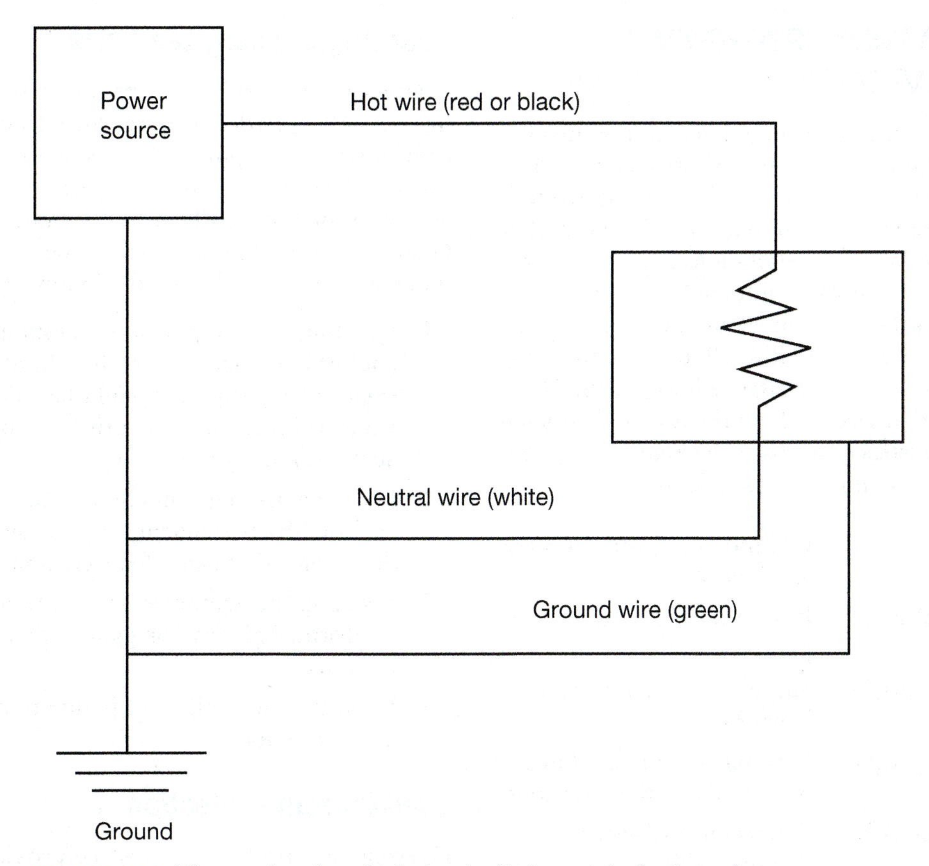

FIGURE 15–2 Three-wire electrical circuit, listing typical wire colors.

highest current level at which a person in contact with the conductor can release the grasp of the conductor. Figure 15–3 shows the relationship between amperage and danger with typical domestic 60-cycle AC.

The severity of injury with electrical shock depends not only on the dosage of current, as shown in Figure 15–3, but also on the path taken through the body by the current. The path is influenced by the resistance of various parts of the body at the time of contact with the conductor. The skin is the major form of resistance to current flow. Current paths through the heart, brain, or trunk are generally much more injurious than paths through extremities.

Electrical Dose (mA)	Effect on Human Body
<1	No sensation, no perceptible effect.
1	Shock perceptible, reflex action to jump away. No direct danger from shock, but sudden motion may cause accident.
>3	Painful shock.
6	Let-go current for women.*
9	Let-go current for men.*
10–15	Local muscle contractions. Freezing to the conductor for 2.5% of the population.
30–50	Local muscle contractions. Freezing to the conductor for 50% of the population.
50–100	Prolonged contact may cause collapse and unconsciousness. Death may occur after 3 minutes of contact due to paralysis of the respiratory muscles.
100–200	Contact for 0.25 second may cause ventricular fibrillation of the heart and death. AC continuing for more than one heart cycle may cause fibrillation.
>200	Clamps and stops the heart as long as the current flows. Heart beating and circulation may resume when current ceases. High levels of current can produce respiratory paralysis, which can be reversed with immediate CPR. Severe burns to the skin and internal organs may result in irreparable damage to body.

FIGURE 15–3 Effects of electrical current on the human body (60-cycle AC [alternating current]).

*Difference between men and women is based on the relative body mass of the "average" man and woman.

INSTALLATION SAFETY REQUIREMENTS

There are specific requirements related to the installation of electrical equipment used to provide power at construction sites.[2] These requirements apply to both permanent and temporary installations. However, they do not apply to permanent installations made before the new construction project was initiated. A construction company can comply with most of the requirements related to electricity in 29 CFR 1926 by performing work in accordance with the latest edition of the NEC. This is the code that electrical contractors and subcontractors use on construction jobs. The remaining sections of 29 CFR 1926 that apply are as follows:

- 1926.404(b)(1) — Ground-fault protection for employees
- 1926.405(a)(2)(ii)(E) — Protection of lamps on temporary wiring
- 1926.405(a)(2)(ii)(F) — Suspension of temporary lights by cords
- 1926.405(a)(2)(ii)(G) — Portable lighting used in wet or conductive locations
- 1926.405(a)(2)(ii)(J) — Extension cord sets and flexible cords

Approval, Examination, Installation, and Use of Equipment

All electrical conductors and equipment used by contractors and their subcontractors must meet the approved specifications set forth in the NEC. Before installing or using equipment, contractors must ensure that electrical equipment is approved and that it is free of identifiable hazards that might cause injuries or death. The following factors should be checked before proceeding with the installation and use of electrical equipment:

1. The listing, labeling, specifications, or certification that a given piece of equipment is suitable for the purpose for which it is to be installed and that it conforms to all applicable requirements of the standard.
2. Adequate protection provided by guards and other parts designed to enclose or otherwise protect equipment in terms of strength and durability.
3. Adequacy of all electrical insulation.
4. Effects of the heat produced by the equipment under normal operating conditions.
5. Classification of the equipment in question (e.g., voltage, current capacity, intended use, type, size).
6. Other characteristics of the equipment that are supposed to eliminate or reduce the potential hazards associated with it.

Guarding of Energized Parts

Contact with energized electrical parts is the principle hazard to be guarded against when dealing with electrically powered equipment. Consequently, energized parts carrying 50 or more volts must be guarded in such a way as to prevent a worker from accidentally making contact. Guarding of energized parts can be accomplished in a number of ways, including the following:

1. Locating the equipment in question in a protected, enclosed environment that limits access. This enclosed environment might be a designated room, vault, cabinet, or some other enclosure that limits access (Figure 15–4).
2. Locating the equipment in question in an area enclosed by permanent barriers, such as partitions or screens, that can effectively limit access.
3. Locating the equipment in question on an elevated platform, balcony, or gallery that can effectively limit access.
4. Elevating the equipment in question 8 feet or more above the floor.

Over-Current Protection

Over-current protection must be provided for circuits rated at 600 volts nominal or less. The requirements for over-current protection are as follows:

1. Conductors of electricity and electrical equipment must have over-current protection that is sufficient in terms of their ability to conduct current.
2. Conductors must have sufficient capacity to carry the required load.
3. Except in the case of a motor running overload protection, over-current devices must not interrupt the continuity of the grounded conductor, except when all conductors of the circuit are opened at the same time.

FIGURE 15–4 Designated electrical closet.
Source: ZStoimenov/Fotolia

4. Over-current devices must be located so that they are: (a) easily accessible, (b) protected from physical damage, and (c) clear of any combustible or inflammable material.
5. Fuses and circuit breakers must be either located or protected to prevent burn or arcing injuries to workers.

Grounding of Equipment Connected by a Cord or Plug

Under specified conditions, proper grounding must be provided for metal parts of equipment when the equipment is connected by a cord or plug and it is possible that these parts might become energized. These conditions are as follows:

1. When the equipment is located in a hazardous (classified) location.
2. When the equipment is operated at more than 150 volts to ground. Exceptions to this are guarded motors and the metal frames of electric appliances when the frames are properly and permanently insulated from grounding.
3. When the equipment is one of the following types: (a) motor-operated hand tools; (b) equipment used in the presence of moisture; (c) equipment used by workers standing on metal floors; (d) equipment used by workers inside of metal tanks, boilers, or other metal containers; (e) portable X-ray equipment and peripherals; (f) any tools likely to be used in the presence of moisture; and (g) portable hand lamps.

There are cases in which equipment connected by a plug or cord do not have to be grounded. These exceptions include (1) equipment that receives power through an isolating transformer with an ungrounded secondary of 50 volts or less and (2) properly labeled portable tools equipped with a system of double insulation.

Ground-Fault Protection at Construction Sites

Part of Subpart K—29 CFR 1926.404(b)—deals specifically with ground-fault protection at construction sites. The two principal ways of preventing electric shock are insulation and grounding. Properly done, both can be effective injury-prevention strategies. However, neither is perfect. Insulation can become worn and deteriorated. A break or fault in a grounding system can occur for a variety of reasons without a worker realizing it, until it is too late. Consequently, ground-fault protection is important.

Such protection can be provided by a ground-fault circuit interrupter (GFCI). A GFCI is a device that is able to sense even small levels of current leakage to ground. When leakage causes an imbalance in the circuit, the GFCI is able to act as a circuit breaker and very quickly shut down the electricity. This device is effective in protecting workers against the ground fault, which is one of the most common sources of electrical hazards in the workplace.

SAFETY-RELATED WORK PRACTICES

Subpart K (29 CFR 1926.416 and 1926.417) contains requirements related to work practices.[3] These requirements are presented in three broad categories as follows:

- 1926.416(a) Protection of employees
- 1926.416(b) Passageways and open spaces
- 1926.417 Lockout and tagging of circuits

Protection of Employees

Employers must not allow workers to perform their duties near any process, equipment, or other device that might expose them to an energized electric power circuit unless the worker is properly protected. The necessary protection can be provided by grounding, insulation, guarding, or other appropriate means. When workers are using hand tools, jackhammers, or other devices that might come in contact with underground electric powerlines, and the location of the lines is unknown, they must be provided with properly insulated gloves sufficiently rated to protect them in the event of accidental contact. In addition to providing the gloves, employers must provide the supervision necessary to ensure that they are worn.

Before allowing workers to begin their tasks in or near an area where electric power circuits might be located, employers must conduct a survey to determine the location of the circuits. This can be accomplished by observation, use of the appropriate instruments, or asking responsible personnel who are familiar with the circuitry (such as representatives of the power company). Once the electric power circuits have been located, employers must make workers aware of their presence and what precautions are to be taken to protect against injury.

Passages and Open Spaces

Open spaces in which electric power circuits might be located must not be allowed to become passageways for workers moving from one area of a construction site to another. Employers must use barriers and guards to prevent such areas from becoming de facto passageways. In addition, areas that are used as passageways, such as walkways and hallways, must be kept free of electrical cords.

Lockout and Tagging of Circuits

When working on electrical equipment, circuits, or tools, controls should be tagged to prevent accidental energizing. Equipment or circuits that are energized must be mechanically deactivated, made inoperable, and have tags attached at all points where the equipment or circuit could be energized.

SAFETY-RELATED MAINTENANCE AND ENVIRONMENTAL CONSIDERATIONS

Subpart K (29 CFR 1926.431 and 1926.432) contains requirements for safety-related maintenance and environmental considerations in the following broad categories of information[4]: (1) maintenance of equipment and (2) environmental deterioration of equipment.

Maintenance of Equipment

Maintenance of electrical equipment is important at construction sites and when stored or used in any other hazardous locations. Employers must ensure that such equipment is properly maintained and stored in a dust-tight, ignition-proof, or explosion-proof condition. The equipment should be free of loose or missing screws, gaskets, threaded connections, or seals or of other impairments to a tight, secure condition.

Environmental Deterioration of Equipment

Certain types of electrical equipment are designed specifically for safe use in hazardous operating environments. All electrical equipment used in these conditions must be specifically designed for such use. These hazardous conditions include the following: (1) damp or wet conditions; (2) environments in which gases, fumes, vapors, liquids, or other substances are present that might cause deterioration of the electrical components of the equipment; and (3) extreme temperatures (hot or cold).

Electrical equipment that is designed for use in dry environments must be properly protected during construction. Any electrical equipment that might be exposed to a corrosive environment must be made of a material that is appropriate to the environment in question.

SAFETY REQUIREMENTS FOR SPECIAL EQUIPMENT

Subpart K (29 CFR 1926.441) contains the requirements for special equipment, such as batteries.[5] Unsealed batteries must be housed in ventilated enclosures or in well-ventilated rooms to guard against the contamination of the air by fumes, gases, or electrolyte spray in other areas, Figure 15–5. Ventilation sufficient to diffuse gases from batteries is required; this not only guards against air contamination, it prevents the buildup of an explosive mixture as gases escape from the batteries. Racks and trays that hold batteries should be treated to make them resistant to the corrosive effects of the electrolyte spray. Floors in areas that contain batteries should be constructed of acid-resistant material unless provisions are made for protecting the floor from the accumulation of acid. Workers who handle batteries should be provided and required to wear the appropriate personal protective equipment (e.g., face shields, aprons, rubber gloves), Figure 15–6. Employers must provide facilities for quick drenching of the eyes and body in the vicinity (within 25 feet) of areas in which batteries are stored or handled.

Batteries should be charged in segregated areas designated only for battery charging. During the charging process, vent caps should be maintained in functioning condition to prevent electrolyte spray. Battery-charging equipment must be protected from damage by moving equipment, industrial trucks, and other hazards.

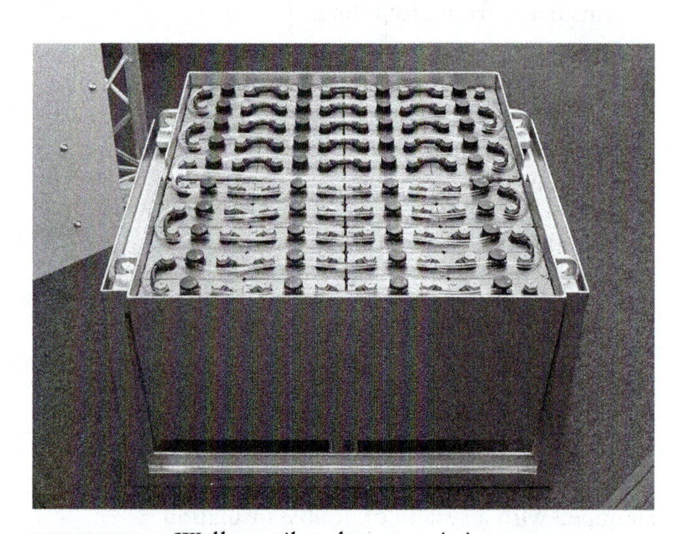

FIGURE 15–5 Well-ventilated storage is important.
Source: Markobe/Fotolia

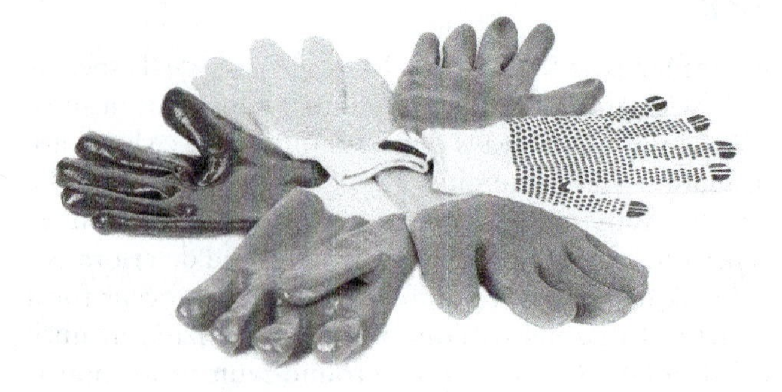

FIGURE 15–6 Workers should wear rubber gloves when handling batteries.
Source: Indigolotos/Fotolia

SUBPART L: SCAFFOLDING REQUIREMENTS AND RELATED SAFETY PRACTICES

Subpart L covers all of OSHA's requirements for the erection and use of scaffolding for performing elevated work. This is an extensive subpart covering the requirements related to general issues, specific types of scaffolds, and aerial lifts. Subpart L contains the following sections:

- 1926.450 Scope, application, and definitions
- 1926.451 General requirements
- 1926.452 Additional requirements applicable to specific types of scaffolds
- 1926.453 Aerial lifts
- 1926.454 Training
- Appendix A–E Scaffolds

RATIONALE FOR THE SCAFFOLDING REQUIREMENTS

Working on scaffolding is one of the most dangerous tasks in construction.[6] The primary cause of injuries associated with scaffolding is falling. In fact, of the more than 900 deaths every year in construction, almost 9 percent, or approximately 80 deaths, involve scaffolding. Of the almost 10,000 scaffolding-related injuries suffered every year by construction workers, most are caused by one of the following: (1) planking failure, (2) support failure, (3) slipping, or (4) impact of a falling object. In the accidents that led to these injuries, only 33 percent of the scaffolds involved were properly equipped with **guardrails**.

GENERAL REQUIREMENTS FOR SCAFFOLDING

A **scaffold** is any one of a variety of types of temporary elevated platforms and its corresponding structure (e.g., supports, means of suspension, planking) used for supporting workers and materials at a distance above the ground.[7] Scaffolding is a field within construction with a language of its own. Figure 15–7 contains the definitions for some of the more widely used types of scaffolds and other associated terms with which construction professionals need to be familiar.

The general requirements related to scaffolding as set forth by OSHA in 29 CFR 1926.451 fall into the following broad categories: capacity, platform construction, supported scaffolds, suspension scaffolds, scaffold access, scaffold use, fall protection, and falling object protection.

Scaffolding Terms

Body belt (safety belt). A strap with means for securing it about the waist and attaching it to a lanyard, lifeline, or deceleration device.
Body harness. A system of straps that may be secured about the employee in a manner that distributes the fall arrest forces and that has means for attaching it to other components of a personal fall arrest system.
Fabricated frame scaffold. A scaffold that consists of a platform supported on fabricated end frames with integral posts, horizontal bearers, and intermediate members.
Ladder jack scaffold. A supported scaffold that consists of a platform resting on brackets attached to ladders.
Lifeline. A component that consists of a flexible line that connects to an anchorage at one end to hang vertically (vertical lifeline) or to anchorages at both ends to stretch horizontally (horizontal lifeline), which serves as a means for connecting other components of a personal fall arrest system to the anchorage.
Maximum intended load. The total load of all persons, equipment, tools, materials, transmitted loads, and other loads reasonably anticipated to be applied to a scaffold or scaffold component at any one time.
Mobile scaffold. A powered or unpowered, portable, caster- or wheel-mounted supported scaffold.
Personal fall arrest system. A system used to arrest an employee's fall, which consists of an anchorage, connectors, a body belt, or body harness and may include a lanyard, deceleration device, lifeline, or combinations of these.
Pole scaffolds. A supported scaffold that consists of a platform resting on cross beams supported by ledgers and a double row of uprights independent of support (except ties, guys, and braces) from any structure.
Power-operated hoist. A hoist that is powered by energy other than human.
Pump jack scaffold. A supported scaffold consisting of a platform supported by vertical poles and movable support brackets.
Rated load. The manufacturer's specified maximum load to be lifted by a hoist or to be applied to a scaffold or scaffold component.
Roof bracket scaffold. A rooftop-supported scaffold consisting of a platform resting on angle-shaped supports against a pitched roof.
Scaffold. Any temporary elevated platform (supported or suspended) and its supporting structure (including points of anchorage) used for supporting employees or materials, or both.
Tube and coupler scaffold. A supported or suspended scaffold that consists of a platform supported by tubing, erected with coupling devices that connect uprights, braces, bearers, and runners.

FIGURE 15–7 Language of scaffolding.

SAFETY FACTS & FINES

Constant monitoring and regular inspections are important, no matter how experienced a company is at working with electricity. Even electric utility companies are not immune to electrical hazards and the injuries they can cause. An electric utility in Providence, Rhode Island, was fined $80,000 when two workers received severe burns when they made contact with energized electrical equipment that was not properly grounded. To complicate matters, one of the employees was wearing a nylon shirt that was burned into his skin. The company was cited for the following violations: (1) failing to properly ground electrical equipment, (2) failing to ensure that electrical equipment was not energized, and (3) failing to monitor the clothing worn by employees (clothing that might increase the severity of an injury).

Capacity

Scaffolds, no matter the type, must be able to support their own weight plus an additional load of at least four times the **maximum intended load**. In the case of suspension scaffolds, the ropes must be able to support the weight of the scaffold and at least six times the maximum intended load. The stall load of any scaffold hoist cannot be allowed to exceed three times its **rated load**.

Platform Construction

All platforms must be fully planked, and the planking must be placed so that gaps between units are no more than 1-inch. The 1-inch requirement cannot be exceeded unless an employer can establish that a larger gap is necessary in order to fit around supporting structures or other similar circumstances. When the 1-inch rule is exceeded, individual planking units must be kept as close together as possible and may not exceed 9½ inches between the platform and upright supports or guardrails. Figures 15–8 and 15–9 are examples of scaffolds used in construction.

Scaffold Width. Scaffold platforms and walkways must be at least 18 inches wide, except for ladder jack, top plate bracket, and pump jack scaffolds, which may be 12 inches wide. In areas too narrow to accommodate an 18-inch wide platform or plank, the walking surface must be as wide as possible, and workers must be protected by either guardrails or personnel fall arrest systems.

Platform Front Edges. The space between the front edge of a scaffold and the face of the structure should not exceed 14 inches unless the appropriate precautions have been taken. These precautions must be either guardrails or the provision of a fall arrest system. This requirement does not apply in the case of outrigger scaffolds that must be no more than 3 inches from the work surface or scaffolds for plastering and lathing. Plastering and lathing scaffolds are limited to 18 inches from the work surface to give workers more room to use their specialized tools.

Platform Ends. There are minimum and maximum distances that apply to the ends of scaffolds. The end of a platform unit for a scaffold must extend at least 6 inches beyond the centerline of its support, unless the platform is cleated or otherwise restrained by an appropriate means. Cleats, hooks, and other appropriate devices can be used to satisfy the 6-inch requirement. Maximum distances apply unless platforms are specially

FIGURE 15–8 Example of a scaffold commonly used in construction.
Source: Etien/Fotolia

FIGURE 15–9 Example of a scaffold commonly used in construction.
Source: Moonrise/Fotolia

Installation Requirements for Guys, Ties, and Braces for Supported Scaffolds
Installed where horizontal members support inner and outer legs.
Installed according to the manufacturer's specifications.
Installed at the closest horizontal member, according to the 4:1 ratio.
Repeated at horizontal members every 20 feet or less for scaffolds 3 feet wide or less.
Repeated at horizontal members every 26 feet or less for scaffolds more than 3 feet wide.
The top support (guy, tie, and brace) must be placed more than the 4:1 ratio from the top.
Installed at each end of the scaffold and at horizontal intervals of 30 feet or less.

FIGURE 15–10 Checklist of installation requirements for scaffold supports.

designed so that cantilevered ends can support workers and materials or guardrails have been installed. Maximum distances are as follows: (1) 12 inches of extension for platform units 10 feet long or less, and (2) 18 inches of extension for platform units longer than 10 feet.

Abutted, Overlapped, and Directional Changing Platforms. Long scaffolds can be created by placing platforms end to end. When this is done, each platform must rest on its own supports. Overlapped platforms must be set up so that the overlap occurs only at supports, and it must be at least 12 inches (unless the platforms are nailed to each other). When a scaffold changes direction, platforms should rest at right angles on the bearers. When a right angle cannot be accomplished, the off-angle platform should be put in place first.

Platform Surface Finishes. Platforms must not be finished with materials that might conceal a structural defect. Opaque finishes may not be used on any part of a platform except the edges. Edges are exempted from the requirement to allow for identification, grading, and other necessary markings. Wood preservatives, fire retardants, and slip-prevention finishes may be used, as long as they do not obscure the top or bottom surfaces of the platforms.

Platform Components. Mixing of platform components is prohibited unless a **competent person** has determined that the mix does not create a hazard to workers. Mixing of components applies to components made of different materials and components produced by different manufacturers. A competent person may allow the components from two different manufacturers to be mixed when the mixing can be accomplished without forcing parts together or decreasing the scaffold's structural integrity in any other way. No mixing or adaptation is allowed that threatens, in any way, the structural integrity of the scaffold.

Supported Scaffolds

Supported scaffolds consist of one or more platforms that are supported by outrigger beams, brackets, poles, legs, uprights, posts, frames, or other types of rigid support systems. Critical concerns related to supported scaffolds include guys, ties, braces, poles, legs, posts, frames, and uprights.

Guys, Ties, and Braces. With supported scaffolds, tipping is always a potential hazard. Consequently, OSHA established a 4:1 ratio as a rule for governing when guys, ties, braces, or some other means of support must be used to ensure stability. The 4:1 ratio means that with freestanding scaffolds, the height cannot exceed four times the width of the minimum base dimension without using guys, ties, or braces. For example, a scaffold that is 5 feet wide, 60 feet tall, and 75 feet long would require special support (guys, ties, or braces) because the 4:1 ratio is exceeded. The requirements for installing supports are summarized in Figure 15–10.

Poles, Legs, Posts, Frames, and Uprights. Scaffolds must be placed on a firm footing that will neither shift nor give way when subjected to the maximum allowable load. Base plates, mudsills, or other appropriate devices designed for this use are required. The devices used must be level, rigid, and capable of enduring the maximum allowable load for the scaffold. No object or device that might shift or give way may be used.

Construction professionals should avoid the temptation to use front-end loaders or **forklifts** as the base for a scaffold unless the equipment has been especially designed for this purpose. The only exception to this rule is that forklifts may be used if the following conditions are fully met: (1) the entire scaffold platform is securely

SAFETY FACTS & FINES

Because falls account for such a high percentage of the accidents in construction, the Occupational Safety and Health Administration (OSHA) watches the use of scaffolding closely. A company in Pharr, Texas, found out just how closely when it was fined $299,300 for failing to provide workers with the necessary fall protection while working on scaffolding. OSHA had targeted construction sites at which scaffolding might be used and found workers at this company's site working more than 10 feet above the ground without the necessary fall protection.

attached to the fork, and (2) the forklift is not moved horizontally while workers are on the scaffold.

Suspension Scaffolds

Suspension scaffolds operate on the same principle as suspension bridges. They are suspended from an overhead support structure by ropes, cables, or other nonrigid devices. Support devices include rigger beams, cornice hooks, parapet clamps, and other similar devices. OSHA mandates that all such support devices rest on surfaces that meet one of the following requirements: (1) can support at least four times the operating load of the scaffold, or (2) can support 1.5 times the scaffold load when operating at the stall load of the hoist. The greater of these two requirements is the one that applies.

Scaffold Access

OSHA sets forth requirements for access to scaffolds that apply to workers who are using, erecting, or dismantling them. The requirements apply any time a scaffold platform is more than 2 feet above or below an access point. OSHA emphasizes that cross-braces may not be used for access to a scaffold. Acceptable methods for accessing scaffolds are listed in Figure 15–11.

Scaffold Use

Once a scaffold has been properly erected, it is critical that workers use it properly and safely. OSHA sets forth the following requirements for safe scaffold use:

1. Never load a scaffold in excess of the maximum intended load. The maximum intended load includes all workers, tools, equipment, and materials that the scaffold is expected to hold at one time.
2. Never load a scaffold in excess of its rated capacity. This is the maximum load a hoist is allowed to lift.
3. Never use shore or lean-to scaffolds.
4. Before a scaffold is used—at least at the beginning of every shift—it should be inspected by a competent person for visible defects. In addition, a thorough inspection should take place any time an incident occurs that might threaten the structural integrity of a scaffold.
5. Scaffolds should never be moved in a horizontal direction while occupied by workers unless they were especially designed for such movement.
6. Scaffolds must not be allowed to come within specified distances of energized power lines. The distances are summarized in Figure 15–12. The requirements in Figure 15–12 may be waived if the construction company is able to have the electric utility company in question do one of the following: (a) de-energize the lines, (b) relocate the lines, or (c) install coverings to protect workers from accidental contact with energized lines.
7. When moving, dismantling, erecting, or altering scaffolds, construction companies must ensure that a competent person is present to supervise. Further, the work involved must be accomplished by individuals with the proper training and experience.
8. Workers must not be allowed to perform any task on a scaffold that is covered with snow, ice, or any other type of slipping hazard.
9. Appropriate precautions, such as tag lines, must be used when swinging loads are to be hoisted onto a scaffold or onto a destination near a scaffold.
10. Suspension ropes or cables used to support adjustable suspension scaffolds must be of a diameter large enough to provide sufficient surface area to accommodate brake and hoist devices. Suspension ropes and cables must be protected from heat-producing processes and corrosive elements.
11. Workers must not be allowed to perform tasks on scaffolds during periods of high wind or in storm conditions. Even if the competent person determines that storm or wind conditions are not sufficient to shut down work on scaffolds, employees must be protected by a fall arrest system and a windscreen.

Acceptable Methods for Scaffold Access
Portable, hook-on, and attachable ladders
Stairway-type ladders
Stairtowers (scaffold stairway or towers)
Ramps and walkways
Integral prefabricated scaffold access frames

FIGURE 15–11 Workers must have safe access to scaffolds.

Safe Distance Requirements: Scaffolds and Power Lines		
Insulated Lines		
Voltage	**Minimum Distance**	**Alternatives**
<300 volts	3 feet (0.9 m)	–
300 volts to 50 kV	10 feet (3.1 m)	–
>50 kV	10 feet (3.1 m) plus 0.4 inches (1.0 cm) for each 1 kV over 50 kV	2 times the length of the line insulator, but never <10 feet (3.1 m)
Uninsulated Lines		
Voltage	**Minimum Distance**	**Alternatives**
<50 kV	10 feet (3.1 m)	–
>50 kV	10 feet (3.1 m) plus 0.4 inches (1.0 cm) for each 1 kV over 20 kV	2 times the length of the line insulator, but never <10 feet (3.1 m)

FIGURE 15–12 Required safe distances between scaffolds and energized power lines.

If using a windscreen, the scaffold must first be secured against the wind forces to which it will be exposed.

12. Boxes, barrels, loose blocks, and other temporary expedients may not be used on scaffolds for lifting workers to a higher level. Even ladders should not be used on scaffolds except in the case of large area scaffolds where the ladder legs can be secured to prevent slippage. Platform units are secured to the scaffold to prevent their movement, and the scaffold is secured to guard against the sideward thrust force exerted on it when a worker climbs the ladder.
13. Platforms on scaffolds shall not deflect more than one-sixtieth of their span when fully loaded.
14. When a worker is welding while standing on a scaffold, appropriate precautions must be taken to guard against the possibility of welding current arcing through suspension cables or metal scaffold members. Precautions should include proper grounding and insulation of suspension cables and metal members.

Fall Protection

The most obvious hazard associated with scaffolds is falling. Consequently, workers on scaffolds more than 10 feet above the next lower level must be protected from falling to that level. Generally speaking, workers must be protected by personal fall arrest systems or guardrail systems.

Fall arrest systems must meet the requirements set forth in 29 CFR 1926.502(d). In addition, personal fall arrest systems used in conjunction with scaffolds must be attached by lanyard to a vertical or horizontal **lifeline** or a structural component of the scaffold. Guardrails must be installed along all open sides of a platform as well as at the ends of platforms. Guardrails must be installed before the scaffold is occupied by workers, except in the case of erecting and dismantling scaffolds.

Falling Object Protection

Objects falling from scaffolds and onto scaffolds both represent significant hazards on construction sites. To protect workers on a scaffold from falling objects, construction companies should take the following precautions: (1) require the use of hard hats; (2) install toeboards, screens, debris nets, catch platforms, or canopy structures to contain or deflect falling objects; and (3) make sure that objects that might fall on workers are kept away from the edges of higher-level surfaces and are properly secured to prevent their falling.

To protect workers below the level of a scaffold, construction companies should take the following precautions: (1) barricade the area below the scaffold, (2) erect toeboards along the edge of any scaffold that is higher than 10 feet above the next lower surface, (3) use paneling or screening erected from the toeboard or platform to the top of the guardrail to prevent objects from falling off the scaffold, (4) erect guardrails with openings too small to allow objects to pass through and fall to a lower level, and (5) erect canopy structures, debris nets, or catch platforms to catch or deflect objects falling from the scaffold.

Checklist of Specialty Scaffolds

- Multi-level suspended scaffolds
- Plasterers', decorators', and large-area scaffolds
- Bricklayers' square scaffolds
- Horse scaffolds
- Form scaffolds and carpenters' bracket scaffolds
- Outrigger scaffolds
- Window jack scaffolds
- Crawling boards (chicken ladders)
- Step, platform, and trestle scaffolds
- Single-point adjustable suspension scaffolds
- Two-point adjustable suspension scaffolds
- Multipoint adjustable suspension scaffolds, stonesetters' multipoint adjustable suspension scaffolds
- Catenary scaffolds
- Float (ship) scaffolds
- Interior hung scaffolds
- Needle beam scaffolds
- Pole scaffolds
- Tube and couples scaffolds
- Fabricated (tubular welded) frame scaffolds
- Roof bracket scaffolds
- Pump jack scaffolds
- Ladder jack scaffolds
- Mobile scaffolds
- Repair bracket scaffolds

FIGURE 15–13 Types of specialty scaffolds. For specific requirements related to such scaffolds, see 29 CFR 1926.452.

Requirements for Specific Types of Scaffolds

The requirements set forth in this section and in 29 CFR 1926.451 apply to all scaffolds. There are additional requirements that apply to specific types of specialized scaffolds. Figure 15–13 contains a list of specialized scaffolds that are sometimes used in construction. For specific requirements related to these scaffolds, students are encouraged to refer to 29 CFR 1926.452.

AERIAL LIFTS

Aerial lifts are vehicle-mounted mechanisms used to elevate workers to locations above the ground that would otherwise be inaccessible (Figure 15–14). Figure 15–15 is a list of the most widely used aerial lifts in the construction industry. OSHA sets forth specific requirements for

FIGURE 15–14 Example of an aerial lift.
Source: R. Roth/Fotolia

extensible and articulating boom platforms. There are also requirements for electrical tests and the bursting safety factor.

Requirements for Extensible and Articulating Boom Platforms

Only individuals who have the proper training and experience to operate aerial lifts may be authorized to do so. These individuals must wear a **body belt** and a lanyard attached to either the boom or the basket when working from an aerial lift. Belting to an adjacent pole, structure, or piece of equipment is not acceptable. Before beginning the operation of a lift, all lift controls should be tested to ensure that they are in proper working condition.

Workers in the basket of an aerial lift should always stand firmly and squarely on the floor of the basket.

Aerial Lifts
Extensible boom platforms
Aerial ladders
Articulating boom platforms
Vertical towers

FIGURE 15–15 Common types of aerial lifts used in construction.

Climbing or sitting on the walls of the basket is forbidden. Construction companies must ensure that workers adhere strictly to the manufacturer's load and boom specifications. Wheel chocks must be in place when using an aerial lift at an angle (such as on a hill). In addition, brakes must be set. When outriggers are used, they must be positioned on a solid surface or on pads.

Aerial lift trucks must remain stationary when the boom is elevated. Articulating and extensible boom platforms that are used primarily to carry people must have controls on the ground and on the platform. These controls must be plainly marked to indicate their functions.

Electrical Tests

Electrical tests conducted on an extensible or articulating boom must conform to the requirements set forth in ANSI A92.2, Section 5. Acceptable equivalent tests may be used. For example, DC voltage tests approved and specified by the equipment manufacturer may be used to meet the requirements for electrical tests.

Bursting Safety Factor

ANSI A92.2, Section 4.9 applies to the critical hydraulic and pneumatic components of extensible and articulating booms. By this standard, a critical component is one that, if it fails, could cause a free fall or free rotation of the boom. Critical components must have a bursting safety factor of at least 2:1.

TRAINING REQUIREMENTS RELATED TO SCAFFOLDING

OSHA sets forth its training requirements for construction companies in 29 CFR 1926.21(b)(2). In addition to these general requirements that mandate instruction to help workers recognize, avoid, eliminate, and control hazardous conditions in the work environment, in 29 CFR 1926.454, OSHA sets forth requirements related specifically to scaffolds. Training must be provided by a "qualified" individual. A "qualified" person has a higher level of training and knowledge than a "competent" person. Training is to be provided in the following areas:

1. Electrical, fall, and falling object hazards.
2. Proper procedures for handling electrical hazards and for erecting and disassembling fall protection and falling object protection systems.
3. Safe use of a scaffold and materials on a scaffold.
4. Load and carrying capacities of scaffolds.
5. Other pertinent requirements of the scaffold rule.

In addition to providing training that qualifies workers to safely use scaffolds, construction companies are required to provide retraining when it is needed.

Retraining is required when any one of the following conditions exist: (1) changes at the work site create conditions for which the worker has not been trained, (2) changes in scaffolding, related equipment, or fall protection systems create conditions for which the worker has not been trained, and (3) inadequacies become apparent in an employee who needs further training.

SUBPART M: FALL PROTECTION REQUIREMENTS AND RELATED SAFETY PRACTICES

Falls are the leading cause of workplace deaths in the construction industry in the United States. Because of this, OSHA developed a comprehensive and detailed standard that covers both the equipment and the human aspects of fall protection. Overall, OSHA requires construction companies to do the following:

1. Select and provide the proper fall protection systems for every situation in which such systems are necessary.
2. Properly construct and install safety systems for fall protection.
3. Train workers in the need for and proper selection, use, and maintenance of fall protection systems.
4. Provide competent supervision for workers exposed to fall-related hazards.
5. Train workers to know, understand, and apply safe work procedures.

OSHA's fall protection standard is not prescriptive. It tells construction companies what they are required to do but does not prescribe how it must be done. The standard covers two aspects of fall protection: (1) protection from falling off, onto, or through working levels, and (2) protection from being struck by falling objects.

OSHA's fall protection standard identifies several different situations in which fall protection is required, including the following:

- Ramps, runways, and walkways
- Excavations more than 6 feet deep
- Hoist areas
- Holes
- Form work and reinforcing steel
- Leading edge work
- Unprotected sides and edges
- Overhand bricklaying and related work
- Roofing work
- Precast concrete erection (steel erection has its own standard, see Subpart R)
- Wall openings
- Residential construction
- Other walking and working surfaces

Exceptions to OSHA's fall protection standard are made for construction-related workers involved in inspecting, investigating, or assessing workplace conditions before the start of work or after all work has been completed. In addition, several other parts of OSHA's construction standard (29 CFR 1926) are exempted from the requirements in Subpart M because they contain their own requirements, unless situations arise that are not covered in the subpart in question. When this occurs, Subpart M should be considered the applicable standard. These subparts are as follows:

- Subpart L: Scaffolds
- Subpart N: Cranes and derricks
- Subpart R: Steel erection
- Subpart V: Electric transmission and distribution
- Subpart X: Stairways and ladders

Subpart M contains the following sections and appendices that list the requirements for fall protection on construction sites:

- 1926.500 Scope, application, and definitions
- 1926.501 Duty to have fall protection
- 1926.502 Fall protection systems and criteria
- 1926.503 Training requirements

Figure 15–16 contains the more commonly used terms in the language of fall protection.

FALL PROTECTION DIRECTIVE

In the past OSHA tried to build *compliance flexibility* into the fall protection requirements of 29 CFR 1926.500 as these requirements applied to residential construction. However, because falls continued to account for a high proportion of deaths in construction, including residential construction, OSHA rescinded its *compliance flexibility* for residential construction and issued its Fall Protection Directive (STD 03-11-002). This directive requires residential construction companies to comply with the requirements of 29 VFR 1926.500 (covered in the sections which follow). In practical terms, the directive means that residential construction companies are required to use guardrails, safety nets, or personal fall arrest systems in most cases. The requirements of OSHA's Fall Protection Standard are summarized in the following sections. These requirements are followed by a comprehensive summary of the fall protection requirements of the ANSI/ASSE Z359 Fall Protection Code.

Language of Fall Protection

Fall protection has a language of its own. To understand OSHA regulations and other fall protection guidelines, it is necessary to know the following terms:

Anchorage. A secure point of attachment for lifelines, lanyards, or deceleration devices.

Body belt. A strap with means both for securing it about the waist and for attaching it to a lanyard, lifeline, or deceleration device.

Body harness. Straps that may be secured about the person in a manner that distributes the fall-arrest forces over at least the thighs, pelvis, waist, chest, and shoulders, with a means for attaching the harness to other components of a personal fall arrest system.

Connector. A device used to couple (connect) parts of a personal fall arrest system or positioning device system together.

Hole. A void or gap of 2 inches (5.1 cm) or more in the least dimension in a floor, roof, or other walking or working surface.

Lanyard. A flexible line of rope, wire rope, or strap that generally has a connector at each end for connecting the body belt or body harness to a deceleration device, lifeline, or anchorage.

Lifeline. A component consisting of a flexible line for connection to an anchorage at one end to hang vertically (vertical lifeline) or for connection to anchorages at both ends to stretch horizontally (horizontal lifeline). Serves as a means for connecting other components of a personal fall arrest system to the anchorage.

Low-slope roof. A roof with a slope ≤4:12 (vertical to horizontal).

Opening. A gap or void of 30 inches (76 cm) or more high and 18 inches (46 cm) or more wide, in a wall or partition, through which employees can fall to a lower level.

Personal fall arrest system. A system including but not limited to an anchorage, connectors, and a body belt or body harness; used to arrest an employee in a fall from a working level. The use of a body belt alone for fall arrest is prohibited.

Platform. A working space for persons, elevated above the surrounding floor or ground.

Positioning device system. A body belt or body harness system rigged to allow an employee to be supported on an elevated vertical surface, such as a wall, and work with both hands free while leaning backwards.

Rope grab. A deceleration device that travels on a lifeline and automatically, by friction, engages the lifeline and locks to arrest a fall.

Self-retracting lifeline or lanyard. A deceleration device containing a drumwound line that can be slowly extracted from, or retracted onto, the drum under minimal tension during normal employee movement, but which—after onset of a fall—automatically locks the drum and arrests the fall.

FIGURE 15–16 Important fall protection terms for construction professionals.

Complying with OSHA's Fall Protection Directive

An easy way for construction professionals to understand what must be done to comply with OSHA's Fall Protection Directive is to apply the *ABCD rule*. This rule of thumb is explained as follows:

- *Anchorage.* Make sure that all roof anchors used can support the intended loads with a safety factor added in.
- *Body Support.* Use full-body harnesses for the best results. Make sure the harness used is the right size for the worker in question, that it is worn properly, and that it is adjusted snuggly.
- *Connector.* Make sure that shock-absorbing lanyards or self-retracting lifelines connect properly to the worker's harness and the anchor.
- *Descent/Recuse.* Make sure devices needed to lower or raise a fallen worker are readily available and in proper working order.

REQUIREMENTS OF THE STANDARD

OSHA's current fall protection standard sets the **trigger height** at 6 feet. This means that any construction employee working higher than 6 feet off the ground must use a fall protection device, such as a safety harness and line (Figure 15–17).

This trigger height means that virtually every small residential builder and roofing contractor is subject to the standard. Because most residential builders and roofing contractors are small businesses, Subpart M of 29 CFR 1926 is a source of much controversy.

OSHA dealt with this controversy with a plain language revision to Instruction STD 3.1. This instruction modifies the fall protection requirements set forth in 29 CFR 1926.501(b)(13) for contractors engaged in residential construction. The modification allows companies involved in four groups of activities to use alternative procedures instead of the regularly required fall protection strategies. Companies engaged in the following types of residential construction activities may develop a fall protection plan and apply it in lieu of applying the normally required fall protection methods:

FIGURE 15–17 Personal fall arrest harness.
Source: Sailom/Fotolia

- *Group 1:* Installation of floor joists, floor sheathing, and roof sheathing; erection of exterior walls; and setting and bracing of roof trusses and rafters.
- *Group 2:* Working on concrete and masonry block foundation walls and related formwork.
- *Group 3:* The following types of work in attics and on roofs: installation of drywall, insulation, electrical systems, HVAC systems, alarm systems, telecommunications systems, plumbing, and carpentry.
- *Group 4:* Installation, repair, or removal of roofing materials, such as shingles, tar paper, and tile.

OSHA officials argue that the 6-foot trigger height saves up to 80 lives per year and prevents more than 56,900 injuries. The rationale is that 6 percent of all lost-time fall injuries in the construction industry are caused by falls from less than 10 feet. Opponents counter that the cost of complying with the standard is almost $300 million annually. Commercial contractors, whose employees typically work much higher than the 6- to 16-foot range, are less concerned about the height controversy.

Requirements from 29 CFR 1926 that apply specifically to fall protection in scaffolding work are as follows:

- 1926.451(g)(2) reads, "The employer shall have a competent person determine the feasibility and safety of providing fall protection for employees erecting or dismantling supported scaffolds. Employers are required to provide fall protection for employees erecting or dismantling supported scaffolds where the installation and use of such protection is feasible and does not create a greater hazard."[8]
- 1926.502(d)(15) reads, "Anchorages used for attachment of personal fall arrest equipment shall be independent of any anchorage being used to support or suspend platforms and capable of supporting at least 5,000 pounds per employee attached, or shall be designed, installed, and used as follows:
  - (i) as part of a complete personal fall arrest system which maintains a safety factor of at least two and
  - (ii) under the supervision of a qualified person."[9]
- 1926.502(d)(16) reads, "Personal fall arrest systems, when stopping a fall, shall:
  - (i) limit maximum arresting force on an employee to 900 pounds (4 kg) when used with a body belt;
  - (ii) limit maximum arresting force on an employee to 1,800 pounds (8 kg) when used with a body harness;
  - (iii) be rigged such that an employee can neither free-fall more than 6 feet (1.8 m) nor contact any lower level;
  - (iv) bring an employee to a complete stop and limit maximum deceleration distance an employee travels to 3.5 feet (1.07 m); and
  - (v) have sufficient strength to withstand twice the potential impact energy of an employee free-falling a distance of 6 feet (1.8 m), or free-fall distance permitted by the system, whichever is less."[10]

FALL PROTECTION SYSTEMS

Fall protection systems widely used in construction include guardrail, safety net, personal fall arrest, positioning device, warning line, controlled access, and safety monitoring systems. Requirements and practices for each of these systems are explained in the following paragraphs.

Guardrail Systems

The top of guardrails must be 42 inches (plus or minus 3 inches) above the walking level of the scaffold. Top rails of guardrails may exceed 45 inches when conditions warrant the additional height, provided all other requirements of the standard are met. Intermediate structural members must be installed between the top edge of the guardrail system and the walking surface when the wall is at least 21 inches high. Guardrails must be able to withstand a force of at least 200 pounds applied in any outward or downward direction within 2 inches of the top edge.

Safety Net Systems

Safety nets must be installed as close as possible under the walking surface of the scaffold (never more than 30 feet below). The fall area between the walking surface and the net should be unobstructed. Figure 15–18 summarizes the outward extension requirements for safety nets. In addition to providing the necessary outward extension on safety nets, it is also important to install nets with sufficient clearance to prevent contact with the surface should a worker fall into the net. Safety nets must have a border rope for webbing with a breaking strength of at least 5,000 pounds.

Outward Extension Requirements for Safety Nets

Vertical	Horizontal
≤5 feet	8 feet
>5 feet and <10 feet	10 feet
>10 feet	13 feet

FIGURE 15–18 Safety nets must extend a safe distance out from the edge of the walking surface of the scaffold.

SAFETY FACTS & FINES

OSHA citations do not necessarily lead to fines. A company in Dayton, Ohio, had a citation overturned because it had adequate procedures in place to detect the problem. The citation was written by an OSHA inspector who, while driving by, noticed two workers on a scaffold that had no guardrails. The supervisor claimed that his employees had been trained to assemble the scaffold properly and that he was preparing to conduct his regularly scheduled inspection. The supervisor had documentation of the training and claimed he would have caught the violation during his regularly scheduled inspection. The OSHA inspector issued a citation anyway, but the OSHA Review Commission accepted the supervisor's explanation and overturned the citation.

Personal Fall Arrest Systems

Before January 1998, body belts could be used as part of a **personal fall arrest system**. But this is no longer the case. Personal fall arrest systems now require a **body harness** with components that meet the following requirements:

- Connectors must be drop forged, pressed, or formed steel or be made of equivalent materials. The connectors must have a corrosion-resistant finish, smooth surfaces, and smooth edges to prevent damage to other components of the system.
- Dee-rings and snaphooks must have a tensile strength of at least 5,000 pounds and must be proof tested to at least 3,600 pounds. Only locking snaphooks may be used.
- Lanyards and vertical lifelines must have a breaking strength of at least 5,000 pounds.
- Anchorages used for attaching personal fall arrest equipment must be capable of supporting at least 5,000 pounds per employee who is attached.

Positioning Device Systems

Positioning device systems must be rigged to ensure that a worker can free-fall no more than 2 feet. They must be secured to an anchorage capable of withstanding at least twice the load of a worker's fall or 3,000 pounds, whichever is greater. Connectors, dee-rings, and snaphooks used with positioning device systems must meet the following requirements:

- Connectors must be drop forged, pressed, or formed steel or be made of equivalent material. They must have a corrosion-resistant finish, smooth edges, and smooth surfaces to prevent damage to other components of the system. Connecting assemblies must have a tensile strength of at least 5,000 pounds.
- Dee-rings and snaphooks must be proof tested to a tensile strength of at least 3,600 pounds. Snaphooks must be the locking type.

Warning Line Systems

When work is to be done on a roof or the top of a structure, warning line systems are sometimes used. Such systems must meet the following requirements:

- Warning lines must be erected no less than 6 feet from the edge of the roof. Points of access must be formed by two warning lines that connect it to the work area. When the access path is not in use, it must be blocked by a rope, wire, chain, or other barricade.
- Warning lines (ropes, wire, and chains) must be flagged at least every 6 feet. Lines must be rigged so that their lowest point does not exceed 34 inches and their highest point does not exceed 39 inches.
- The rope, wire, or chain used for warning lines must have a tensile strength of at least 500 pounds.

Controlled Access Zones

When leading edge work is to be done, an effective safety strategy is the **controlled access zone**. When this strategy is used, the zone must be marked by a control line that meets the following requirements:

- Control lines must be erected at least 6 feet from the leading edge, but not more than 25 feet from the edge (except when erecting precast concrete). When erecting precast concrete, the control line must be at least 6 feet from the leading edge, but no more than 60 feet or half the length of the precast member being erected (whichever is greater).
- Control lines must extend along the entire length of the access zone and must be approximately parallel to the leading edge. Lines must be connected at both ends to a guardrail or wall.
- Control lines must be flagged at least every 6 feet and have a minimum breaking strength of 200 pounds.

Safety Monitoring Systems

When work is to be done in settings where fall hazards are present, construction companies should designate a competent person to monitor the safety of employees. This is called having a safety monitoring system. The competent person must meet the following requirements:

- Competent to recognize fall hazards and to warn employees of the hazards.
- Be on the same walking surface as the workers being monitored and within visual distance and oral communication distance of the workers.

- Have no other duties at the time of monitoring that might distract from the monitoring duties.

Rescuing a Fallen Worker

Even the best fall protection system is not infallible. Falls will sometimes still occur in spite of the best efforts of committed professionals. Consequently, it is important for construction professionals to understand how to respond quickly and effectively when a fall occurs. What follows are recommendations for responding to a fall:

- *Have a comprehensive rescue plan already in place and readily accessible and train all workers in the procedures it contains.* Planning for falls is critical. After the fall, it is too late. In fact, OSHA requires that companies have a comprehensive rescue plan in writing, that all workers be trained in the procedures contained in the plan, and that the plan be readily accessible. The plan should contain procedures for: (1) preventing prolonged suspension, (2) identifying orthostatic intolerance signs and symptoms, and (3) rapid rescue and treatment procedures. All workers should be trained concerning the rescue plan so that they know how to properly and immediately respond in the event of a fall.
- *Respond quickly to the fall.* OSHA requires that medical treatment be provided to fallen workers within four to six minutes (promptness is especially important if workers are suspended or unconscious).
- *Monitor and communicate continually.* Monitor fallen workers constantly for signs of medical distress (i.e., dizziness, nausea, difficulty breathing, impaired vision, low heart rate, etc.). Get professional medical attention to the worker at the first sign of distress. For suspended workers, communicate continually so they know they are being rescued. The goal of this communication is to prevent further injuries caused by panic.
- *Stick to the rescue plan.* Workers should have been trained on all applicable procedures in the rescue plan and should, therefore, know what to do when a fall occurs. Stick to the plan. Do not confuse workers by going outside of the plan unless improvisation is absolutely necessary to deal with circumstances not anticipated in the plan.
- *Get the fallen worker to safe ground and accessible to medical care.* Get the fallen worker to safe ground quickly. If nothing more than first aid is required, administer it on the spot. However, if medical treatment is needed, make sure that emergency personnel have access to the injured worker for the quickest possible evacuation.
- *Collect all fall protection equipment that was in use during the fall.* All fall protection equipment in use when the fall occurred should be collected and protected from tampering. It should be taken out of commission until it can be inspected to determine if equipment failure was a factor in the fall.

PROTECTION FROM FALLING OBJECTS

The fall hazard that gets the most attention in scaffold work is height. However, this is not the only fall hazard associated with scaffold work. There is also the hazard of objects, such as tools or building materials, falling on workers below the scaffold. When this happens, there is a substantial risk of acceleration and impact injuries. Four principal methods are used to protect employees from objects that might fall from scaffolding. These methods are toeboards, guardrails, proper storage and removal of materials, and canopies.

Toeboards, Guardrails, and Canopies

Toeboards used for protection against falling objects must be erected along the edge of the walking surface for as long a distance as necessary to protect workers below. The toeboards must be able to withstand a downward or outward force of at least 50 pounds, and they must extend at least 3½ inches in height above the walking surface. Toeboards should be constructed of solid material or have openings that are no more than 1 inch in their greatest dimension. The openings in guardrails must be small enough to prevent the passage of tools or other objects that might fall on workers below. Canopies used as protection against falling objects must be sufficiently strong to withstand the force of objects that may fall on them without failing by collapse or penetration.

Proper Storage and Removal of Materials

Masonry and roofing work by their nature introduce falling object hazards. To guard against these hazards, masonry materials and equipment must be stored at least 4 feet back from the working edge. This distance increases to 6 feet in the case of roofing work (unless guardrails are installed). Waste and excess masonry materials should be removed regularly from the work area. In the case of roofing work, materials that are stacked, grouped, or piled near the edge of the roof must be stable and self-supporting.

FALL PROTECTION PLAN

In the rare case when using conventional fall protection methods actually makes the work more hazardous or when conventional methods are not feasible, contractors undertaking leading edge work, precast concrete erection, or residential construction may elect to use this option. A fall protection plan used in lieu of conventional fall protection methods must meet the following requirements:

1. The plan must be site specific and prepared by a qualified individual. It should be kept up-to-date, and any changes to it must be approved by the qualified person.
2. A copy of the latest edition of the plan must be available at the job site.
3. Implementation of the plan must be supervised by a competent person.
4. The plan must document why conventional fall protection methods are not feasible.
5. The plan must contain an explanation of the methods that will be used to provide fall protection instead of the conventional methods that would normally be used.
6. The plan must identify specific zones at the job site in which conventional fall protection methods cannot be used and designate these zones as controlled access zones. Names of employees allowed access to controlled access zones must be listed in the plan.
7. The plan must contain provisions for safety monitoring in accordance with 29 CFR 1926.502(h).
8. The plan must be subject to revision (e.g., new practices, procedures) if an employee falls or a near miss occurs.

FALL PROTECTION TRAINING

Employers are required by OSHA to provide whatever safety and health training is necessary to ensure that workers are able to perform their duties safely. The general requirements with regard to training are as follows: (1) establish and supervise programs of education and training in recognition, avoidance, and prevention of unsafe employment and (2) instruct employees in the recognition and avoidance of unsafe conditions and in the recognition of the regulations applicable to their work environment to control or eliminate hazards there.

More specifically, the fall protection portion of the standard requires that the following types of training be provided by an individual who is competent in all of these areas:

1. Nature of fall hazards in the workplace.
2. Proper procedures for erecting, maintaining, disassembling, and inspecting the types of fall protection systems to be used.
3. Proper use and operation of guardrail, personal fall arrest, safety net, warning line, safety monitoring, controlled access systems, and any other systems that are used.
4. Role of each employee in the safety monitoring system when such a system is to be used.
5. Limitations and proper use of mechanical equipment during the performance of work on low-sloped roofs.
6. Proper procedures for handling and storing materials and for the erection of overhead protection.
7. Role of employees in fall protection plans.
8. Standards contained in Subpart M of 29 CFR 1926.

Certification of Training

Certification of training is critical. Contractors are required to verify that the necessary training has been provided for workers. This is accomplished by maintaining a written certification record containing the following information: (1) name of the worker who received the training, (2) dates of the training, and (3) signature of either the employer or the competent person who provided the training. This signature certifies that the training was provided and completed.

Retraining

Retraining is not required for workers who were trained by other employers as long as they demonstrate an understanding of the various subjects that are to be covered in a training program. A competent person or employer may verify that training was provided by another employer for certification purposes. If the date of the training is not known, the date when it was determined that the previous training was adequate may be used. Fall protection retraining is required only when one of the following conditions exists:

1. Changes in conditions at the job site render the previous training obsolete.
2. Changes in fall protection systems or equipment render the previous training obsolete.
3. Inadequacies in a worker's understanding of fall protection methods indicate the need for additional training.

ANSI Z359: FALL PROTECTION CODE

Widely recognized fall protection standards for the construction industry include OSHA's standard (29 CFR 1926, Subpart M) as well as two ANSI standards: ANSI Z359 – Fall Protection Code and ANSI A10.32 – Construction Fall Protection. Construction companies must comply with OSHA's fall protection requirements set forth in 29 CFR 1926, Subpart M, Appendix C. The ANSI fall protection standards are not mandatory, but they can be helpful for companies that are establishing comprehensive fall protection programs. The ANSI standards—particularly ANSI Z359—are important enough that construction professionals should be well versed in their requirements. The critical components of ANSI Z359 are as follows:

- ANSI Z359.0: Definitions and nomenclature used for fall protection and fall arrest.
- ANSI Z359.1: Safety requirements for personal fall arrest systems, subsystems, and components.
- ANSI Z359.2: Minimum requirements for a comprehensive managed fall protection program.
- ANSI Z359.3: Safety requirements for positioning and travel restraint systems.
- ANSI Z359.4: Safety requirements for assisted-rescue and self-rescue systems, subsystems, and components.
- ANSI Z359.6: Specifications and design requirements for active fall protection systems.
- ANSI Z359.7: Qualification and verification testing of fall protection products.
- ANSI Z359.12: Connecting components for personal fall arrest systems.
- ANSI Z359.13: Personal energy absorbers and energy absorbing lanyards.
- ANSI Z359.14: Self-retracting devices for personal fall arrest and rescue systems.

ANSI Z359.2: Managed Fall Protection Program

Several of the components of ANSI Z359 provide guidance for designers, manufacturers, and testers of fall protection systems, subsystems, and components. However, ANSI Z359.2 applies directly to construction companies in that it provides guidance for establishing a comprehensive fall protection program. ANSI Z359.2 provides guidelines for appropriate policies, responsibilities/duties, training, fall protection procedures, rescue procedures, incident investigations, and program evaluation. What follows is a sample outline for an ANSI Z359 approved fall protection program:

- Develop a fall protection policy that established the scope of the program.
- Develop specific goals for the program.
- Conduct a comprehensive hazard analysis and record all known fall hazards.
- Develop measures for eliminating, preventing, mitigating, or controlling all known fall hazards identified.
- Plan and implement fall protection training.
- Develop job-site specific plans for all job sites (these plans are comprised of the safety measures that are specific to the job site in question).
- Establish and implement procedures maintaining and regularly inspecting fall protection equipment.
- Develop a comprehensive rescue plan and include it in the training provided for workers.
- Monitor the program continually and conduct periodic audits to ensure its viability. Make corrections, adjustments, and improvements the moment the need for them is identified. Revise the written program accordingly.

Other Applicable Standards

In addition to 29 CFR 1926, Subpart M and ANSI Z359, there are other fall protection related standards that construction professionals may find useful. These include:

- ANSI A10: Safety requirements for construction and demolition operations
- ANSI Z117: Confined spaces
- ANSI Z535: Safety signs and colors

SUBPART N: HOISTS, ELEVATORS, AND CONVEYORS AND RELATED SAFETY PRACTICES

Subpart N covers the requirements for construction companies that use cranes, derricks, hoists, helicopters, conveyors, and aerial lifts. This section sets forth the requirements for all of this equipment and emphasizes the importance of adhering to the manufacturer's specifications concerning loading capacities, speed limits, special hazards, and unique equipment characteristics. A competent person is required to inspect cranes and derricks before daily use. Comprehensive records must be maintained showing that inspections actually occur as required. Subpart N contains the following sections:

- 1926.550 (Reserved)
- 1926.551 Helicopters
- 1926.552 Material hoists, personnel hoists, and elevators
- 1926.553 Base-mounted drum hoists
- 1926.554 Overhead hoists
- 1926.555 Conveyors
- 1926.556 Aerial lifts

Hoists and elevators used in the construction industry are classified as material hoists, personnel hoists, base-mounted drum hoists, and overhead hoists. There are specific requirements and safety practices related to each type of hoist. Requirements related to conveyors have to do primarily with having the appropriate "on/off" mechanism and providing an appropriate audible warning system before start-up.

Material Hoists

When using **material hoists**, it is important to comply with the specifications and limitations provided by the manufacturer. If manufacturer's specifications are not available, the construction company must obtain specifications from a competent engineer. In addition, critical information must be posted in clearly visible locations on cars and platforms. This information includes rated load capacities, recommended operating speeds, special hazard warnings, and any special instructions that may apply.

Ensuring that the wire rope used with material hoists is properly installed and capable of carrying the expected maximum load is important. Manufacturer's specifications must be followed for installation and load requirements. Wire rope must be removed and replaced when even one of the following conditions exists:

1. Six randomly distributed broken wires occur in one rope lay, or three broken wires in one strand in one rope lay.
2. One third or more of the original diameter of outside wires is lost due to abrasion, scrubbing, flattening, preening, or any other type of wear.
3. Ropes are damaged by either heat or electricity.
4. Reduction in the nominal diameter of wire ropes is as follows: (a) more than $\frac{3}{64}$ inch for ropes with a diameter up to $\frac{3}{4}$ inch, (b) $\frac{1}{16}$ inch for ropes with a diameter ranging from $\frac{7}{8}$ inch to $1\frac{1}{8}$ inches, and (c) $\frac{3}{32}$ inch for ropes with a diameter ranging from $1\frac{1}{4}$ inches to $1\frac{1}{2}$ inches.

Entrance openings to hoists must be guarded by substantial gates or bars that cover the entire width of the opening. Overhead protection must be provided on the top of all material hoist cages or platforms (2-inch planking, $\frac{3}{4}$-inch plywood, or equivalent material). Car arresting devices must be installed to automatically function in the case of rope failure. Enclosed hoist towers must be enclosed on all sides for the entire height of the hoist with a screen of at least $\frac{1}{2}$-inch mesh, No. 18 U.S. gauge wire, or equivalent. These requirements and practices apply to all of the hoists covered in this section. In addition, there are other requirements and practices that apply to each specific type of hoist.

Personnel Hoists

Towers for **personnel hoists** outside the structure must be enclosed on all sides. In addition, towers must be anchored to the structure at least every 25 feet. Hoist cars are to be enclosed on all sides and the top (except the gates or doorways). Doors or gates must be at least 6 feet, 6 inches high and must be equipped with mechanical locks that cannot be operated from the landing side.

Traction hoists must have a minimum of three ropes, and drum-hoists must have a minimum of two. Personnel hoists must be constructed of materials and components that meet the specification set forth in ANSI A10.4, "Safety Requirements for Workmen's Hoists."

Overhead Hoists

Overhead hoists are to be used only in locations that allow the operator to stand clear when the hoist is in operation. Air-powered hoists must be connected to an air supply with sufficient capacity to safely drive the hoist under the anticipated load. Air hoses must be connected in a way that prevents their disconnection during operation.

Base-Mounted Drum Hoists

Appropriate guards must be provided for **base-mounted drum hoists** that are exposed to moving parts (e.g., gears, chains, cables). All controls must be within easy reach of the operator's station. A remotely operated hoist stop must be available so that it can be activated if any other control fails. Hoists operated by electric motors must have a disconnect device that can be activated when the electric power fails.

Conveyors

Conveyors must be equipped with an "on/off" switch that is accessible to workers and an emergency cutoff switch. There must be an audible warning system that alerts personnel in the vicinity that the conveyor is about to start up. Workers must not be allowed to ride conveyors. In addition, conveyors should be equipped with appropriate guards to shield workers from moving parts. Conveyors used above ground should be equipped with overhead protection. All conveyors should be blocked off by barricades to prevent workers from trying to pass, step, or jump over them.

HELICOPTERS

Using helicopters for lifting purposes in construction is a highly specialized operation that requires personnel with the appropriate training. Any helicopter used in construction must comply not just with the applicable OSHA regulations, but with the Federal Aviation Administration (FAA) regulations too. Before every operation involving a helicopter, a plan of operation must be developed, and the pilot and ground crew must be briefed on the plan.

Workers on the ground crew must wear the personal protective equipment that is called for in the situation. This might change somewhat from situation to situation, but should always include eye protection and a hard hat equipped with a chinstrap. Loose clothing and anything else that might flap in the breeze should be avoided. The landing or loading site should be cleared of anything that might be blown around by the downwash of the helicopter. The clearance zone should extend at least 100 feet beyond the perimeter of the landing or loading site.

Ground-crew personnel should work under a hovering helicopter only when absolutely necessary to hook or unhook loads. When such work is necessary, a safe means must be provided for workers to quickly and easily reach the hoist line hook and to engage or disengage the cargo slings. A means of dissipating static electricity must be provided and used before ground crew are allowed to touch a load suspended from a hovering helicopter. When such a static dissipation device is not available, rubber gloves must be worn by any ground crew personnel who might touch the load.

Visibility is an important factor when unloading a helicopter. Appropriate precautions must be taken to minimize visibility problems that might be caused by the downwash. Ground crew members should be especially careful to stay clear of the helicopter's rotors when visibility is reduced. Construction personnel not involved in the ground crew's operations should stay clear of the landing or loading site by at least 50 feet.

There should be continual communication between the pilot and a designated member of the ground crew. Communication is achieved by the use of hand and arm signals (Figure 15–19). The signalperson must be easily recognizable by all personnel, including the pilot, and distinguishable from other members of the ground crew.

AERIAL LIFTS

In addition to meeting the requirements set forth in Subpart N, construction companies that use aerial lifts must use only lifts that satisfy the guidelines contained in ANSI 92.2, "Vehicle-Mounted, Elevated, and Rotating Platforms." Aerial lifts (Figure 15–20) used in construction fall into one of the following categories: (1) extensible boom platforms, (2) aerial ladders, (3) articulating boom platforms, or (4) vertical towers.

It is important to adhere to all manufacturers' specifications concerning load limits and equipment operation. In addition, the boom and ladder must be secured and locked before movement of a lift. Lifts that transport workers must be equipped with complete lower and upper controls. The lower controls must be able to override the upper controls. Both sets of controls should be labeled to show their functions, and they should be tested every day.

Personnel who work in the basket of a boom should wear a body harness with a lanyard that is securely fastened to either the boom or the basket. Lanyards should never be attached to other equipment or pools.

SUBPART O: MOTOR VEHICLES, MECHANIZED EQUIPMENT, AND MARINE OPERATIONS AND RELATED SAFETY PRACTICES

Subpart O covers the requirements for motor vehicles, including industrial trucks (forklifts), that operate at off-the-highway job sites. In addition, the requirements related to marine operations are covered in this subpart. Subpart O contains the following sections:

- 1926.600 Equipment
- 1926.601 Motor vehicles
- 1926.602 Material handling equipment
- 1926.603 Pile driving equipment
- 1926.604 Site clearing
- 1926.605 Marine operations and equipment
- 1926.606 Definitions applicable to this subpart

MOTOR VEHICLES AND MECHANIZED EQUIPMENT: REQUIREMENTS AND RELATED SAFETY PRACTICES

There are general requirements for motor vehicles in Subpart O and additional requirements specifically for industrial trucks (forklifts and **forked trucks**). All motor vehicles must have a system of operating, emergency, and parking brakes. They must also have two headlights, two taillights, brake lights, an audible warning device for the use of the operator (such as a horn), and an audible alarm system for backing up. An observer can be used for backing up if the motor vehicle is not equipped with a backup alarm. The cabs of motor vehicles must be equipped with safety glass, windshield wipers, and a system for defogging the windows. Motor vehicles that transport workers must be equipped with seats and safety belts for every passenger.

Requirements for Industrial Trucks

What construction professionals typically call a forklift, OSHA calls a powered **industrial truck**, Figure 15–21.[13] OSHA's requirements for industrial trucks are contained in 29 CFR 1910.178 and 29 CFR 1926.602. The general requirements of this standard are as follows:

- *Truck operations, 29 CFR 1910.178(m):* There are numerous specific requirements concerning truck operations. Those listed herein are provided as representative examples. Construction professionals and students should refer to 29 CFR 1910.178(m) for a complete list of all requirements. Examples of

MOVE RIGHT

Left arm extended horizontally; right arm sweeps upward to position over head.

The signal "Hold" is executed by placing arms over head with clenched fists.

Right arm extended horizontally; left arm sweeps upward to position over head.

Right hand behind back; left hand pointing up.

Combination of arm and hand movement in a collecting motion pulling toward body.

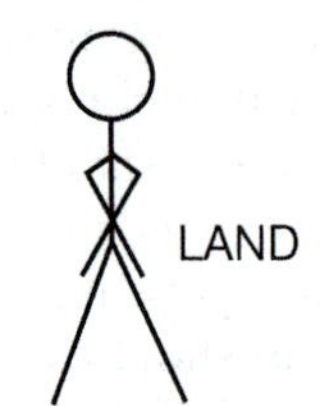

Arms crossed in front of body and pointing downward.

Hands above arm, palms out using a noticeable shoving motion.

Arms extended, palms up; arms sweeping up.

Left arm held down away from body. Right arm cuts across left arm in a slashing movement from above.

MOVE DOWNWARD

Arms extended, palms down; arms sweeping down.

FIGURE 15–19 Hand and arm signals used with helicopters in construction.
Source: 29 CFR 1926.551.

the types of requirements pertaining to *truck operation* are: (a) Trucks may not be driven up to anyone standing in front of a bench or other fixed object; (b) no person is allowed to stand or pass under the elevated portion of any truck, whether loaded or unloaded; and (c) unauthorized personnel are not allowed to ride on a truck.

- *Traveling, 29 CFR 1910.178(n):* There are numerous specific requirements concerning industrial trucks that are traveling. Those listed herein are provided as representative examples. Construction professionals and students should refer to 29 CFR 1910.178(n) for a complete list of all requirements. Examples of the types of requirements pertaining

FIGURE 15–20 Aerial lift commonly used in construction.
Source: Hipercom/Fotolia

to *traveling* are: (a) All traffic regulations must be observed including authorized plant (or site) speed limits. A safe distance shall be maintained approximately three truck lengths from the truck ahead, and the truck shall be kept under control at all times; (b) right of way shall be yielded to ambulances, fire trucks, or other vehicles in emergency situations; and (c) other trucks traveling in the same direction at intersections, blind spots, or dangerous locations shall not be passed.

- *Loading, 29 CFR 1910.178(o):* There are several specific requirements concerning truck loading. Those listed herein are provided as representative examples. Construction professionals and students should refer to 29 CFR 1910.178(o) for a complete list of all requirements. Examples of the types of requirements pertaining to truck *loading* are: (a) only stable or safely arranged loads shall be maintained. Caution will be observed when handling off-center loads that cannot be centered; (b) only loads within the rated capacity of the truck shall be handled; and (c) a load engaging means shall be placed under the load as far as possible; the mast shall be carefully tilted backward to stabilize the load.
- *Operation of the truck, 29 CFR 1910.178 (p):* There are several specific requirements concerning operation of industrial trucks. Those listed herein are provided as representative examples. Construction professionals and students should refer to 29 CFR 1910.178(p) for a complete list of all requirements. Examples of the types of requirements pertaining to the operation of industrial trucks are: (a) if at any time a powered industrial truck is found to be in need of repair, defective, or in any way unsafe, the truck must be taken out of service until it has been completely restored to safe operating condition; (b) fuel tanks must not be filled while the engine is running. Spillage of fuel shall be avoided; (c) no truck shall be operated with a leak in the fuel system until the leak has been corrected.
- *Maintenance of industrial trucks, 29 CFR 1910.178(q):* There are numerous specific requirements concerning maintenance of industrial trucks. Those listed herein are provided as representative examples. Construction professionals and students should refer to 29 CFR 1910.178(q) for a complete list of all requirements. Examples of the types of requirements concerning maintenance of industrial trucks are: (a) any power-operated industrial truck not in safe working condition must be removed from service and all repairs must be made by authorized personnel; (b) no repairs shall be made in Class I, II, and III locations; and (c) repairs to fuel and injection systems which involve fire hazards shall be conducted only in locations designated for such repairs.

FIGURE 15–21 Forked truck.
Source: Ftfoxfoto/Fotolia

Forked Truck Training Requirements

All persons who operate a forked truck in a construction setting must receive the training specified in 29 CFR 1926.602(d).[14] Employers are required to develop a complete training program. Training for operators of forked trucks must consist of classroom training (concepts and principles) and practical training (hands-on application). The training must cover hazards of operating a forked truck in a construction setting, proper operation of a forked truck, and OSHA's requirements for the safe operation of a forked truck.

Operators who have completed the required training must be evaluated in a live setting while they operate a forked truck. The evaluation is based on the following criteria:

1. Did the operator complete the necessary training? In what formats (classroom and practical)?
2. Was the training provided by an individual who has the competence (i.e., training, knowledge, and experience) to train and evaluate forked truck operators?

3. Were all of the required topics covered during the training (Figure 15–22)?
4. Did the operator receive training in the operating instructions, warnings, and precautions explained in the operator's manual for the type of forked truck the operator is being trained to use (including operator restraint systems and seat belts)?
5. Does the employee properly and safely operate the forked truck in question in a live setting?
6. Has the employer certified that the proper training has been provided and completed?

Once operators have completed their initial training, they are to be evaluated at least every three years and must receive refresher training whenever there is a need. The requirements for refresher training are covered in the next section.

Forked Truck Evaluation and Refresher Requirements

Industrial truck operators must be evaluated at least once every three years to ensure that their knowledge and skills are up-to-date.[15] If the evaluation reveals the need, refresher training is required. There are also several other circumstances that, when present, mandate that refresher training be provided. The various "triggers" for refresher training are as follows:

- Operator is involved in an accident or a near miss.
- Operator is observed operating the industrial truck in a manner that is reckless, negligent, or otherwise unsafe.
- Evaluation reveals the need for refresher training.
- Work environment has changed in a way or ways that could affect the operation of the industrial truck.
- Operator is assigned to a different type of industrial truck.

Checklist of OSHA Training Requirement for Forklift Operators
Truck-Related Requirements
Operating instructions and precautions for the specific type of forklift
Differences between driving forklifts and automobiles
Controls and instrumentation for the forklift
Engine or motor operation
Steering and maneuvering
Visibility (including restriction caused by loading)
Operation, adaptation, and limitations of the fork and attachments
Capacity and stability of the forklift
Vehicle maintenance and inspection
Batteries (charging, recharging, and refueling)
Operating limitations for the forklift
Any other pertinent information from the operator's manual
Workplace-Related Requirements
Surface conditions where the forklift will be operated
Load stability and makeup
Stacking, unstacking, and load manipulation
Pedestrian traffic in the vicinity
Narrow isles and other areas of restricted operation
Hazardous areas of operation
Ramps and other sloped surfaces
Operating in closed and unventilated environments
Other hazardous conditions in the operating environment

FIGURE 15–22 Forklift operators must undergo OSHA-mandated training.

Requirements for Dump-Body Trucks

Trucks with dump bodies, hereafter referred to as dump trucks, must be equipped with a system to prevent accidental lowering of the body during maintenance or inspection work.[16] Operating levers and dumping devices must be equipped with a system to prevent an accidental tripping of the dumping mechanism. Trip handles on the tailgates must be located so that the operator is clear when activating the handle. Rubber-tired vehicles must have fenders unless they are not designed by the manufacturer for fenders. In such cases, they must have mud flaps.

Mechanized Equipment

Equipment that is left unattended at night, whether near a highway or on a construction site, must be equipped with lights, reflectors, or barricades to clearly identify its location.[17] Equipment must be inspected by competent supervisory personnel and deemed to be in safe operating condition before it is used.

Load capacity rating and safe operation rules for mechanized equipment must be conspicuously posted at the operator's station for each piece of equipment. A portable fire extinguisher with a rating of 5 BC or more must be available and easily accessible at the stations of all equipment operators.

MARINE OPERATIONS AND EQUIPMENT: REQUIREMENTS AND RELATED SAFETY PRACTICES

Marine operations, in the context of OSHA's construction standard, include the loading, unloading, handling, or moving of construction materials, equipment, and supplies into a vessel.[18] When conducting marine operations, a safe ramp or walkway must be provided, unless it is safe to board the vessel without them. When a Jacob's ladder is used, it must be of the double-rung or flat-tread type, and it must hang without slack from its lashings or be pulled up entirely.

When the upper end of the ramp or walkway is flush with the bulwark or rests on it, a handrail at least 33 inches high must be provided between the top of the bulwark and the deck. Substantial steps that are properly secured are also required. The ramp or walkway must be kept free of obstructions, and it must be illuminated for its entire length. The ramp or walkway should be located so that the load it carries does not pass over workers.

Workers on barges must not be allowed to walk along the sides or around deckloads fore or aft, unless there is safe passage (e.g., walkway, grabrail, taut handline). Each barge in use for marine operations must be equipped with at least one 30-inch life ring (U.S. Coast Guard approved) with 90 or more feet of line attached. In addition, each barge in use must be equipped with at least one ladder (portable or permanent) that will reach from the top of the apron to the surface of the water. Workers on unguarded decks must wear U.S. Coast Guard–approved life jackets.

Summary

Subpart K of OSHA's construction standard covers the temporary and permanent installation and the ongoing operation of electric power and electrically powered devices and equipment in construction projects. Subpart K does not apply to permanently installed electrical power put in place before the current construction project. There are four major areas of emphasis in Subpart K: (1) safety requirements for installation, (2) safety-related work practices, (3) safety-related maintenance and environmental consideration, and (4) safety requirements for special equipment.

Subpart L covers all OSHA requirements for the erection and use of scaffolding for performing elevated work. This subpart covers the requirements as they relate to general issues, specific types of scaffolds, and aerial lifts. It also covers the requirements related to training for personnel who work on scaffolds.

Subpart N covers the requirements related to the safe operation of cranes, derricks, hoists, helicopters, conveyors, and aerial lifts. This subpart sets forth the requirements for all of these various types of equipment and emphasizes the importance of adhering to the manufacturer's specifications concerning loading capacities, speed limits, special hazards, and unique equipment characteristics. The requirements for having a competent person inspect cranes and derricks before daily use and for keeping comprehensive records of inspections are also explained.

Subpart O covers the requirements for motor vehicles that operate at off-the-highway locations. The term *motor vehicle* includes industrial trucks (also called forklifts or forked trucks). The requirements for construction companies that perform maritime operations are also covered in Subpart O. The requirements in this subpart deal with motor vehicles, material handling equipment, pile driving equipment, site clearing, and marine operations and equipment.

Key Terms and Concepts

Amperes
Annual inspections
Base-mounted drum hoists
Body belt
Body harness
Certification of training
Competent person
Conductors
Controlled access zone
Conveyors
Electrical hazards
Electricity
Electrons
Equipment knowledge
Fall protection systems
Forked trucks
Forklift
Freeze
Frequent inspections
Grounded conductor
Guardrails
Industrial trucks
Let-go current
Lifeline
Material hoists
Maximum intended load
Monthly inspections
Neutrons
Ohms
Overhead hoists
Periodic inspections
Personal fall arrest system
Personnel hoists
Protons
Rated load
Resistance
Retraining
Safety monitoring systems
Scaffold
Semiconductors
Short circuit
Static electricity
Toeboards
Trigger height
Voltage
Z359

Review Questions

1. What are the four major areas of emphasis in Subpart K?
2. Define the term *electricity*.
3. What is a conductor? A semiconductor?
4. What is *static electricity,* and what hazards can it pose in a work setting?
5. Define the term *electrical hazard*.
6. What is *frequency* as it relates to electricity, and how is it measured?
7. List four of the major sources of electrical hazards in the workplace.
8. List three things that should be checked before proceeding with the installation and use of electrical equipment.
9. Explain two methods that can be used for guarding energized parts.

10. There are cases in which equipment connected by a plug or cord does not have to be grounded. What are these exceptions?
11. Certain types of electrical equipment are designed for safe use in hazardous conditions. What are these conditions?
12. Describe the safe way to charge batteries at a job site.
13. What is a scaffold?
14. Describe the requirements for platform construction for scaffolds.
15. What are the requirements for the proper width of scaffold platforms?
16. What are the requirements that must be met with suspension scaffolds?
17. List four practices for safe scaffold use.
18. List the precautions construction personnel should take to protect employees working below scaffolding.
19. What is an aerial lift?
20. Describe the safe practices for workers in the basket of an aerial lift.
21. List the five areas that must be covered in training programs for employees who work on scaffolds.
22. Because falls are a leading cause of death in the construction industry, OSHA's requirements related to fall protection are very specific. What are the five things construction companies must do to protect their workers from fall hazards?
23. Construction companies engaged in certain types of work may develop a fall protection plan in lieu of applying the normal fall protection methods. What are these types of work?
24. What are the requirements that a fall protection plan must meet?
25. What are the major areas that fall protection training must cover?
26. Fall protection retraining is required only in certain circumstances. What are they?
27. Summarize briefly the general requirements and related safety practices for the operation of cranes and derricks.
28. When certain conditions exist, wire rope used with hoists must be replaced. What are these conditions?
29. Summarize the requirements and related safety practices for conveyors in construction.
30. When working with a helicopter in construction, what are the hand and arm signals for the following (sketch your answers): move right, move upward, release sling load.

Critical Thinking and Discussion Activities

1. You are the superintendent for job site 31 for your construction company. You are attending a meeting this morning of all of the company's site superintendents and will not get out to job site 31 until noon at the earliest. You receive a call from one of your supervisors at the job site who says, "The supervisor who normally checks our electrical equipment before we start it up each morning just called in sick. You are the only other person qualified to conduct the preoperation inspections. We really need this equipment, or I'm going to have a crew standing around doing nothing until you get here at noon. Can we just skip the inspections this time?" What would you do if you were this job-site superintendent? Assume that the meeting you are attending is critical, and you cannot skip it.
2. You have just taken over as the safety director for Jones Construction Company (JCC), a company that hired you because it has a poor safety record and just received a serious citation and substantial fine from OSHA (an employee was killed on a JCC job site). It is your first day on the job, and you are astounded at the unsafe behavior you see on scaffolds. In spite of its poor safety record, JCC has never had a scaffold-related accident. You are trying to decide if you should stop all work on the scaffolds until OSHA's requirements for safe scaffolding are met or let the work continue while you take each situation you find one at a time. What do you think should be done here? Assume the job in question is already behind schedule.
3. You are a recent college graduate, and this is your first real job. You are the assistant director of safety for Rayster Construction, Inc. (RCI). Your job today is to conduct preoperation inspections of the company's two cranes that are being used to construct a bridge across a small bay. As you begin your first inspection, the crane operator—who has worked for RCI for 25 years—brushes past you and starts the crane's engine. He says, "Get out of the way, kid. This crane is just fine. It doesn't need inspecting. You are holding up my work." This crane operator has an excellent safety record, as does RCI in general. How should you handle this situation?
4. The new forklift operator at your job site concerns you. He has had the proper training, but you wouldn't know it by his actions. You are the job-site supervisor on this project, and this employee worries you. He has no interest in retraining and does not think he needs it. What should you do in this situation?

Application Activities

1. Get permission to visit a construction site in your community. Conduct a walk-through, paying special attention to electrical equipment. Can you identify any situations in which electrical equipment is not properly guarded? Write your findings without naming the company in question.
2. Repeat Activity 1, paying special attention to scaffolding.
3. Identify a construction company that has developed a fall protection plan in lieu of using the normal fall protection methods. Get permission to review the plan. How does the company protect its employees from fall-related hazards?
4. Get permission to visit a job site where a crane is being used. When the crane is not in operation, get permission to enter the cab. Are safe operating instructions and checklists properly posted? Write your findings without naming the company.
5. Get permission to conduct a walk-through at a job site where hoists are used. Examine the wire ropes on the hoist. Should any of the ropes be replaced? Are the hoist towers properly enclosed? Are the entrances to hoists properly guarded? Write your findings without naming the construction company.
6. Identify a construction company that uses forked trucks. Ask to review their safety training program for operators. Critique the program in writing without naming the company.

Endnotes

1. Fish, Raymond M. "Conduction of Electrical Current to and Through the Human Body: A Review." Retrieved from http://www.ncbi.nlm.nih.gov/pmc/articles/PMC2763825/ on January 10, 2015.
2. 29 CFR 1926.403
3. 29 CFR 1926.416 and 1926.417
4. 29 CFR 1926.431 and 1926.432
5. 29 CFR 1926.441
6. OSHA 3150 (Revised). "Scaffold Use in the Construction Industry." Retrieved from www.osha.gov on January 10, 2015.
7. Ibid., 9–29.
8. 20 CFR 1926.451(g)(2)
9. 29 CFR 1926.502(d)(15)
10. 29 CFR 1926.451(d)(16)
11. 29 CFR 1926.550
12. 29 CFR 1926.32
13. 29 CFR 1926.600 and 1926.602(c)
14. 29 CFR 1926.600 and 1926.602(d)
15. Ibid.
16. 29 CFR 1926.601
17. Ibid.
18. 29 CFR 1926.605

CHAPTER SIXTEEN

SUBPARTS P THROUGH U AND RELATED SAFETY PRACTICES

LEARNING OBJECTIVES

- Explain the main provisions of Subpart P.
- Summarize the main provisions of Subpart Q.
- Explain the main provisions of Subpart R.
- Describe the main provisions contained in Subpart S.
- Summarize the main provisions of Subpart T.
- Summarize the main provisions of Subpart U.

Subparts P through U contain many critical requirements of OSHA's construction safety and health standard (29 CFR 1926). Included in this chapter are the requirements and corresponding safety practices related to the following topics: excavations; concrete and masonry work; **tunnels and shafts**, caissons, cofferdams, and compressed air; and demolition.

SUBPART P: EXCAVATIONS AND RELATED SAFETY PRACTICES

Excavation work is dangerous. Every year in the United States, about 200 construction workers engaged in trench work die. Soil weighs approximately 100 pounds per cubic foot. Consequently, a cave-in can easily crush a human, breaking bones and damaging organs. An even greater danger is that of suffocation. Even partial engulfment of the body can be enough to crush and suffocate a worker. Rescue operations in excavation work are also dangerous. Workers who have not been trained in excavation rescue operations often do more harm than good. Buried workers have been injured more severely and even killed when untrained rescuers tried to unearth them using improper procedures. On other occasions, the rescuers themselves have become victims.

The most common hazards of excavation accidents include cave-ins, contact with energized power sources or conductors, toxic atmospheres, loose rocks, rising water, and the collapse of nearby structures or equipment. Subpart P covers the general requirements for excavations and specific requirements for cave-ins, protective support systems, and other related hazards. Subpart P contains the following sections:

- 1926.650 Scope, application, and definitions
- 1926.651 Specific excavation requirements
- 1926.652 Requirements for protective systems

EXCAVATIONS: REQUIREMENTS AND RELATED SAFETY PRACTICES

An *excavation* is a man-made cut, cavity, or trench in the ground made by removing earth. Subpart P applies to open excavations except surface mines and certain house foundation excavations. Mining excavations are covered by the Mine Safety and Health Administration.[1] The most common form of excavation in construction is the trench, which is a narrow excavation that is deeper than its wide and is no wider than 15 feet. Subpart P does *not* apply to house foundation or basement excavations when all of the following conditions exist:

- Excavation is less than 7½ feet deep.
- Excavation is benched for at least 2 feet horizontally for every 5 feet (or less) of depth.
- Minimum horizontal width at the bottom of the excavation is no less than 2 feet (or wider, if possible).
- No water, surface cracks, or other conditions exist that might reduce the stability of the excavation.
- No heavy equipment or machinery operate in the vicinity that might destabilize the excavation.
- All loose soil, equipment, and material are kept back from the edge of the excavation for a distance at least equal to the depth of the excavation.
- The number of workers in the excavation is the minimum needed to perform the work.
- Excavation work is carried out in such a way to minimize the time spent in the excavation.

Planning Requirements

Proper planning can help minimize the hazards associated with excavation work. Before undertaking such work, construction professionals should consider all of the following factors:

1. Traffic in the vicinity of the excavation (heavy or light? cars? trucks? heavy equipment?)

2. Proximity of structures or equipment to the excavation, condition of structures, type of equipment (does it cause vibration?)
3. Soil factors (moist or dry? packed or loose?)
4. Surface and ground water; depth at which the excavation will begin to accumulate water
5. Utilities (overhead and underground)
6. Weather (is rain likely?)

To answer the questions raised by these considerations, it is necessary to conduct onsite observations of the excavation site, take test borings for soil type and condition, and seek input from gas, electric, water, and telecommunications providers. Doing these things before beginning excavation can save time, trouble, and even lives. Using the information gathered, construction professionals should develop a plan for safely undertaking the excavation in question and share that plan with all workers who will participate.

Onsite Inspections

The Occupational Safety and Health Administration (OSHA) requires daily inspections by a competent person to detect possible problems or hazards in excavations and in the vicinity of excavations. Inspections must also be made following heavy rains or any man-made activities that might destabilize the excavation (e.g., blasting, operating vibrating equipment). In large operations, construction companies should assign a full-time safety professional as the competent person. In smaller operations, a supervisor can be assigned responsibility for conducting inspections.

CAVE-INS: REQUIREMENTS AND RELATED SAFETY PRACTICES

The most common hazard when conducting excavation work is the **cave-in**. To protect workers from this hazard, OSHA has set forth specific requirements and safety practices covering support systems, specific safety precautions to be taken, installation and removal of protective systems, and materials and equipment.[2]

Soil Types

Type A: The most stable of the three types of soil. Cohesive soil with an unconfined compressive strength of 1.5 tons per square foot or more. Examples of type A soil include clay, silty clay, sandy clay, and clay loam. Cemented soils, such as caliche and hardpan, are also considered type A.

Type B: This soil type is moderately stable. Cohesive soil with an unconfined compressive strength greater than 0.5 ton per square foot, but less than 1.5 tons per square foot. Examples of type B soil include angular gravel, silt, silt loam, and sandy loam.

Type C: This is the least stable of the three types of soil. Cohesive soil with an unconfined strength of 0.5 ton per square foot or less. Examples of type C soil include gravel, sand, and some sandy loams.

FIGURE 16–1 Soil types.

Support Systems

OSHA requires that workers in excavations be protected by one of the following methods: (1) sloping or benching the sides of the excavation, (2) supporting the sides of the excavation, or (3) placing a shield between the side of the excavation and the work area in the excavation. The requirements and practices for ensuring the safety of workers in an excavation depend on the type of soil at the site. Figure 16–1 contains an explanation of the three types of soils as set forth in Appendix A to Subpart P. Appendix B to Subpart P contains a variety of configurations for sloping and benching. Figure 16–2 illustrates the concept of sloping. Figure 16–3 illustrates the concept of benching. When sloping, the bottom-line rule to remember is that excavations up to 20 feet deep meet the requirements of safe practice as long as the sides slope at least 1.5:1 horizontal to vertical (see Figure 16–2).

Trench shields or trench boxes designed or approved by a registered professional engineer may also be used to protect workers in excavations (Figure 16–4 and Figure 16–5). Employers have some latitude with regard to the design of trench shields and the materials used in their

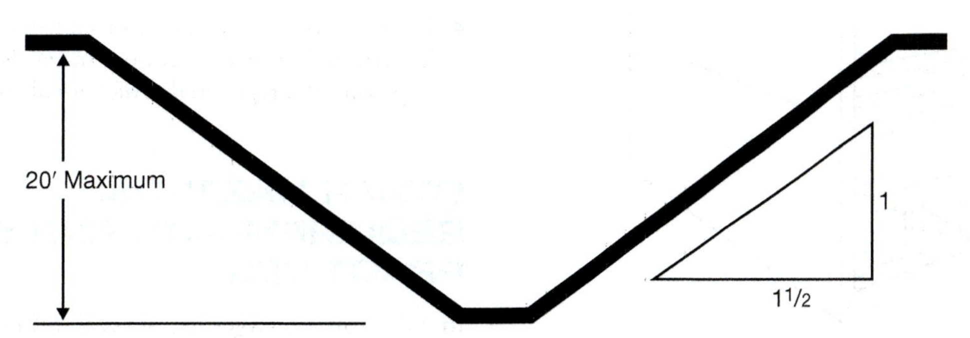

FIGURE 16–2 Sloping the sides of an excavation no less than 1.5 (horizontal):1 (vertical) helps to ensure worker safety in excavations up to 20 feet deep (even in type C soil).

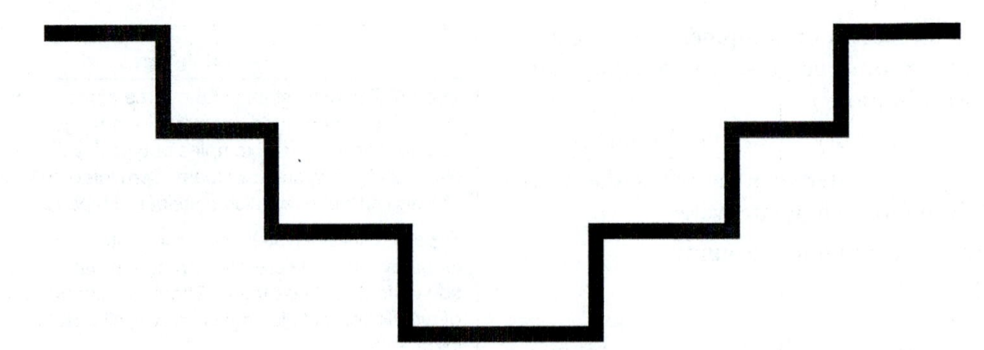

FIGURE 16–3 Excavations dug in this configuration are called *benched*.

construction. These structures must provide protection that is at least equivalent to or exceeds the protection of the shoring system that otherwise would have been used.

Specific Safety Precautions

Construction companies that undertake excavation work are required to provide support systems, such as shoring, bracing, or underpinning, to adequately stabilize structures in the vicinity of an excavation. These structures include walls, buildings, towers, sidewalks, and pavement. Excavation below the level of a foundation or footing of a nearby structure is prohibited, unless one of the following conditions exists or is provided:

1. Proper support system is constructed to ensure the stability of the structure.
2. Excavation is being made in stable rock.
3. Registered professional engineer determines and certifies that the structure is far enough away from the excavation in question that it poses no threat.

Installation and Removal of Protective Systems

Installing protective systems can be just as dangerous as working in an excavation. Consequently, OSHA requires the following safe practices when installing protective systems:

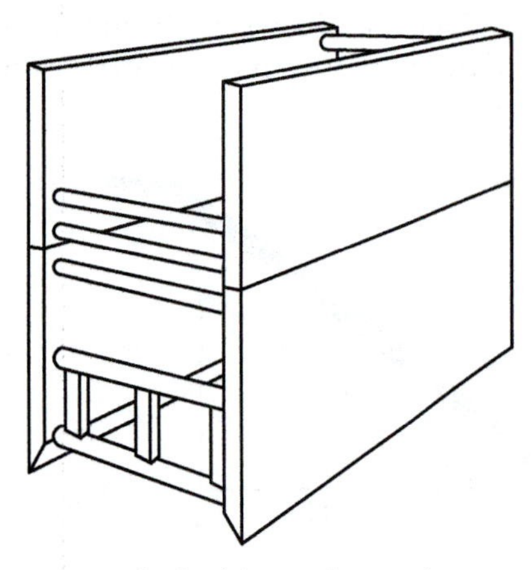

FIGURE 16–4 Trench shields can be used to protect workers from cave-ins.

- Make sure that all structural members of the support system are properly and securely connected.
- Make sure that no structural component of a support system is overloaded.
- When temporary removal of a structural component of a support system is necessary, first install another to take its place.
- Coordinate the installation of support systems closely with the excavation work.

When removing support systems, dismantling should occur from the bottom up with each component removed carefully. Once the dismantling process is completed, the excavation should immediately be backfilled.

Materials and Equipment

The materials and equipment used for constructing, installing, and dismantling protective systems are the responsibility of the construction company using them. Defective materials or equipment can cause a protective system to fail. To guard against this, OSHA has established the following safe practices and requirements:

- Companies must use materials and equipment that are free of defects.
- Manufactured materials and equipment must be used in strict accordance with the manufacturer's specifications.
- Immediately remove from service any material or equipment that is deemed to be unsafe by a competent person.
- Materials or equipment removed from service cannot be returned to service without inspection by and approval of a registered professional engineer.

OTHER HAZARDS: REQUIREMENTS AND SAFETY PRACTICES

In addition to cave-ins, several other hazards are commonly associated with excavation work.[3] These are fall and equipment hazards, **water accumulation hazards,**

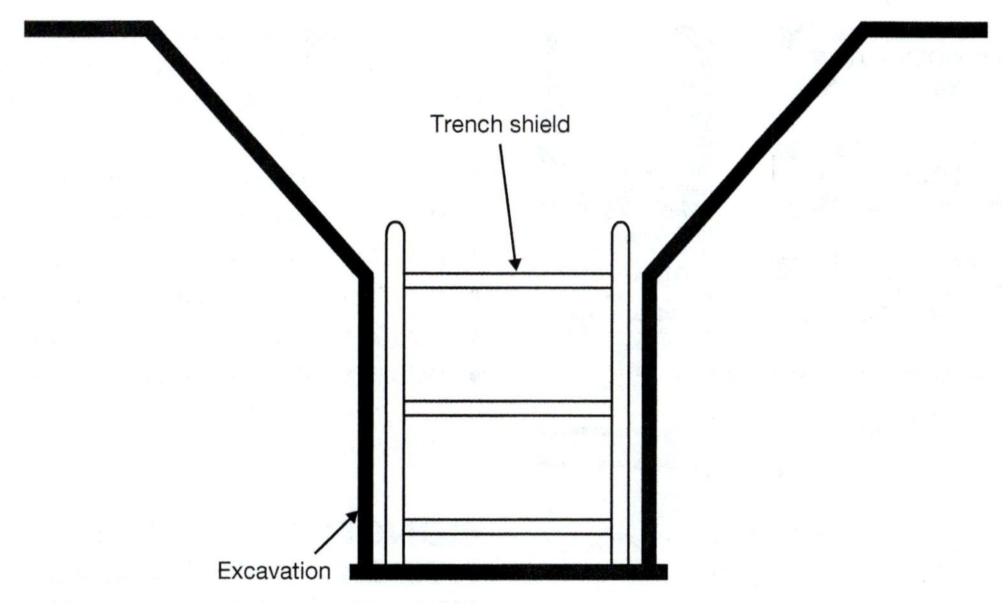

FIGURE 16–5 Trench shield placed in an excavation.

hazardous or toxic atmospheres, and access and egress hazards. Specific requirements and safe practices related to these hazards are explained in the following sections.

Fall, Load, and Equipment Hazards

When conducting excavation work, construction professionals must take the precautions necessary to prevent workers, equipment, and loads from falling into the excavation. OSHA's requirements and safe practices with regard to **falls, loads,** and equipment are as follows:

- Construct retaining devices around excavations or keep all equipment, workers not in the excavation, and loads back from the edge at least 2 feet. The preferred method is to combine both of these precautions.
- Provide warning systems that alert workers and operators of equipment that they are coming too near the edge of the excavation (Figure 16–6).
- Provide protective barricades or other equivalent means to guard against falling rock, soil, or other materials.
- Do not allow people to work on sloped or benched excavation walls at levels above other workers in the excavation.
- Do not allow workers to be underneath loads that are being handled by other workers or mechanical equipment.

Water Accumulation Hazards

Workers should not be allowed to perform their duties in an excavation in which water is accumulating, unless the appropriate precautions have been taken to remove the water. These precautions are as follows:

- Properly functioning water removal equipment is being used to prevent the accumulation of water. While in use, this equipment must be monitored by a competent person.
- Diversion ditches, dikes, or other suitable means are used to keep water from running into the excavation, especially excavations subject to runoff from heavy rain.
- Excavations that are subject to runoff, or that are prone for any other reason to accumulate water, must be inspected by a competent person and determined to be safe before workers are allowed to enter the excavation.

Hazardous or Toxic Atmospheres

Before allowing a worker to enter an excavation that is 4 feet or more deep or that could reasonably be expected to have a hazardous atmosphere, a competent person must test the atmosphere. If a hazardous condition is determined to exist, workers must use the appropriate respiratory protection devices, and ventilation of the excavation must be undertaken. When toxic substances are present in the atmosphere, appropriate controls must be used to minimize the concentration of these substances. The controls must be monitored and tested regularly.

Emergency rescue equipment must be made readily available if a hazardous atmosphere might exist or develop in the excavation. This equipment should include such items as respiratory protection devices, a safety harness and line, and a basket stretcher. Workers who enter confined excavations, such as bell-bottomed piers, must wear safety harnesses and lifelines. There must be an observer present at all times when the worker is in the confined space. This observer must maintain communication with the worker and ensure that the lifeline is working properly.

FIGURE 16–6 Warning signs should be used to alert workers that they are approaching an excavation.
Source: Booka/Fotolia

Access and Egress Hazards

It is important to ensure that workers have a safe means of entering and exiting excavations. For trenches 4 feet or more deep, OSHA regulations and safe practice require that workers be provided with appropriate means of **access** and **egress**, such as ladders, steps, ramps, or other equivalent means. If ramps are used, they must be designed or approved by a competent person. If the ramp is to be used for vehicular access, a person qualified in the area of structural design must design or approve the ramp. In addition, the structural members used in ramps must be fastened together in such a way as to avoid the introduction of tripping hazards.

SUBPART Q: CONCRETE AND MASONRY CONSTRUCTION AND RELATED SAFETY PRACTICES

Subpart Q covers the requirements and related safety practices for concrete construction and masonry work. The requirements and practices are arranged according to the following broad areas of concern: equipment and tools, cast-in-place concrete, precast concrete, lift-slab concrete, and masonry construction. Subpart Q contains the following sections:

- 1926.700 Scope, application, and definitions
- 1926.701 General requirements
- 1926.702 Requirements for equipment and tools
- 1926.703 Requirements for cast-in-place concrete
- 1926.704 Requirements for precast concrete
- 1926.705 Requirements for lift-slab construction operations
- 1926.706 Requirements for masonry construction

CONCRETE WORK: REQUIREMENTS AND RELATED SAFETY PRACTICES

Most construction projects involve some form of **concrete work**. Consequently, OSHA specifies requirements and safe practices that apply to all types of concrete work.[4] These general requirements are summarized in this section.

- Before any type of load can be placed on a concrete structure or any concrete portion of a structure, an individual qualified in the field of structural design must determine that the structure or the portion of a structure being used can support the load.
- Workers involved in tying or placing reinforcing steel at heights of 6 feet or more above any working surface must wear the appropriate fall protection devices.
- Protruding reinforcing bars must be guarded to prevent impalement if a worker falls on one (Figure 16–7).
- When reinforcing strands are tensioned at the job site, workers should not be allowed behind the jack during the tensioning process. A steel reinforcing strand that snaps under high tension creates a whiplash hazard that can injure or even kill a nearby worker. Access to the tensioning area should be clearly marked with signs and blocked off with barriers.
- Concrete buckets must be equipped with safety latches to prevent accidental or premature dumping.

SAFETY FACTS & FINES

Trenches, like all confined spaces, can subject workers to **hazardous atmospheres**. A common hazard is lack of oxygen. This is why it is so important for the atmosphere to be tested by a competent person before workers are allowed access to a trench or an excavation of any type that is more than 4 feet deep. A company in Shrewsbury, Massachusetts, lost an employee who died when he was allowed access to a confined space that lacked sufficient oxygen. The company was cited for failure to (1) implement appropriate procedures to protect workers in confined spaces, (2) test the atmosphere in a confined space before allowing workers access, (3) train workers in recognizing and handling confined space hazards, and (4) provide workers with the proper ventilation, monitoring, and testing equipment for use while in the confined space.

SAFETY FACTS & FINES

It is critical for construction companies to consult with local utilities before beginning to excavate on a job site. Gas, electricity, telecommunications, and other utility lines are often buried underground and can easily be damaged by excavation equipment. A construction company in Plummer, Minnesota, was fined $50,000 by the Environmental Protection Agency when it began an excavation project without first checking with utility companies to locate underground lines. The company's excavation equipment ruptured an underground pipeline, resulting in a spill of more than 200,000 gallons of crude oil, which found its way into a nearby river. Not only did the spill violate the Clean Water Act, it also created a fire hazard that forced a nearby town to be evacuated.

Workers should not be allowed to work underneath concrete buckets when they are being lifted or elevated (Figure 16–8).

Concrete Equipment and Tools

In addition to the general requirements and safety practices related to all concrete work, there are general requirements for the tools and equipment used in concrete work.[5] The more critical of these requirements and safe practices are summarized in this section.

- Equipment, such as saws, mixers, screens, and pumps, must be properly locked out and tagged out before workers are allowed to perform routine maintenance duties on them (Figure 16–9).
- Concrete mixers with loading skips of 1 cubic yard or larger must be equipped with a mechanical means of clearing the skip of material. Guardrails are required on each side of the skip.
- Electrically powered, rotating trowel machines must be equipped with an automatic shutoff switch that activates if the operator releases the handles.
- Concrete pumping systems that use pipes for discharging the concrete must be equipped with pipe supports that are rated for at least 100 percent overload. Compressed air hoses used on concrete pumping systems must be equipped with fail-safe joint connectors.
- Workers who use a pneumatic hose to apply cement, sand, or water mixture must wear appropriate head and face protection.
- Masonry saws, such as the one shown in Figure 16–9, must be equipped with a guard over the blade and a method for retaining blade fragments.

FIGURE 16–7 Protruding rebar should be guarded to protect against impalement and other hazards.
Source: Hoda Bogdan/Fotolia

CAST-IN-PLACE CONCRETE: REQUIREMENTS AND RELATED SAFETY PRACTICES

The term *cast-in-place* means that the concrete is poured (cast) at the job site (in place) into forms that are either constructed or assembled at the site.[6] **Cast-in-place concrete** work involves the following activities: (1) building

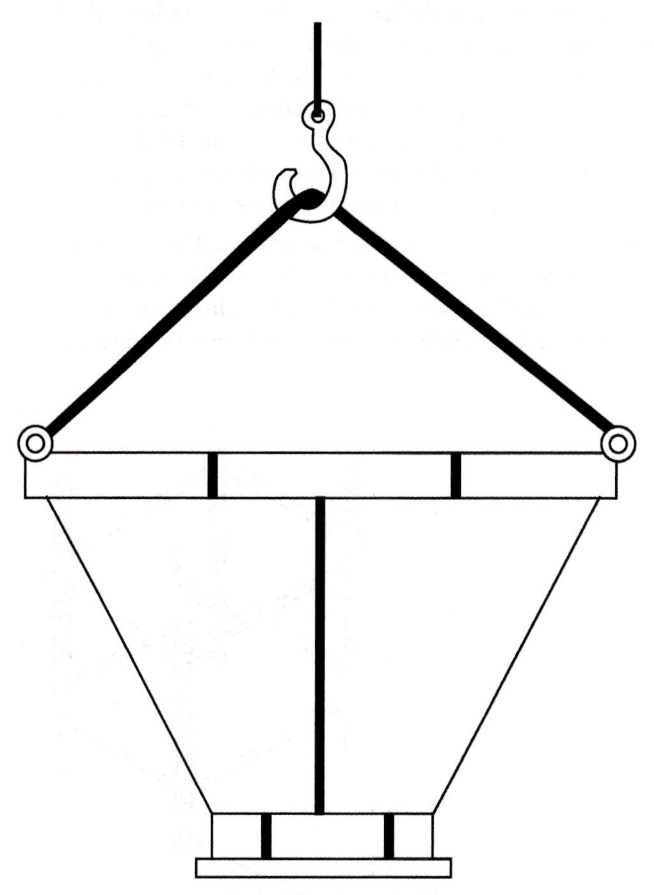

FIGURE 16–8 Concrete buckets should be lifted away from workers to prevent injuries from falling material.

FIGURE 16–9 Equipment, such as this masonry saw, must be properly locked out and tagged out before routine maintenance may be performed.
Source: Vallefrias/Fotolia

FIGURE 16–10 Cast-in-place concrete form.
Source: Peuceta/Fotolia

or assembling forms at the job site, (2) placing steel reinforcing bars (typically called "rebar") into the forms, (3) pouring concrete into the forms (Figure 16–10), and (4) treating the surface of the concrete in some manner, such as applying broom strokes (for a rough finish) or using a trowel (for a smooth finish).

When undertaking cast-in-place concrete work, the following requirements and safety practices apply:

- Forms must be designed and built to withstand all vertical and lateral loads that might be applied to them. Further, forms must be built in accordance with design specifications that set forth everything that is necessary to properly build the forms, including drawings, shoring equipment, working decks, scaffolds, and other equipment or materials.
- Shoring equipment must be inspected before erection to ensure that it meets the specifications set forth in the engineering drawings. Inspections should be completed immediately before, during, and immediately after concrete is poured into the forms. Shoring equipment that is found to be damaged or weakened in any way must be reinforced immediately.
- When single post shores are tiered (placed one on top of the other), the following additional requirements and safety practices apply: (1) shoring must be designed by a qualified designer and inspected after it is put in place (tiered) by a qualified structural engineer, (2) single post shores must be aligned vertically and spliced to prevent misalignment, and (3) single post shores must be properly braced in two mutually perpendicular directions at the splice level. Each tier must be diagonally braced in the same directions.
- Reinforcing bars for vertical components of the structure (columns, walls, piers) must be properly supported to prevent overturning or collapse (Figure 16–11). When wire mesh is unrolled in a form, measures must be taken to prevent recoiling.

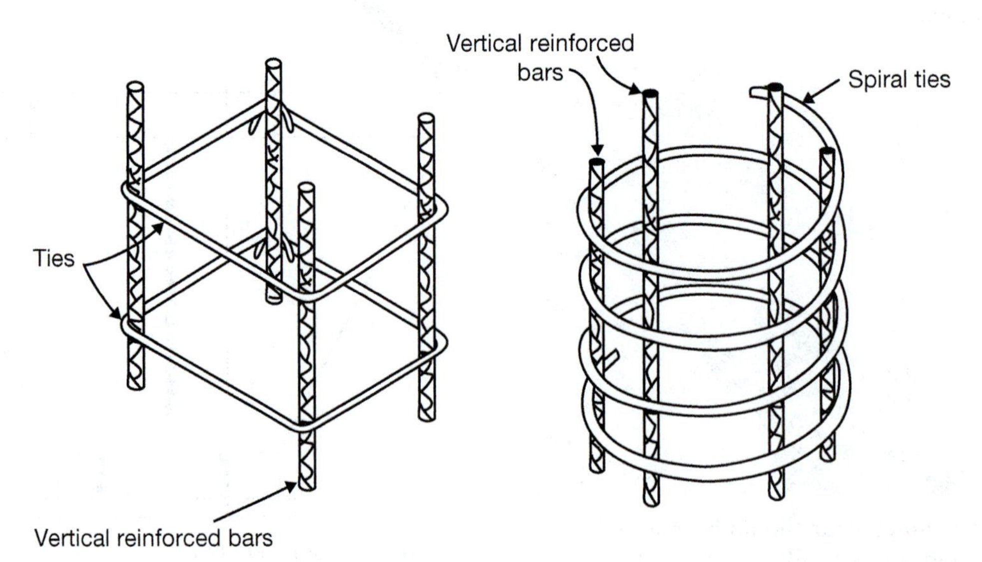

FIGURE 16–11 Reinforcing bar cages for cast-in-place concrete columns.

- Forms and shores must remain in place until it has been determined by a standard test approved by the American Society for Testing of Materials (ASTM) that the concrete has cured sufficiently to support its own weight and that of any superimposed loads. If the plans and specifications provide guidance as to when the forms may be removed, that guidance should be followed.

Any forms used for cast-in-place concrete that were designed, built, maintained, and erected in accordance with Sections 6 and 7 of ANSI A10.9: "American National Standard for Construction and Demolition Operations Concrete and Masonry Work," also meet the requirements and safe practices specified in 29 CFR 1926.703.

PRECAST CONCRETE: REQUIREMENTS AND RELATED SAFETY PRACTICES

Precast concrete is poured in forms at a location off the construction site, allowed to cure until it attains the specified strength, shipped to the job site, and erected.[7] Precast concrete contains all of the reinforcing materials needed to withstand the loads it will have to carry once erected, as well as all of the bolts, plates, and other connectors needed to transport and erect it. Not always, but often, precast concrete components (e.g., columns, beams, wall members, floor members) are fabricated by a subcontractor that specializes in this type of work. Figures 16–12 and 16–13 contain cross-sectional configurations of some of the more commonly used precast concrete components. The most common uses of precast concrete construction are bridges, parking garages, commercial buildings, and multistory residential buildings, such as high-rise condominiums.

Regardless of whether they are erected by the general contractor or a specialized subcontractor, precast concrete components must be properly supported until they are permanently attached (typically welded or bolted connections) to prevent overturning or collapse. Lifting inserts embedded in the concrete for use when erecting precast members must be capable of supporting at least four times the anticipated load. Lifting mechanisms must be capable of supporting at least five times the anticipated load. Workers should not be allowed to work under precast components that are in the process of being erected.

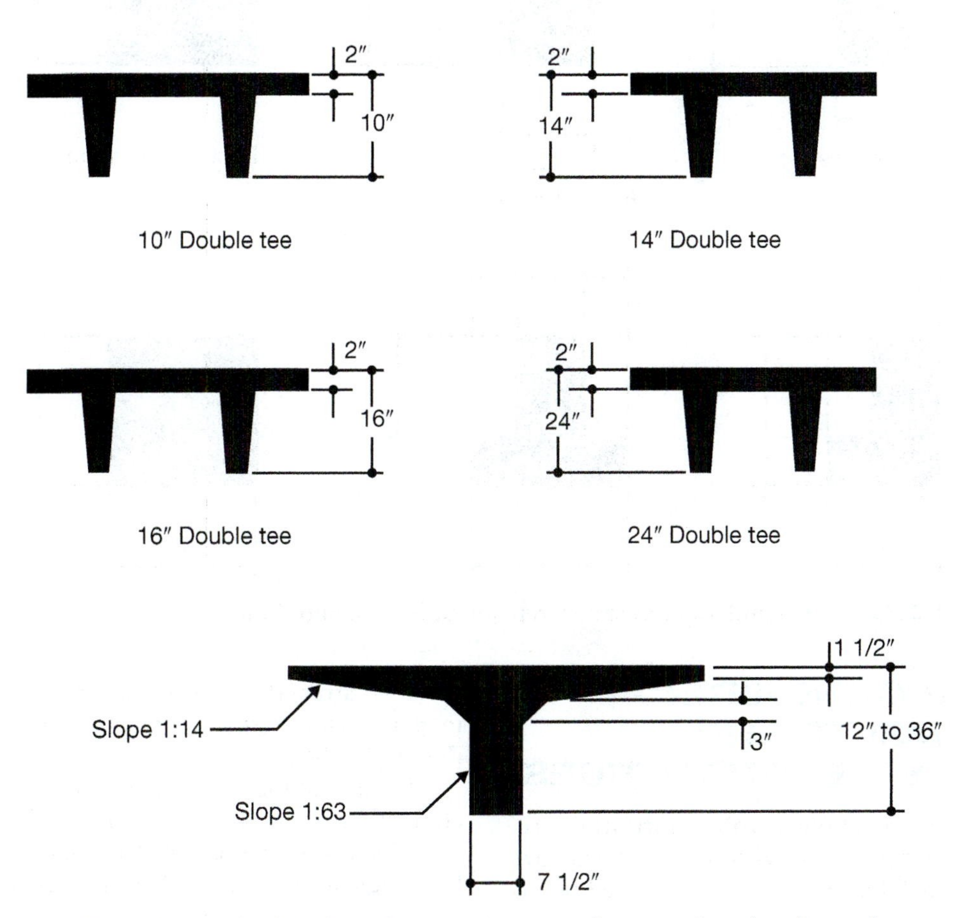

FIGURE 16–12 Standard T-shaped precast concrete floor, roof, and wall members.

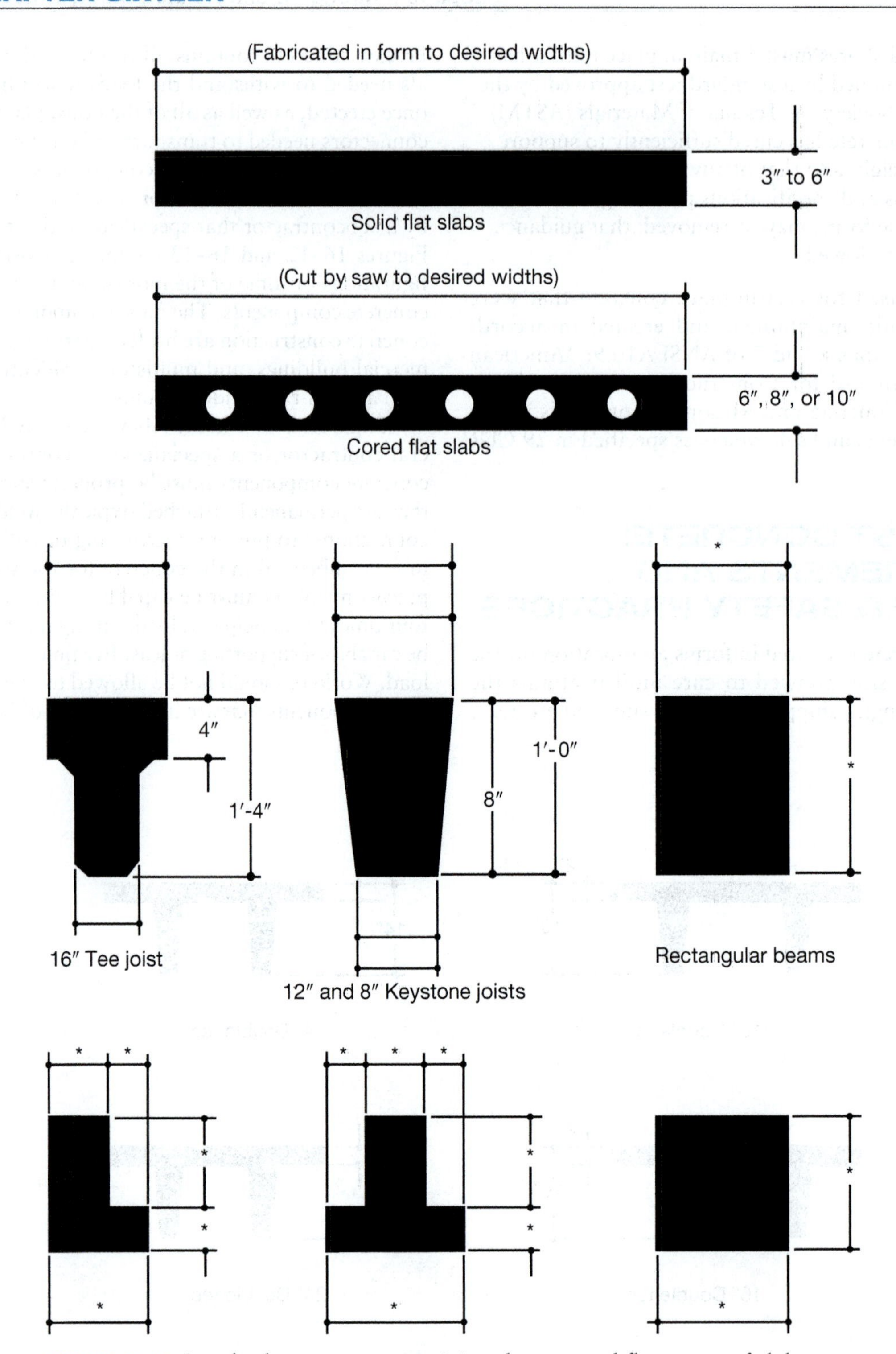

FIGURE 16–13 Standard precast concrete joists, beams, and floor or roof slabs.

LIFT-SLAB CONCRETE: REQUIREMENTS AND RELATED SAFETY PRACTICES

Lift-slab concrete operations involve fabricating precast concrete components with lifting hardware embedded in them.[8] Once the slabs have cured to the specified strength, the slabs are tilted up and secured in place using either bolted or welded connections (Figure 16–14).

Because of the inherent danger in lifting heavy concrete slabs into place and securing them with bolted or welded connections, OSHA requires that such slabs be designed by a licensed professional engineer with

FIGURE 16–14 Lift-slab components are precast and lifted into place. They are secured using either welded or bolted connections.

expertise in lift-slab construction. The plans and specifications developed by the engineer must give the contractor clear guidance concerning how the slabs are to be erected and how the lateral stability of the structure is to be ensured during the lifting and connecting operations.

Jacks and other equipment used to lift the slabs must be clearly marked to show the manufacturer's specifications for lift capacity. Jacking and lifting equipment (e.g., threaded rods, lifting attachments, lifting nuts, hook-up collars) must be able to support at least 2½ times the anticipated load and must not be used beyond their rated capacity. Jacks and lifting equipment must be equipped with a safety mechanism that can support the load in any position in the event of a malfunction.

The maximum number of lifting units in a slab is 14, but the preferred number is the minimum necessary to allow the operator to keep the slab level to within prescribed tolerances. Unless the structure in question has been appropriately reinforced, no workers other than those directly involved in the lift-slab operations are to be allowed in the structure during operations. A structure is "appropriately reinforced" only if an independent, qualified professional engineer—other than the one who designed and planned the lift operations—certifies that failure of a lifting unit in any part of the structure will be confined to that part and will not threaten the stability of the overall structure. No worker who is not essential to the lifting operation should be allowed under the slab while it is being lifted. Welding on temporary and permanent connections must be completed by a certified welder who is familiar with lift-slab operations and the plans and specifications of the project.

MASONRY CONSTRUCTION: REQUIREMENTS AND RELATED SAFETY PRACTICES

Masonry involves the use of bricks or blocks as the primary building material (Figure 16–15 and Figure 16–16).[9] Stability is an issue with **masonry construction**. Consequently, masonry walls more than 8 feet in height must be supported to prevent collapse. Bracing used to support a masonry wall under construction must remain in place until permanent supporting elements of the structure are in place.

FIGURE 16–15 A principal building material in masonry construction is bricks such as these.
Source: Dja65/Fotolia

In addition to bracing, a limited access zone must be established wherever a masonry wall is being constructed. The limited access zone must meet the following requirements:

- It must be established before beginning construction of the masonry wall.
- It must be at least 4 feet wider than the height of the masonry wall and must run its entire length.
- It must be located on the side of the masonry wall that will not be scaffolded.
- Only employees actively engaged in constructing the masonry wall may be allowed entry into it.
- It must remain in place until the masonry wall is adequately supported to prevent collapse.

FIGURE 16–16 Stability is always a concern with masonry construction.
Source: Vladislav Gajic/Fotolia

SUBPART R: STEEL ERECTION AND RELATED SAFETY PRACTICES

Subpart R covers the requirements and related safety practices for structural steel construction and assembly.[10] Subpart R contains the following sections as updated in the latest **steel erection** standard, which became effective in July 2001:

- 1926.750 Scope
- 1926.751 Definitions
- 1926.752 Site layout, site-specific erection plan, and construction sequence
- 1926.753 Hoisting and rigging
- 1926.754 Structural steel assembly
- 1926.755 Column anchorage
- 1926.756 Beams and columns
- 1926.757 Open web steel joists
- 1926.758 Systems-engineered metal buildings
- 1926.759 Falling object protection
- 1926.760 Fall protection
- 1926.761 Training

STRUCTURAL STEEL CONSTRUCTION

Steel construction is a specialized field within the broader field of construction. It has its own unique hazards, which result in an average of 35 deaths and 2,300 lost-time injuries every year. Because of the inherent dangers in erecting structural steel, OSHA established the requirements set forth in Subpart R and then, in July 2001, increased the requirements with passage of a new "Steel Erection Rule."

Structural steel is manufactured in rolling mills in numerous standard shapes (Figure 16–17). These standard structural members are cut to length and prepared for erection through the attaching of various plates and other connection devices. Figure 16–18 shows

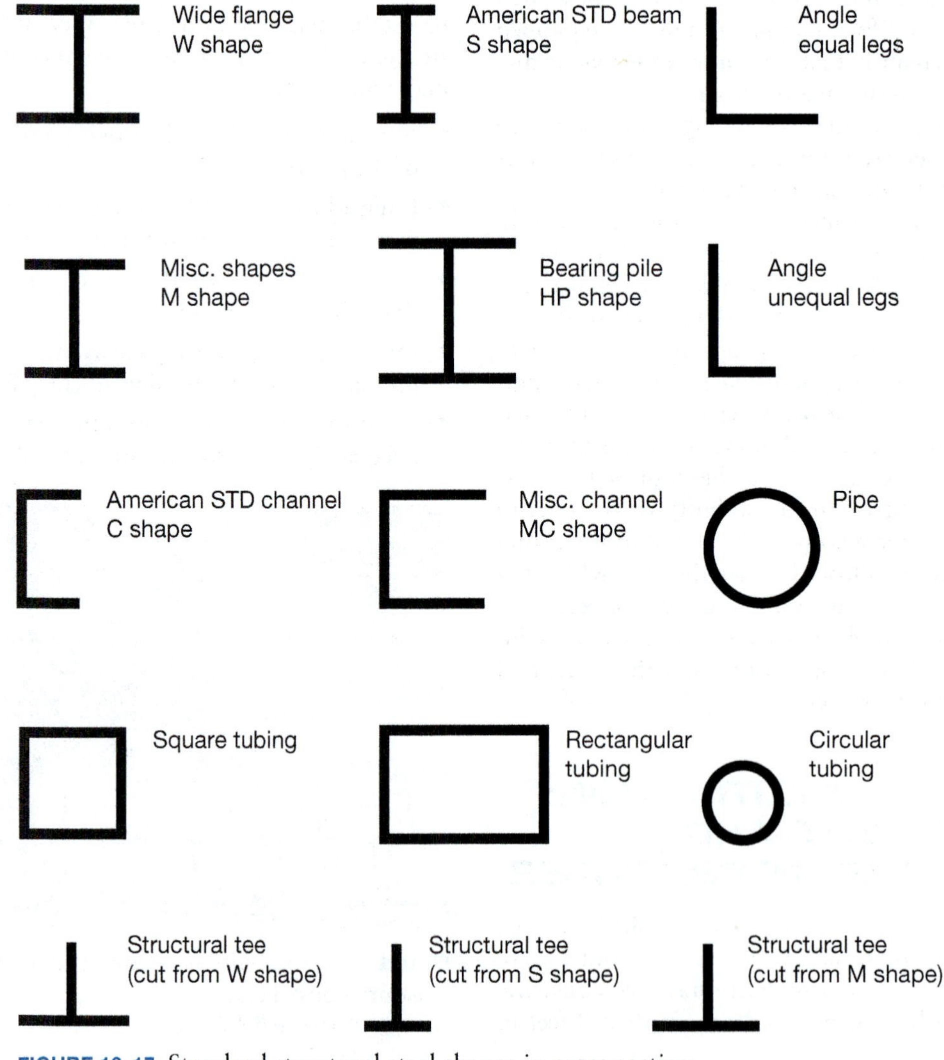

FIGURE 16–17 Standard structural steel shapes in cross section.

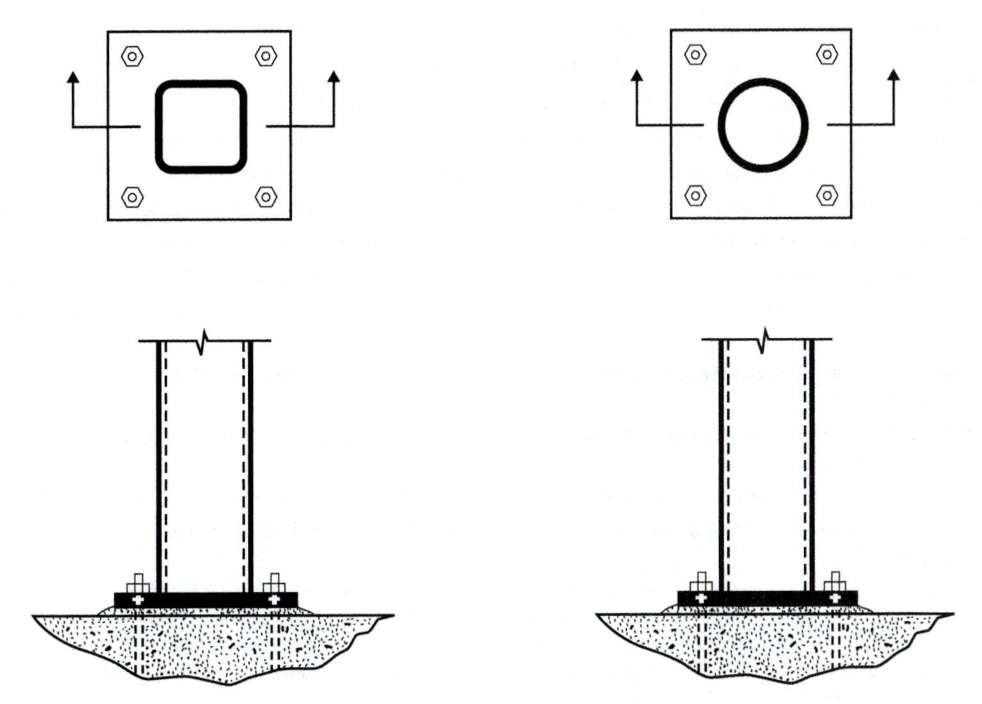

FIGURE 16–18 Connection details for structural steel columns; rectangular tubing (left) and circular tubing (right).

connection details for two structural steel columns that will be bolted to a concrete footing.

The hazards most commonly associated with steel erection work include working under loads; hoisting, landing, and placing decking and steel joists; column stability; double connections; and falling to lower levels. Figure 16–19 contains a summary of terms commonly used in steel construction.

SITE LAYOUT, SITE-SPECIFIC ERECTION PLAN, AND CONSTRUCTION SEQUENCE: REQUIREMENTS AND RELATED SAFETY PRACTICES

Construction sequence is important. Steel erection cannot begin until the controlling contractor authorizes it in writing.[11] This written authorization must notify the steel erection contractor that concrete used in footings has either: (1) cured to at least 75 percent of its intended minimum compressive design strength or (2) cured sufficiently to support the loads that will be imposed during erection. The written authorization must also notify the steel erection contractor that repairs, replacements, and modifications to anchor bolts were made in accordance with 29 CFR 1926.755(b). There must be a **site-specific erection plan**.

The controlling contractor is also required to provide the steel erector with a **site layout** plan, including the preplanning routes for hoisting loads. There must be adequate roads into and through the construction site to accommodate the safe delivery and movement of all equipment needed to erect the steel and the steel itself. There must also be sufficient room for operation of all erection equipment and storage of all steel and related materials until erected. Before steel erection can begin, there must be preplanning to develop the sequence for erecting individual steel members. The details of coordination between the controlling contractor and the steel erector must also be worked out during preplanning.

HOISTING AND RIGGING: REQUIREMENTS AND RELATED SAFETY PRACTICES

The **hoisting and rigging** requirements already set forth in 29 CFR 1926.550 also apply to steel erection, except those in 29 CFR 1926.550(g)(2).[12] Cranes used in steel erection must undergo a thorough visual inspection before each shift to ensure that all controls and mechanisms are in proper and safe working order. The operator is responsible for all tasks under his control. If there is any doubt as to the safety of a task, the operator has the authority to stop operations and refuse to handle loads until safety can be assured.

Responsible personnel must minimize the work that is done under a load that is being hoisted. This is achieved through preplanning of routes that workers will use in the vicinity of the erection work. When it is absolutely necessary to work underneath a load, the steel and other materials being hoisted must be properly rigged by a qualified rigger to prevent unintentional displacement.

Steel Construction Terms

Bolted diagonal bridging: DDD Diagonal bridging that is bolted to a steel joist or joists.

Bridging clip: AA device that is attached to the steel joist to allow the bolting of the bridging to the steel joist.

Choker: A wire rope or synthetic fiber rigging assembly that is used to attach a load to a hoisting device.

Cold forming: The process of using press brakes, rolls, or other methods to shape steel into desired cross-sections at room temperature.

Column: A load-carrying vertical member that is part of the primary skeletal framing system. Columns do not include posts.

Connector: An employee who, working with hoisting equipment, is placing and connecting structural members or components.

Derrick floor: An elevated floor of a building or structure that has been designated to receive hoisted pieces of steel before final placement.

Double connection: An attachment method in which the connection point is intended for two pieces of steel that share common bolts on either side of a central piece.

Double connection seat: A structural attachment that, during the installation of a double connection, supports the first member while the second member is connected.

Erection bridging: Bolted diagonal bridging that is required to be installed before releasing the hoisting cables from the steel joists.

Girt (in systems-engineered metal buildings): A Z- or C-shaped member formed from sheet steel, spanning between primary framing and supporting wall material.

Hoisting equipment: Commercially manufactured lifting equipment designed to lift and position a load of known weight to a location at some known elevation and horizontal distance from the equipment's center of rotation. Hoisting equipment includes, but is not limited to, cranes, derricks, tower cranes, barge-mounted derricks or cranes, gin poles, and gantry hoist systems. A come-along (a mechanical device, usually consisting of a chain or cable attached at each end, that is used to facilitate movement of materials through leverage) is not considered hoisting equipment.

Leading edge: The unprotected side and edge of a floor, roof, or formwork for a floor or other walking or working surface (such as decking), which changes location as additional floor, roof, decking, or formwork sections are placed, formed, or constructed.

Metal decking: A commercially manufactured, structural grade, cold-rolled metal panel formed into a series of parallel ribs; for this subpart, this includes metal floor and roof decks, standing seam metal roofs, other metal roof systems, and other products, such as bar gratings, checker plate, expanded metal panels, and similar products. After installation and proper fastening, these decking materials serve a combination of functions including, but not limited to: a structural element designed in combination with the structure to resist, distribute, and transfer loads and to stiffen the structure and provide a diaphragm action; a walking or working surface; a form for concrete slabs; a support for roofing systems; and a finished floor or roof.

Multiple lifting rigging: A rigging assembly manufactured by wire rope rigging suppliers that facilitates the attachment of up to five independent loads to the hoist rigging of a crane.

Permanent floor: A structurally completed floor at any level or elevation (including slab on grade).

Positioning device system: A body belt or body harness rigged to allow an employee to be supported on an elevated, vertical surface, such as a wall or column, and to work with both hands free while leaning.

Post: A structural member with a longitudinal axis that is essentially vertical and that (1) weighs 300 pounds or less and is axially loaded (a load presses down on the top end) or (2) is not axially loaded but is laterally restrained by the above member. Posts typically support stair landings, wall framing, mezzanines, and other substructures.

Purlin (in systems-engineered metal buildings): A Z- or C-shaped member formed from sheet steel, spanning between primary framing and supporting roof material.

Safety deck attachment: An initial attachment that is used to secure an initially placed sheet of decking to keep proper alignment and bearing with structural support members.

Shear connector: Headed steel studs, steel bars, steel lugs, and similar devices that are attached to a structural member for the purpose of achieving composite action with concrete.

Steel erection: The construction, alteration, or repair of steel buildings, bridges, and other structures, including the installation of metal decking and all planking used during the process of erection.

Steel joist girder: An open web, primary load–carrying member, designed by the manufacturer, used for the support of floors and roofs, does not include structural steel trusses.

Steel truss: An open web member designed of structural steel components by the project structural engineer of record. For the purposes of Subpart 1926.751, a steel truss is considered equivalent to a solid web structural member.

Structural steel: A steel member, or a member made of a substitute material (such as, but not limited to, fiberglass, aluminum, or composite members). These members include, but are not limited to, steel joists, joist girders, purlins, columns, beams, trusses, splices, seats, metal decking, girts, and all bridging, and cold-formed metal framing, which is integrated with the structural steel framing of a building.

Systems-engineered metal building: A metal, field-assembled building system, consisting of framing, roof, and wall coverings. Typically, many of these components are cold-formed shapes. These individual parts are fabricated in one or more manufacturing facilities and shipped to the job site for assembly into the final structure. The engineering design of the system is normally the responsibility of the systems-engineered metal building manufacturer.

FIGURE 16–19 Steel construction terms (from 29 CFR 1926.751).

Source: From the U.S. Department of Labor. Published by the U.S. Department of Labor.

SAFETY FACTS & FINES

A steelworker was standing on a ladder while attempting to connect a 24-foot steel beam to a 21-foot high column. During the process, the column broke loose from its foundation bolts, knocking the worker off the ladder and killing him. The column was not connected by four bolts as is required in the steel erection standard.

A worker was walking along a steel beam more than 23 feet in the air, bolting a steel purlin to a rafter. He lost his balance and fell to the ground. The fall was fatal. His death could probably have been prevented had he been wearing the fall protection gear required in the steel erection standard.

Hooks with self-closing safety latches or their equivalent are to be used to prevent slippage or displacement of hoisted materials.

Multiple-lifting procedures (sometimes referred to as "Christmas-treeing") are to be undertaken in ways that reduce overhead exposure and operator fatigue and that eliminate improper, unsafe work practices. All workers involved in the multiple lift must have been trained in proper procedures in accordance with 29 CFR 1926.761 (discussed later in this chapter). Components of the multiple-lift rigging assembly must be designed and assembled with a maximum capacity for total assembly and for each individual attachment point. This capacity must have a safety factor of at least 5:1.

STRUCTURAL STEEL ERECTION: REQUIREMENTS AND RELATED SAFETY PRACTICES

When working on multistory structures, the permanent floors planned for the structure must be installed as the erection of structural members progresses.[13] At no time during the process should there be more than eight stories between the floor currently being erected and the highest permanent floor. In addition, there must be no more than 48 feet or four floors (whichever is less) of unfinished connections above the foundation or the highest permanently secured floor. Either safety nets or a planked or decked floor must be maintained within two stories or 30 feet (whichever is less) directly under any erection work being performed. In addition, the contractor must take special precautions to eliminate or at least minimize tripping and slipping hazards.

COLUMN ANCHORAGE: REQUIREMENTS AND RELATED SAFETY PRACTICES

Column anchorage is important. One of the main causes of collapsing columns and the resulting injuries to workers is insufficient anchorage.[14] Consequently, OSHA's steel erection standard requires that all columns be anchored by a minimum of four anchor bolts (Figure 16–20). The columns must also be set on level finished surfaces with pregrouted leveling plates, leveling nuts, or shim packs that are sufficient to transfer the column loads (Figure 16–21). In addition, anchor bolts that are to be repaired or modified in any way in the field must be approved by the structural steel engineer of record for the project in question. Before steel erection can begin, the controlling contractor must notify the steel erector in writing if anchor bolts or rods have been repaired or modified in any way.

BEAMS AND COLUMNS: REQUIREMENTS AND RELATED SAFETY PRACTICES

Beams are vertical structural members and columns are horizontal structural members. When beams are being secured to columns, they must not be released from the hoisting lines until secured by at least two bolts per connection that have been drawn up wrench-tight as specified by the structural engineer for the project (Figure 16–22). Solid web structural members used as bracing must be secured by at least one bolt per connection drawn up wrench-tight as specified by the structural engineer for the project.[15]

One of the more hazardous tasks for steel erection workers is making double connections. Such connections are made when two structural members (e.g., beams) on opposite sides of a column web are connected by sharing

FIGURE 16–20 Steel columns must be anchored by at least four bolts.
Source: Maximnorby/Fotolia

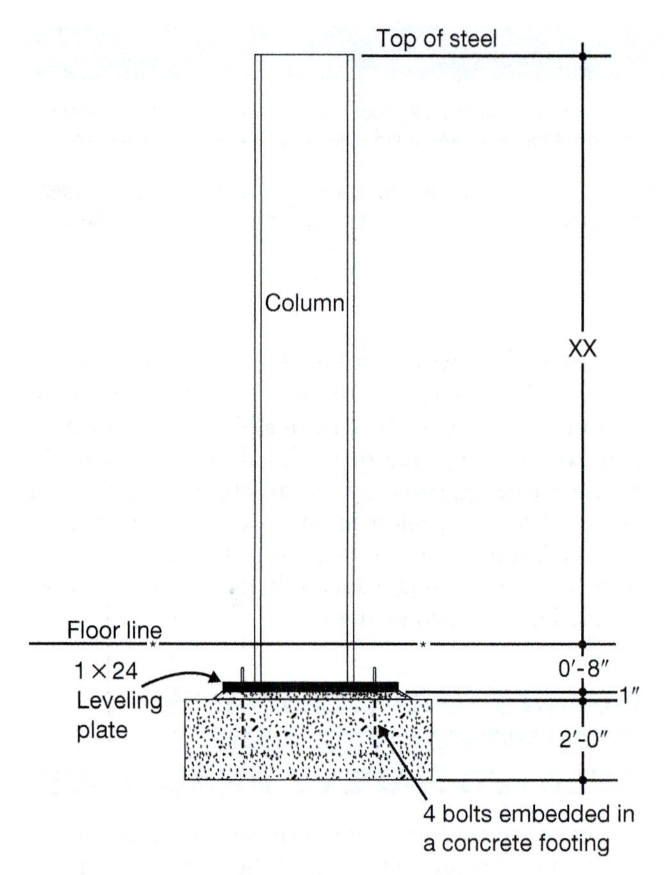

FIGURE 16–21 Steel columns require four bolts and a leveling plate for anchorage.

common connection holes for the connection bolts. Consequently, when making a double connection, at least one bolt with a wrench-tight nut must remain connected to the first member to secure that member and to prevent the column from being displaced. An equivalent device,

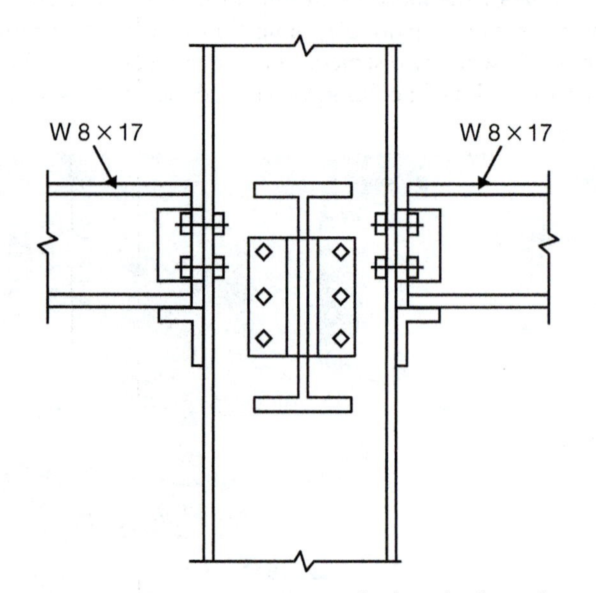

FIGURE 16–22 Beams remain attached to the hoist lines until two bolts per connection have been drawn up wrench-tight.

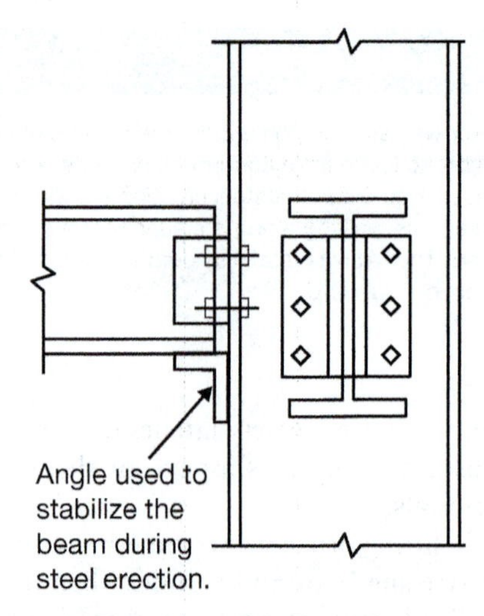

FIGURE 16–23 An angle, such as this one welded to the column, is an acceptable method when making double connections.

such as a securely attached seat on which to rest the beam, may be used instead of the one-bolt method when making double connections (Figure 16–23). However, the seat or other equivalent device must be properly designed to carry the loads imposed during the erection process.

OPEN WEB STEEL JOISTS: REQUIREMENTS AND RELATED SAFETY PRACTICES

When working with lightweight steel joists, known as **open web steel joists**, collapse is always a concern.[16] To minimize this hazard, steel joists must be field bolted at the column to provide lateral stability. A vertical stabilizer plate is to be provided at each column. Hoisting cables must not be released until the seat at each end of the steel joist is field-bolted and each end of the bottom chord is restrained by the column stabilizer plate. OSHA's steel erection standard provides comprehensive tables of joist configurations and corresponding lengths. When steel joists with spans longer than those contained in the tables are used, erection bridging is required. Loads placed on joists must not be allowed to exceed their carrying capacity. No construction loads are allowed on steel joists until all bridging is installed and anchored and all joist-bearing ends are attached.

SYSTEMS-ENGINEERED METAL BUILDINGS: REQUIREMENTS AND RELATED SAFETY PRACTICES

Pre-engineered metal buildings, also known as systems-engineered buildings, are common in modern construction (Figures 16–24, 16–25, and 16–26).[17] Most of the

FIGURE 16–24 Pre-engineered metal buildings are common in modern construction.
Source: Fallesen/Fotolia

FIGURE 16–25 Most of the requirements and safety practices set forth in the Steel Erection Standard apply to pre-engineered metal buildings.
Source: Wittybear/Fotolia

FIGURE 16–26 Displacement and collapse are always concerns when erecting metal buildings such as the one shown here.
Source: Wittybear/Fotolia

various requirements and related safety practices set forth in the steel erection standard apply to metal buildings. Structural columns must be anchored by a minimum of four anchor bolts. Before hoisting equipment is released during the erection process, at least half of the connection bolts, or the number specified by the manufacturer (whichever number is greater), must be tightened on both sides of the web adjacent to each flange. In girt and eave strut-to-frame connections, when girts or eave struts share common connection holes, steps must be taken to guard against displacement. These steps include making sure either that at least one bolt with its wrench-tight nut is connected or a field-attached seat or similar equivalent device is used.

FALLING OBJECT PROTECTION: REQUIREMENTS AND RELATED SAFETY PRACTICES

Falling object protection is important. Objects falling from a higher level and injuring employees working at a lower level is a concern in steel erection.[18] Consequently, the steel erection standard requires that all materials, equipment, and tools used by workers who are aloft be properly secured against falling. In addition, the controlling contractor is to bar work from taking place under steel that is being erected, unless overhead protection is provided for those workers situated below.

FALL PROTECTION: REQUIREMENTS AND RELATED SAFETY PRACTICES

Falls from higher to lower levels represent the greatest hazard in steel erection. Consequently, OSHA requires **fall protection** for employees working at heights greater than 15 feet.[19] This protection can be in the form of guardrails, safety nets, personal fall arrest systems, positioning devices, or fall restraint systems. When metal decking is being installed and it forms the leading edge of a work area, a controlled decking zone (CDZ) may be established in areas of the structure more than 15 feet above a lower level. A CDZ is a controlled access zone where work can be performed by a limited number of workers without the use of guardrails, safety nets, or personal fall arrest systems. However, these controlled access employees must be protected from falls of more than two stories or 30 feet, whichever is less. CDZs must be clearly marked, and employees who work in CDZs must have completed the training specified in 29 CFR 1926.761.

TRAINING: REQUIREMENTS AND RELATED SAFETY PRACTICES

OSHA requires that personnel who work in steel erection have the necessary training.[20] **Training** must be provided by a person who is qualified in terms of the specific project. Training is required for all employees who are exposed to fall hazards. This training must cover the following topics: (1) recognition of fall hazards; (2) use and proper operation of fall protection systems (e.g., guardrails, personal fall arrest systems, positioning device systems, fall restraint systems); (3) correct procedures for erecting, maintaining, disassembling, and inspecting fall protection systems; (4) procedures for preventing falls through holes and other openings; and (5) requirements of Subpart R. Special training is required for employees who will be involved in the following types of work: multiple-lift rigging, connector procedures, and CDZ procedures.

SUBPART S: TUNNELS, SHAFTS, CAISSONS, COFFERDAMS, AND COMPRESSED AIR AND RELATED SAFETY PRACTICES

Subpart S covers the construction of underground tunnels, shafts, chambers, and passageways. It does not apply to excavations, trenching operations, or underground electrical transmission and distribution lines—all of which are covered in other subparts. Subpart S contains the following sections:

- 1926.800 Underground construction
- 1926.801 Caissons
- 1926.802 Cofferdams
- 1926.803 Compressed air
- 1926.804 Definitions

UNDERGROUND CONSTRUCTION: REQUIREMENTS AND RELATED SAFETY PRACTICES

Underground construction has its own set of unique hazards.[21] The principal hazards include reduced natural ventilation, reduced natural light, limited access and egress, air contaminants, fire, and explosion. Of particular concern are the following issues: safety instruction, access and egress, check-in and check-out procedures, hazardous classifications, potentially gassy operations, gassy operations, air monitoring, ventilation, illumination, fire control, hot work, cranes and hoists, and emergencies.

Safety Instruction

Workers who will be involved in underground construction operations or who will be exposed to the hazards of underground construction in any way must first receive the proper training. Training must focus on hazard recognition and avoidance and should cover the following topics as appropriate for the specific job site:

- Air monitoring.
- Ventilation and illumination.
- Communication.
- Flood control.
- Equipment: mechanical and personal protective.
- Explosives, fire prevention, and fire protection.
- Emergency procedures: check-in, check-out, and evacuation plans.

Comprehensive training records must be maintained for all employees who need the training. These records must be made available to employees upon request.

Access and Egress

Employers are required to provide safe access and egress to underground construction sites (typically tunnels). Unauthorized entry must be both prohibited and prevented. This means it is not sufficient to simply put up "Keep Out" signs. Fences, barricades, or other appropriate devices should be used to ensure against unauthorized access to underground work areas. Completed or unused openings must be clearly marked with "Keep Out" signs and covered, fenced, or barricaded to prevent unauthorized use.

Check-In and Check-Out Procedures

Employers must use an appropriate check-in and check-out procedure to keep track of employees who work underground. An accurate head count of employees working underground must be kept. At least one designated person should be assigned above ground to maintain the head count and call for assistance in the event of an emergency.

Gassy Operations

In underground construction work there are *potentially* gassy operations and confirmed gassy operations. Potentially gassy operations exist when either of the following conditions is present: (1) when air monitoring for a 24-hour period shows the presence of methane or any other flammable gas in concentrations of 10 percent or more of the lower explosive limit measured at 12 inches from the floor, ceiling, or walls of the underground work area or (2) when the geological formation or history of the work area makes it likely that the conditions just explained will exist.

Gassy operations exist when one of the following conditions is present: (1) when air monitoring for three consecutive days shows the presence of methane or any other flammable gas in concentrations of 10 percent or more of the lower explosive limit measured at 12 inches from the floor, ceiling, or walls of the underground work area; (2) when a flammable gas leaking from the strata in the underground work area has actually ignited; or (3) when the underground work area is connected to another work area that is known to be gassy. Safety precautions called for in gassy operations include the following as a minimum:

1. Ventilation.
2. Prohibiting the use of diesel equipment, unless it is specifically approved for gassy operations.
3. Prohibiting smoking or other forms of ignition in or near the work area.
4. Maintaining a fire watch if any hot work is to be performed.
5. Additional air monitoring.
6. Suspending all operations in the work area until all required precautions can be put into place.

Air Monitoring

A competent person must be assigned to conduct air-monitoring courses to ensure that the ventilation available is sufficient and to record measurements of potentially hazardous gases. The atmosphere in all underground work areas must be tested for carbon monoxide, nitrogen dioxide, hydrogen sulfide, methane, and other toxic gases, dusts, vapors, mists, and fumes as often as necessary to ensure that the requirements of 29 CFR 1926.55 are satisfied.

Ventilation

Ventilation is critical in underground construction. Contractors are required to provide an adequate supply of fresh air to all underground work sites. *Adequate* is defined as 200 cubic feet per minute for each employee working in the underground site. Unless natural ventilation provides adequate fresh air, mechanical ventilation with reversible airflow must be provided. In addition, in the presence of blasting, drilling, or similar operations that may introduce harmful vapors, dust, fumes, mists, or gases, the velocity of airflow must be at least 300 feet per minute.

Illumination

Underground construction is similar to aboveground construction in terms of the illumination requirements. Contractors are required to provide proper illumination in all underground sites as specified in 29 CFR 1926.56 (Figure 16–27). When explosives are to be handled underground, a distance of 50 feet must be maintained between any heading and the lighting.

Fire Prevention and Control

The various fire-prevention requirements and practices explained in Chapter 14 (Subpart F) also apply to underground construction. In addition, open flames and fires are prohibited in underground construction sites, except in cases where hot work is to be performed. Smoking should be completely discouraged in underground sites. But if it is allowed, the work area must be free of fire and explosion hazards. When such hazards exist, "No Smoking" signs should be prominently displayed. When piping fuel from aboveground to an underground site, the pipes should remain empty of fuel except during actual pumping; this means that residual fuel left in the line after pumping must be pumped out completely. Gasoline must be prohibited in underground work sites; this

Minimum Illumination Requirements in Foot-Candles

Foot-candles	Area of Operation
5	General construction area lighting
3	General construction areas, concrete placement, excavation and waste areas, accessways, active storage areas, loading platforms, refueling, and field maintenance areas
5	Indoors: warehouses, corridors, hallways, and exitways
5	Tunnels, shafts, and general underground work areas: (Exception: minimum of 10 foot-candles is required at tunnel and shaft heading during drilling, mucking, and scaling. Bureau of Mines–approved cap lights shall be acceptable for use in the tunnel heading.)
10	General construction plant and shops (e.g., batch plants, screening plants, mechanical and electrical equipment rooms, carpenter shops, rigging lofts and active store rooms, barracks or living quarters, locker or dressing rooms, mess halls, and indoor toilets and workrooms)
30	First aid stations, infirmaries, and offices

FIGURE 16–27 Minimum illumination requirements.

means that it is not to be stored, used, or even carried underground. Leaks and spills of flammable substances must be cleaned up immediately.

Hot Work

Hot work in underground settings typically involves welding or cutting. When performing such work underground, noncombustible barriers must be installed in or over a shaft. Only the amount of fuel and oxygen actually needed over a 24-hour period for the hot operation should be allowed in the underground site. When the hot work is completed, fuel and oxygen cylinders should be removed immediately.

Cranes and Hoists

When using cranes and hoists in an underground construction setting, the following precautions should be taken:

- All tools or materials to be raised or lowered must be stacked or secured to prevent shifting or snagging.
- When a load is being moved, flashing warning lights should be used to make sure that employees are aware that a load is being lowered or raised.
- To guard against malfunctioning of operational controls, limit switches. Anti-two-block devices must be used to limit the undesired travel of loads.
- When maintenance and repair work is to be undertaken in a shaft served by a cage, bucket, or skip, employees must be alerted before the work commences.
- When work is being performed in a given shaft, warning signs must be placed in clearly visible locations at each underground landing, at the shaft collar, and at the operator's station.
- Connectors must be used between the hoisting rope and cage skip that are compatible with the wire rope being used to lift the hoist.
- Cage, skip, and load connectors must be used that can withstand the force of the hoist pull, vibration, misalignment, or impact without disengaging.
- All spin connectors must be kept in a properly maintained, clean working condition.
- Proper seating of wire rope wedge sockets must be ensured when they are used.

Emergencies

Rescue teams or services must be provided at construction sites in which 25 or more employees work underground. Two teams of at least five persons each must be provided. One team must be available on site or within a 30-minute travel distance of the site. The other team may be on standby within a two-hour travel distance from the site. When fewer than 25 employees work underground, only one 5-person rescue team is required; this team must be available on site or within a 30-minute travel distance of the site. All members of the rescue team must be qualified in rescue procedures, the use of firefighting equipment, and the use of breathing equipment, such as the various kinds of respirators. Selection, use, and maintenance of respirators must comply with 29 CFR 1926.103 (see Subpart E, Chapter 13). A designated person must be available to call for emergency assistance and to maintain an accurate head count of those working underground.

CAISSONS: REQUIREMENTS AND RELATED SAFETY PRACTICES

A **caisson** is an airtight and watertight structure in which construction work can be done underground or underwater.[22] Caissons that have a diameter or side longer than 10 feet must be equipped with a man lock and shaft for the use of employees. There must be a gauge provided in the locks, on the outer side of each bulkhead, and on the inner side of each bulkhead. These gauges must be properly maintained and accurate. If a caisson is to be suspended at any time when work is being done in it and the bottom of the excavation is 9 feet or more below the deck of the working chamber, a shield must be erected to protect employees.

COFFERDAMS: REQUIREMENTS AND RELATED SAFETY PRACTICES

A **cofferdam** is a watertight enclosure that can be pumped dry so that construction activities, such as the construction of piers, can take place.[23] A means must be provided to prevent flooding if overtopping of the cofferdam by high waters is possible. Signals that warn employees to evacuate the cofferdam must be available, and employees must recognize the signal. Notices informing employees of the signal must be posted. At least two means of rapid exit must be provided with cofferdams. These exits—whether walkways, bridges, or ramps—must be equipped with guardrails.

COMPRESSED AIR: REQUIREMENTS AND RELATED SAFETY PRACTICES

Any time work is being done under **compressed air**, there must be at least one competent person readily available who is familiar with all aspects of working in these conditions.[24] This person serves as the party responsible, on behalf of the employer, for ensuring that all requirements and related safety practices are met.

Medical Issues

A qualified and licensed physician who is competent in all of the medical aspects of working under compressed air and in the treatment of decompression-related illnesses must be retained and available during the entire period of time such work is being done. No employee should be allowed to work under compressed air until examined by the physician and deemed physically capable of undertaking work in these conditions. The physician retained must keep accurate and complete records of decompression-related illnesses. In addition, a fully stocked first aid station must be provided at each tunnel project regardless of the number of people working in it, and an ambulance or equivalent transportation must be available at each project.

Other Critical Issues

Whenever the air pressure in the working chamber is increased to a level above normal atmospheric pressure, a medical lock must be established and maintained. The medical lock must have at least 6 feet of headroom at its center and be divided into at least two compartments. It must be properly heated, ventilated, and sanitized, and it must be readily accessible to employees who will work under compressed air. Identification badges must be issued to employees who work under compressed air that indicate that they work in these conditions, and a permanent record must be kept of all identification badges issued. A communication system (e.g., telephones, bells, whistles) must be established and maintained whenever employees are working under compressed air in all appropriate locations (e.g., the working chamber, first aid station, lock attendant's station, compressor plant, and any other relevant location in which a stakeholder in the operation might be located). Time-of-compression signs must be clearly posted in each man lock (Figure 16–28).

Before allowing a first-time employee into an environment with compressed air, instruction on how to avoid excessive discomfort must be provided. The acclimation to compression of employees is to be gradual and carefully observed. If an employee experiences discomfort, the pressure must be held steady until the employee indicates that the symptoms no longer exist. No employee is to be subjected to more than 50 pounds of pressure per square inch, except during emergency situations. Decompression of employees must be accomplished in accordance with the Worker Decompression Table provided by OSHA (Figure 16–29).

TIME OF DECOMPRESSION FOR THIS LOCK

__________ pounds to __________ pounds in __________ minutes

__________ pounds to __________ pounds in __________ minutes

(Signed by) ____________________
(Superintendent)

FIGURE 16–28 Form for recording the decompression time in a specific lock.

Worker Decompression
Table (Decompression times for various working pressures during the work period in hours)

	Working Period (In hours)							
	½	1	1½	2	3	4	5	6
Work Pressure (psig*)	**Total Decompression Time**							
9–12	3	3	3	3	3	3	3	3
14	6	6	6	6	6	6	6	6
16	7	7	7	7	7	7	17	33
18	7	7	7	8	11	17	48	63
20	7	7	8	15	15	43	63	73
22	9	9	16	24	38	68	93	103

FIGURE 16–29 Employees who work under pressure must undergo an appropriate period of decompression.

*Pounds per square inch gauge.

SUBPART T: DEMOLITION REQUIREMENTS AND RELATED SAFETY PRACTICES

Subpart T covers the requirements and related safety practices for **demolition**. A critical first step before undertaking a demolition operation is to obtain a comprehensive engineering survey of the structure that is to be demolished. This survey helps to guard against premature or unplanned collapse of the structure in question. Subpart T contains the following sections:

- 1926.850 Preparatory operations
- 1926.851 Stairs, passageways, and ladders
- 1926.852 Chutes
- 1926.853 Removal of materials through floor openings
- 1926.854 Removal of walls, masonry sections, and chimneys
- 1926.855 Manual removal of floors
- 1926.856 Removal of walls, floors, and material with equipment
- 1926.857 Storage
- 1926.858 Removal of steel construction
- 1926.859 Mechanical demolition
- 1926.860 Selective demolition by explosives

PREPARATORY OPERATIONS

Preparatory operations are important. A comprehensive engineering survey must be completed on the structure in question before demolition operations can begin.[25] The purpose of the survey is to determine the structural integrity of columns, beams, floors, walls, framing, and roofs. Once the engineering survey has been completed, all utilities (gas, electricity, and other energy sources) are disconnected and turned off. Openings in floors and walls are covered, sealed, or otherwise protected, and structural members are shored or braced.

If the structure to be demolished contains pipes or tanks that hold or once held any kind of hazardous substances, these vessels must be thoroughly purged. Once purged, they must be tested to ensure that all hazardous substances have been eliminated before demolition operations begin. If combustible materials are present in the structure, firefighting equipment, including water tank trucks with pumps and charged hose lines, must be available in nearby proximity to the site. During the actual demolition, a competent person must conduct continual inspections as operations progress to identify and act on any hazards that might be introduced by structural members that have been weakened or materials that have been knocked loose.

OTHER REQUIREMENTS IN DEMOLITION OPERATIONS

Once a structure has been surveyed and properly prepared for demolition, operations may begin.[26] During demolition operations, there are requirements and related safety practices to be observed in the following areas: (1) chutes; (2) removal of material through floor openings; (3) removal of walls, masonry sections, and chimneys; (4) manual removal of floors; (5) removal of walls, floors, and material with equipment; (6) storage; (7) removal of steel construction; (8) mechanical demolition; and (9) selective demolition by explosives. The most critical requirements and safety practices related to these areas of concern are explained in this section.

Material may not be dropped outside of the exterior walls of the structure being demolished, unless the area where it will land has been properly blocked off. Chutes that run down at angles of 45 degrees or more must be enclosed. A gate must be installed near the discharge end of every chute, and a competent person must be assigned to control the gate. The end of the chute into which the debris is dumped must be protected by a guardrail that is at least 42 inches in height (Figure 16–30).

To prevent weakening of the floor, openings for the removal of debris may not exceed 25 percent of the total floor unless sufficient lateral support members for the floor are still in place. If a floor has been weakened, it must be shored sufficiently to carry any loads it may be subjected to during demolition operations.

Masonry walls, chimneys, or other masonry units must not be allowed to fall onto floors during demolition unless it has been determined in the engineering survey that the floors can withstand the force of the fall and the load imposed by the masonry units. Wall sections more than one story tall must not be allowed to stand unless they are properly braced or were originally designed to stand without bracing and still have the structural integrity to stand. No unstable wall shall be left at the end of a work shift. In structural steel-framed buildings, the steel frame must be left in place during the demolition of masonry, but all beams, girders, and other horizontal members are to be kept free of debris. Walls that support earth or adjoining structures must remain in place until the earth or adjoining structure is moved or properly supported (Figure 16–31). Walls must be capable of supporting any debris that might be piled against them.

When manually removing floors, appropriate precautions must be taken to guard against workers falling through areas where flooring has been removed. Openings cut into floors must extend the full span between supports. Walkways must be provided that are at least 18 inches wide and 2 inches thick so that workers do not have to walk on exposed beams. No workers should be allowed in the area below a floor that is being demolished. Floor arches should not be removed until

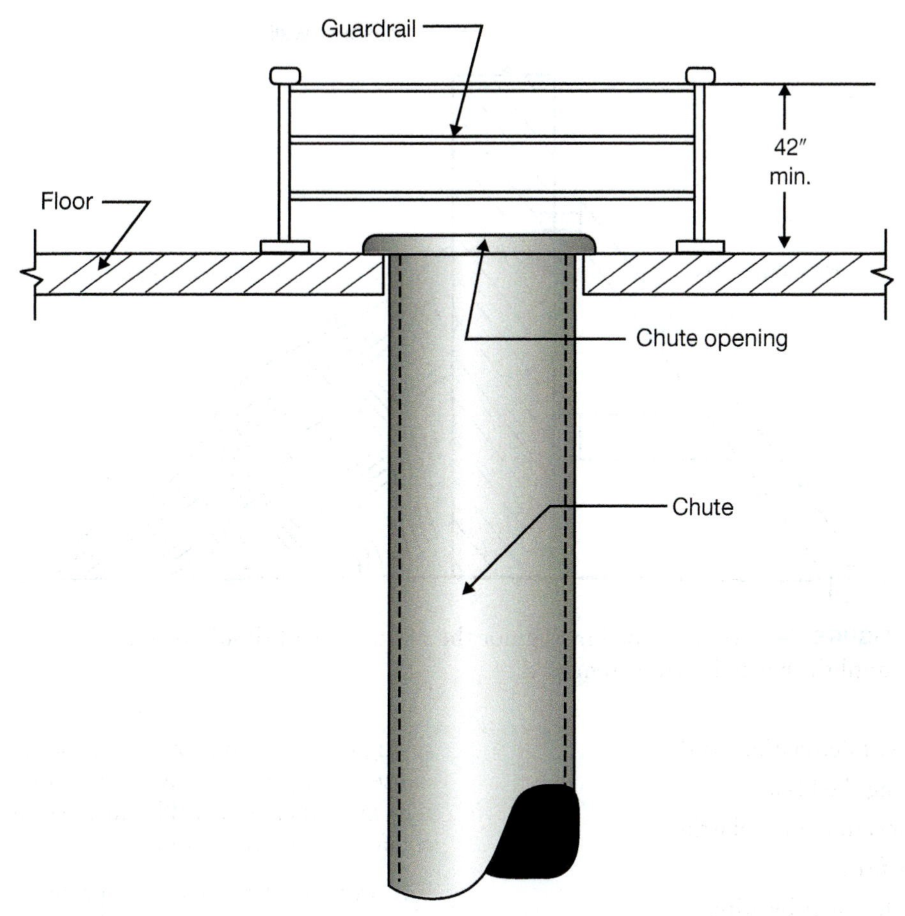

FIGURE 16–30 Guardrails must be provided at the top end of chutes.

the surrounding area for at least 20 feet has been cleared of debris and any other loose material. When using mechanical equipment to remove floors or walls, it must be determined beforehand that all work surfaces have sufficient strength to support the equipment. Equipment must be curbed or otherwise secured to guard against its slipping over the edge of a floor or into an opening.

When removing floors it is important to guard against debris falling from one level to another. It is also important to ensure that floors are capable of carrying the load of any debris stored on them. Storage space where debris is collected must be properly blocked off. Openings through which debris is dumped should be closed when not in use. When removing steel members during a demolition operation, planking should be provided for a work surface after floor arches have been removed. Steel should be removed one column length at a time and tier by tier.

When mechanical demolition operations (balling and clamming) are in progress, workers must be kept clear of the work area. Only essential personnel should be allowed into the demolition area at any time. A demolition ball should not exceed 50 percent of the crane's rated load, nor 25 percent of the nominal break strength of the line from which it is suspended. The load line for a demolition ball should be kept as short as it can possibly be and still accomplish the required task. The ball must be attached to the line with a swivel connector to prevent twisting of the line. Subpart U contains the requirements for using explosives during demolition operations.

SUBPART U: BLASTING AND USE OF EXPLOSIVES AND RELATED SAFETY PRACTICES

Subpart U covers the requirements for blasting and the use of explosives on construction sites. Subpart U contains the following sections:

- 1926.900 General provisions
- 1926.901 Blaster qualifications
- 1926.902 Surface transportation of explosives
- 1926.903 Underground transportation of explosives
- 1926.904 Storage of explosives and blasting agents
- 1926.905 Loading of explosives and blasting agents
- 1926.906 Initiation of explosive charges: electric blasting
- 1926.907 Use of safety fuse

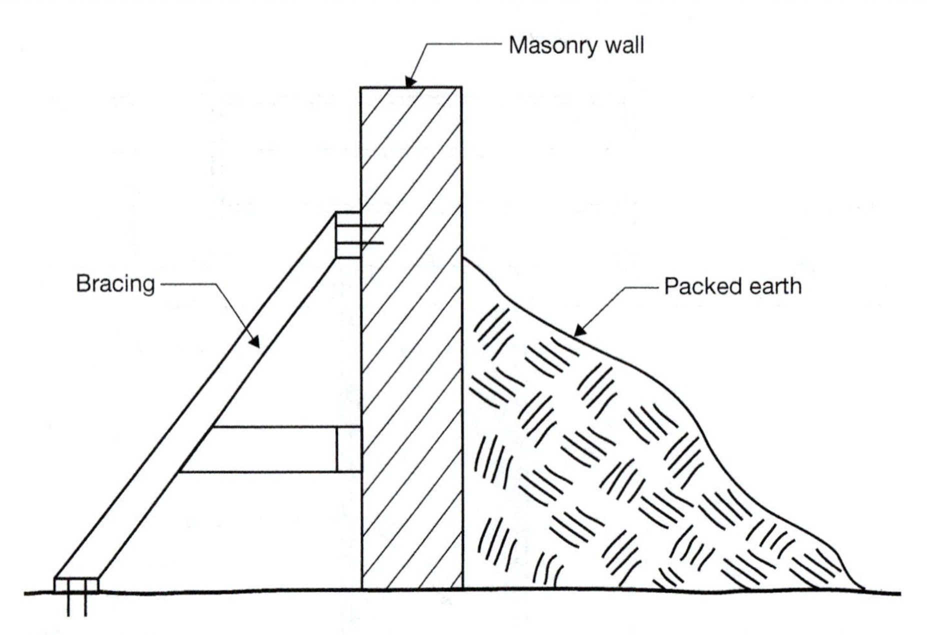

FIGURE 16–31 Neither the bracing nor the masonry wall should be removed until the packed earth is removed.

- 1926.908 Use of detonation cord
- 1926.909 Firing the blast
- 1926.910 Inspection after blasting
- 1926.911 Misfires
- 1926.912 Underwater blasting
- 1926.913 Blasting in excavation work under compressed air
- 1926.914 Definitions applicable to this subpart

BLASTING: GENERAL PROVISIONS AND RELATED SAFETY PRACTICES

Blasting is one of the most dangerous of construction practices. All types of heat- and spark-producing devices are prohibited near explosives.[27] No person may handle explosives while under the influence of intoxicating substances (e.g., alcohol, drugs). Explosives that are not being used must be stored in a locked magazine. All explosives must be accounted for at all times. A comprehensive, up-to-date, and accurate inventory of explosives must be maintained along with accurate records of use. Explosives must never be abandoned, and the appropriate authorities must be notified if explosives are missing for any reason.

Persons who are qualified to use explosives must take all necessary precautions, including the use of signals, flags, and barricades, to protect other employees. Appropriate precautions must be taken to contain the blast when explosives are used in a congested area. Unless it is impossible to do so, blasting operations conducted aboveground should be completed during daylight hours. To protect against accidental discharges caused by lightning, power lines, radar, radio transmitters, and other electrical hazards, contractors should take the following precautions:

1. In holes that have been primed and shunted, detonators should be short-circuited until they are connected into the blasting circuit.
2. In the event of a lightning storm, blasting operations should be suspended, and all personnel should be removed from the blasting area.
3. Warning signs should be prominently displayed to warn against the use of radio transmitters within 1,000 feet of the blasting site.

OTHER REQUIREMENTS IN BLASTING OPERATIONS

When transporting explosives, contractors must familiarize themselves with applicable regulations of the Department of Transportation (49 CFR Parts 146 through 397).[28] These regulations cover the requirements for motor carriers, highways and railroads, water carriers, and pipelines. Because the regulations for transporting explosives are so detailed and so specific, many construction companies choose to contract this work out to specialists. Several key requirements are as follows:

1. Vehicles used to transport explosives must be capable of carrying the load without difficulty.
2. Blasting caps may not be transported in the same vehicle with explosive materials.
3. Vehicles used to transport explosive materials must be adapted on the interior so that spark-producing

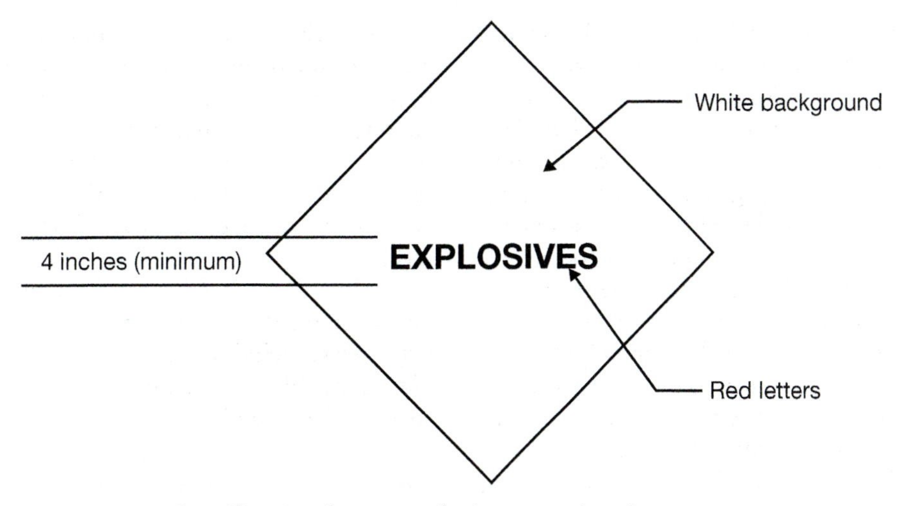

FIGURE 16–32 Specification for an explosives warning sign.

metal areas are covered with wood or another equivalent nonsparking material.

4. Vehicles used to transport explosive materials must be clearly marked on all sides (front, back, and both sides) with the word "Explosives" in red letters at least 4 inches high printed on a white background (Figure 16–32).
5. All vehicles used to transport explosive materials must carry a fire extinguisher with an Underwriter's Laboratory rating of at least 10-ABC, and drivers must be trained in the proper use of it.

Before beginning the preparations for blasting, contractors should post warning signs that explain the **code blasting signals** around the blasting area (Figure 16–33). Initiating devices, such as blasting caps and detonating primers, may not be stored in the same magazine with other explosives or blasting materials. There is to be no smoking within 50 feet of explosives and detonators. When drilling holes for explosive charges, the holes must be sufficiently large to easily accommodate the explosive material. If tamping is necessary, it must be done with a nonsparking implement, such as a wooden dowel. Primers must never be tamped.

Once all holes to be used for the immediate blasting have been filled, all remaining explosives and detonators must be returned to proper storage before detonation. Before beginning the next round of blasting, all holes from the previous blast must be checked for remaining explosive materials before beginning to drill for the next round. Old holes should not be reused. For each new blast, new holes must be drilled. No machines, equipment, personnel (other than those specifically authorized), or activity should be allowed in the blasting area.

Avoid the use of electric blasting caps in any area where extraneous electrical charges might be present that could cause an unplanned detonation. The leg wires on electric blasting caps are to be shunted (short-circuited) until they are wired into the blasting circuit. When using electric blasting caps, it is critical that the manufacturer's specifications be followed explicitly. If extraneous sources of electricity require that safety fuses be used rather than electric blasting caps, the following procedures should be carefully adhered to: (1) dispose of and do not use any safety fuse that has been dented, hammered, or misused in any way; (2) never put a sharp bend in a fuse; (3) when capping a fuse, cut a small amount off the reel each time to ensure a fresh end for every fuse; (4) never use safety fuses less than 30 inches in length; (5) when using the hand-lighting method, have at least two people present; and (6) do not allow one person to light more than 12 fuses when using the hand-lighting method.

Detonating cord should be handled carefully and properly as with any other type of explosive device. The cord should be kept free of sharp kinks, loops, or turns that might direct the cord back toward the line of detonation. Before firing the blast, all connections should be thoroughly inspected. When other devices are used in conjunction with detonating cord, the manufacturer's specifications for those devices must be followed to the letter.

Before initiating detonation, the blaster in charge makes sure that all extra explosives, detonators, and

Warning!
Listen for the Following
Blasting Signals

- **Warning Signal**
  One-minute series of long blasts five minutes before the blast signal.
- **Blast Signal**
  A series of short blasts one minute before the detonation.
- **All-Clear Signal**
  A prolonged blast after inspection of the blast area.

FIGURE 16–33 Blasting signals warning signs.

other extraneous materials have been removed from the blast area and properly stored. This person also makes sure that all equipment, vehicles, and personnel are at safe distance away from the blast area and that flagmen are properly stationed on nearby roadways. Only then does the blaster in charge give the blast "warning signal," followed five minutes later by the "blast signal" (see Figure 16–33). Once the detonation has occurred, the blaster must immediately disconnect the firing line from the blasting machine or device. If power switches are used, they must be locked in the "open" position. Before anyone is allowed to return to the blast area, sufficient time must be allowed for the smoke, dust, and fumes to dissipate. In underground blasting, "sufficient time" is considered to be at least 15 minutes. In addition, the responsible blaster must conduct a thorough inspection of the site to ensure that all explosives detonated completely.

In the case of misfires, the blaster in charge must keep all personnel out of the blast area except those required to clear the misfire. No other work is to proceed until the misfire hazard has been eliminated. *No person* should attempt to clear misfires out of a hole. Instead, the area is once again prepared for a blast, a new primer is put in, and the hole is reblasted. If this is not feasible, the misfired charge may be washed out of the hole with water or blown out with air. In the case of misfires when using a cap or fuse, all personnel must stay back and well away from the hole for at least one hour. No type of work is allowed in the blast area until all misfired charges have been eliminated.

Summary

Subpart P of 29 CFR 1926 contains the requirements and corresponding safety practices for excavations, cave-ins, and related issues, such as falls, loads, equipment, water contamination, access, egress, and hazardous atmospheres. Subpart Q of 29 CFR 1926 contains the requirements and corresponding safety practices for concrete and masonry construction, including concrete work, cast-in-place concrete, precast concrete, lift-slab concrete, and masonry construction.

Subpart R of 29 CFR 1926 contains the requirements and corresponding safety practices for steel erection, including the following: site layout, site-specific erection plan, construction sequence, hoisting and rigging, **structural steel assembly**, column anchorage, **beams and columns**, open web steel joists, **systems-engineered metal buildings**, falling object protection, fall protection, and training.

Subpart S of 29 CFR 1926 contains the requirements and corresponding safety practices for tunnels, shafts, caissons, cofferdams, and compressed air, including the following: underground construction, caissons, cofferdams, and compressed air. Subpart T of 29 CFR 1926 contains the requirements and corresponding safety practices for demolition work, including the following: preparatory operations; stairs; passageways; ladders; chutes; removal of materials through floors; removal of walls, masonry sections, and chimneys; manual removal of floors; removal of walls, floors, and material with equipment; storage; removal of steel construction; and mechanical demolition.

Subpart U of 29 CFR 1926 contains the requirements and related safety practices for blasting and the use of explosives, including the following: general provisions, blaster qualifications, surface transportation of explosives, underground transportation of explosives, storage of explosives and blasting agents, loading of explosives and blasting agents, initiation of explosive charges, use of safety fuses, use of detonation cord, firing the blast, inspection after blasting, misfires, underwater blasting, and blasting in excavations under compressed air.

Key Terms and Concepts

Access
Beams and columns
Blasting
Caissons
Cast-in-place concrete
Cave-ins
Code of blasting signals
Cofferdams
Column anchorage
Compressed air
Concrete work
Construction sequence
Demolition
Egress
Excavation
Falling object protection
Fall protection
Falls
Hazardous atmospheres
Hoisting and rigging
Lift-slab concrete
Loads
Masonry construction
Open web steel joists
Precast concrete
Preparatory operations (in demolition)
Site layout
Site-specific erection plan
Steel erection
Structural steel assembly
Systems-engineered metal buildings
Training
Tunnels and shafts
Underground construction
Water accumulation hazards

Review Questions

1. What is meant by the term *excavation*?
2. What is meant by the term *trench*?
3. Under what circumstances does Subpart P *not* apply to house foundation or basement excavations?
4. List the factors that minimize the hazards associated with excavation work.
5. When installing protective systems as part of an excavation, there are specific requirements that must be observed. List these requirements.
6. List the precautions that can be taken to guard against water accumulation hazards.

7. List the four general requirements and safe practices for concrete work.
8. Define the term *cast-in-place concrete,* and explain what it involves.
9. Explain the general precautions to be taken by contractors when using precast concrete.
10. Explain what is meant by the term *appropriately reinforced* when using lift-slab construction.
11. Limited access zones in masonry construction must meet several requirements. List three of these requirements.
12. In steel erection, work cannot begin without written notification by the controlling contractor. What information must be contained in the written notification?
13. What are the precautions that should be taken during multiple-lifting procedures in steel erection?
14. Explain the column anchorage requirements in steel erection.
15. Explain the precautions to be taken when making "double connections" in steel erection.
16. Training for steel erection workers must cover what topics?
17. Training for workers in underground construction must cover what topics?
18. Potentially gassy operations exist when what conditions are present?
19. What are the minimum safety precautions to be taken in gassy operations?
20. List the five precautions to be taken when using cranes and hoists in underground construction.
21. What is a caisson? What is a cofferdam?
22. Explain the medical requirements when using compressed air.
23. What is the purpose of completing the comprehensive engineering survey before demolition operations can begin on a structure?
24. What are the areas in which safety practices and requirements apply during demolition operations?
25. When using explosives, what are the specific precautions to be taken to prevent accidental discharges by extraneous electrical charges?

Critical Thinking and Discussion Activities

1. You overhear the following conversation between two employees: First employee: "I'm not going to work in a trench. How do I know it won't cave in on me?" Second employee: "Don't be ridiculous. It's just as dangerous on the surface as it is down there." What do you think about this conversation? Is either of these employees right? Wrong? What would you tell these two employees about working in excavations?
2. Your company is fabricating lift-slab members for a warehouse that it is constructing. Your fabrication team has solved an unexpected problem by making a slab that is not in the engineer's design package. It is similar to the other members designed by the consulting engineer your company employed, but slightly wider, taller, and thicker. The fabrication foreman wants to use the same type, size, and number of lifting devices without consulting the engineer (whose office is 75 miles away). According to the foreman, "This slab is almost identical to the ones we've already fabricated and lifted into place. If we consult the engineer, he will just hold us up for two days, put us behind schedule, use the same lifting hardware, and charge us for his time." Your safety director on the other hand is adamant that the engineer be asked to check the design before lifting the slab. According to the safety director, "If that slab falls back on an employee because the lifting hardware isn't strong enough, we are going to wish we had consulted our engineer." What is your opinion in this debate? If the decision were yours, what would you do?
3. You are the safety director for Jones Construction Company. Your blasting team is drilling holes in preparation for placing charges. The blaster in charge wants to use cellular phones to coordinate the preparations and to communicate with the members of his team. He promises to shut his phone off and to ask all of his team members to do the same before initiating the blast, but you are concerned about extraneous electricity. Should you agree to let the blasting team use cellular phones or prohibit their use?

Application Activities

1. Conduct research into the terrorist attack on the World Trade Center towers in New York City on September 11, 2001. The towers were designed to withstand the force of the impact of jetliners like the ones that slammed into them. However, when the steel columns supporting the top floor became so hot from the burning jet fuel that they failed, each successive floor fell on top of the next lower floor, until the entire building had been demolished. Compare what you learn about this tragedy with OSHA's rules concerning demolition of buildings, and write a brief paper that explains the rationale for these rules based on what happened to the World Trade Center.

2. Identify a job site in your community that is using precast or lift-slab construction, and observe the erection of the building. Compare what you observe with the requirements and recommended safety practices for concrete work. Write a summary of what is being done right and anything that is being done in a questionable manner. Write your findings without naming the company.
3. Identify a steel erection company in your community or the nearest one to you and ask to see their training program for employees who are exposed to fall hazards. Does the program contain all of the required elements? Analyze the program. What would you add to it? Does the program include lessons for employees who will be involved in multiple-lift rigging and connector procedures? Write your findings without naming the company.

Endnotes

1. 29 CFR 1926.651
2. 29 CFR 1926.652
3. 29 CFR 1926.651
4. 29 CFR 1926.701
5. 29 CFR 1926.702
6. 29 CFR 1926.703
7. 29 CFR 1926.704
8. 29 CFR 1926.705
9. 29 CFR 1926.706
10. 29 CFR 1926.750
11. 29 CFR 1926.752
12. 29 CFR 1926.753
13. 29 CFR 1926.754
14. 29 CFR 1926.755
15. 29 CFR 1926.756
16. 29 CFR 1926.757
17. 29 CFR 1926.758
18. 29 CFR 1926.759
19. 29 CFR 1926.760
20. 29 CFR 1926.761
21. 29 CFR 1926.800
22. 29 CFR 1926.801
23. 29 CFR 1926.802
24. 29 CFR 1926.803
25. 29 CFR 1926.850
26. 29 CFR 1926.851–1926.860
27. 29 CFR 1926.900
28. 29 CFR 1926.901–1926.911

SUBPARTS V THROUGH CC AND RELATED SAFETY PRACTICES

LEARNING OBJECTIVES

- Explain the main provisions of Subpart V.
- Explain the main provisions of Subpart W.
- Summarize the main provisions of Subpart X.
- Summarize the main provisions of Subpart Y.
- Explain the main provisions of Subpart Z.
- Describe the requirements contained in Subpart CC.

Subparts V through Z contain information on the Occupational Safety and Health Administration's (OSHA) requirements and related safety practices that are vital to construction companies and contractors in the areas of electrical power transmission, rollover protection, stairways and ladders, commercial diving operations, and toxic or hazardous substances.

SUBPART V: POWER TRANSMISSION AND DISTRIBUTION AND RELATED SAFETY PRACTICES

Subpart V contains the requirements and related safety practices for construction of electrical transmission lines, distribution lines, and related electrical equipment. Subpart V contains the following sections:

- 1926.950 General requirements
- 1926.951 Tools and protective equipment
- 1926.952 Mechanical equipment
- 1926.953 Material handling
- 1926.954 Grounding for protection of employees
- 1926.955 Overhead lines
- 1926.956 Underground lines
- 1926.957 Construction in energized substations
- 1926.958 External load helicopters
- 1926.959 Lineman's body belts, safety straps, and lanyards
- 1926.960 Definitions applicable to this subpart

GENERAL REQUIREMENTS

Before beginning construction work involving power transmission, distribution, and related equipment, the tests and inspections necessary to identify hazardous conditions must be completed.[1] Once the inspections have been completed, the following general precautions should be taken:

1. All electrical equipment and lines should be considered energized until it has been determined from tests or inspections that they are not.
2. Operating voltage of all lines and equipment must be determined before beginning work so that the safe distance set forth in Figure 17–1 can be properly maintained.
3. Lines and equipment must be disconnected from their sources of energy, and the connector must be locked out before work begins.
4. Workers must be trained in emergency procedures and first aid.
5. Drown-proofing precautions must be taken when work is done over water.
6. Sufficient lighting must be provided when work is done at night or in any setting that is dark.

Minimum Distances: Working and Clear Hot Stick	
Voltage Range (Phase to Phase) (kV)	Minimum Distance
2.1–15	2 ft. 0 in.
15–35	2 ft. 4 in.
35.1–46	2 ft. 6 in.
46.1–72.5	3 ft. 0 in.
72.6–121	3 ft. 4 in.
138–145	3 ft. 6 in.
161–169	3 ft. 8 in.
230–242	5 ft. 0 in.
345–362	7 ft. 0 in.
500–552	11ft. 0 in.
700–765	15ft. 0 in.

FIGURE 17–1 Selected examples of minimum safe work distances from alternating current for clear hot stick (from OSHA 29 CFR 1926.950).

Source: From Occupational Safety & Health Administration. Published by U.S Department of Labour.

When referring to Figure 17–1, note that the minimum working distance and the minimum clear hot stick distance may be reduced provided the reduced distance is not less than the shortest distance between the energized part and a grounded surface. In other words, never reduce the distance to a length less than the shortest distance between the energized part and a grounded surface.

TOOLS AND PROTECTIVE EQUIPMENT

Tools and protective equipment represent an important area in construction safety. Proper insulation on the handles of tools is important in electrical transmission work.[2] Any tool or object that is a conductor must be kept at a safe distance from energized lines or equipment. Safe distances are set forth in Figure 17–1. Even rubber-protected equipment must be carefully inspected before being used near energized lines or equipment to ensure against worn spots or breakdowns in the rubber coating. Rubber gloves must be air tested to identify holes or thin spots. Before using any tool in a power transmission setting, it should be inspected thoroughly and removed from service if it is not functioning properly. Metal ladders or ladders made of any type of conductive material must be kept away from energized lines and equipment. Double insulation is required on electric hand tools, and non-conductive hoses are required on hydraulic and pneumatic tools.

MECHANICAL EQUIPMENT

Mechanical equipment must be inspected before it is used.[3] If any defects or malfunctioning of mechanisms is noted, the equipment should be taken out of use until fully repaired and properly functioning. Aerial-lift trucks are to be treated in the same manner as energized equipment. This means they must be properly grounded or barricaded. The best approach is to ground *and* barricade. Only properly insulated material may be passed from an aerial lift to a utility pole or other structure. Unless it is certified for work on specific voltages, mechanical equipment should be kept a safe distance from energized lines. Distances considered safe are set forth in Figure 17–2.

Minimum Distances: Live-Line and Bare-Hand Work

Voltage Range	Distance for Maximum Voltage	
(Phase to Phase) (kV)	Phase to Ground	Phase to Phase
2.1–15	2 ft 0 in.	2 ft 0 in.
15.1–35	2 ft 4 in.	2 ft 4 in.
35.1–46	2 ft 6 in.	2 ft 6 in.
46.1–72.5	3 ft 0 in.	3 ft 0 in.
72.6–121	3 ft 4 in.	4 ft 6 in.
138–145	3 ft 6 in.	5 ft 0 in.
161–169	3 ft 8 in.	5 ft 6 in.
230–242	5 ft 0 in.	8 ft 4 in.
345–362	7 ft 0 in.	13 ft 4 in.
500–552	11 ft 0 in.	20 ft 0 in.
700–765	15 ft 0 in.	31 ft 0 in.

FIGURE 17–2 Selected examples of minimum safe distances from alternating current for live-line and bare-hand work (from OSHA 29 CFR 1926.952).
Source: From Occupational Safety & Health Administration. Published by U.S Department of Labour.

MATERIAL HANDLING

Utility poles being transported should be clearly marked by a red flag at the end of the longest pole.[4] Poles and other large, bulky materials should be inspected before they are unloaded to make sure they have not shifted during transport. Taglines should be used when utility poles and other large objects are hoisted, and workers should not be allowed under the hoisted materials. Unless no other option is feasible, materials should not be stored under energized busses or other conductors. In the rare cases when storage of materials under energized conductors is necessary, the safe distance set forth in Figure 17–1 should be carefully observed.

GROUNDING FOR PROTECTION OF EMPLOYEES

Grounding for protection of employees should never be overlooked. Construction personnel should always assume that any conductor or equipment is energized until it is known without doubt, through testing, to be grounded or de-energized.[5] When attaching conductors, the ground end should always be connected first. In addition, properly insulated tools must be used when connecting the "hot" end. When removing grounds, the grounded end also comes off first, using properly insulated tools. Grounds should be placed at the location of the work in question or between the work location and the source of energy. When working on a line section, grounds must be provided at every work location along the line. If, for any reason, it is impossible to provide a ground at a given work location, all work at that location should be considered energized. Grounds may be removed for testing. However, this should be undertaken with great caution. In addition, only sufficiently rated and properly designed tower clamps may be used for ground towers.

OVERHEAD LINES

Overhead lines, poles, and other structures that may not be able to support the weight of a climber must be sufficiently supported through bracing or guy wires.[6] Before they may be used, utility poles, ladders, lifts, and all other elevated structures must be thoroughly inspected for workers' safety. Workers standing on the ground

SAFETY FACTS & FINES

Checking for underground hazards is important. Before beginning any type of digging or excavation, construction professionals should first check for underground cables, pipelines, and other buried hazards. A construction company in Plummer, Minnesota, failed to check for underground hazards before digging and ruptured a crude oil pipeline. Before the ruptured pipeline could be repaired, more than 200,000 gallons of crude oil had leaked into a nearby river. Not only did this failure on the part of the construction company cause an environmental nightmare, it also caused the nearest town to be evacuated. The company was fined $50,000.

should not be allowed to touch any piece of equipment or any other potential conductor located near an energized pole or piece of equipment. Utility poles must be examined thoroughly for structural integrity before they are installed, removed, relocated, or hoisted and moved for any reason. Appropriate precautions should be taken to ensure that utility poles do not contact energized lines when the poles are being hoisted and moved. Poles that are being installed or moved should never be left unattended. Equipment used to install, move, and hoist utility poles should be properly grounded. In addition, even when properly grounded, lifting equipment used near energized lines should be considered energized.

UNDERGROUND LINES

When working underground, it is important to let people aboveground know there are workers below.[7] This means placing barriers or barricades and warning signs around the underground opening (e.g., manhole, vault). It is also important to guard against hazardous atmospheres when working on **underground lines**. Before allowing workers to go underground, the atmosphere of the work environment should be tested and determined to be safe. If the atmosphere is found to be hazardous, appropriate precautions should be taken, such as providing forced-air ventilation, personal protective equipment (PPE), and an attendant by the opening to take immediate action in the case of an emergency.

Before conducting excavation in conjunction with power transmission and distribution work, potentially hazardous underground conditions (e.g., cables, gas lines) should be identified and the appropriate precautions taken to protect underground workers from exposure to them. When splicing underground cables, continuity of the metallic sheath must be maintained by bonding across the opening or by equivalent measures.

CONSTRUCTION IN ENERGIZED SUBSTATIONS

Construction in energized substations is a special case safety professionals must understand how to deal with. Extra caution is called for when working in energized electrical substations (Figure 17–3).[8] No worker should be allowed in a substation without the permission of proper authorities. Then, before workers enter the substation, a checklist for working there should be developed.

FIGURE 17–3 Extra caution is called for when doing construction work in an electrical substation.
Source: Sergbob/Fotolia

This checklist should contain the precautions that workers will be required to take, including any PPE they must wear, barriers or barricades that must be erected, signs that must be posted, grounding of vehicles and heavy equipment that must be accomplished, temporary fences that must be erected when permanent fences are removed to provide access to heavy equipment, and provisions for ensuring that gates are locked and the facility is secured when the substation is unattended (Figure 17–4).

Planning Checklist for Work Performed in Energized Substation

1. Has proper authorization to enter substation been given? By whom?
2. Has an inspection of the substation been completed and all hazardous conditions identified?
3. What personal protection equipment should workers use?
4. Will barriers or barricades be needed? Where?
5. Will warning signs be posted? Where?
6. Who will be designated to work on or near control panels?
7. Who will be designated to control vehicles, gin poles, cranes, and other equipment?
8. Who will ensure that mobile cranes and derricks used in close proximity to energized lines and equipment have been properly grounded?
9. Will permanent fencing have to be removed to provide access for vehicles and for equipment? Who will be responsible for erecting temporary fences and tying them into the remaining permanent fence?
10. Who will be responsible for ensuring that all gates are locked when the substation is left unattended?

FIGURE 17–4

EXTERNAL LOAD HELICOPTERS

The requirements and safety practices for helicopters are explained in Subpart N, Chapter 15 (29 CFR 1926.551).[9] These requirements also apply when performing construction operations in the vicinity of power transmission and distribution lines.

LINEMAN'S BODY BELTS, SAFETY STRAPS, AND LANYARDS

Construction safety professionals should understand how to ensure the safety of workers using **lineman's body belts, safety straps, and lanyards.** When working at elevated locations, employees are required to wear body belts, safety straps, and lanyards that meet the specification set forth in ASTM Standard B-117-64 (unless doing so will create a greater hazard than not wearing them; Figure 17–5).[10] Each item of PPE should be thoroughly inspected before each use. Worn or damaged equipment should be removed from service immediately. If there is a question concerning the condition of PPE, err on the safe side. As of January 1998, body belts are permitted for positioning only.

SUBPART W: ROLLOVER PROTECTIVE STRUCTURES AND OVERHEAD PROTECTION

Subpart W contains the requirements and related safety practices for **rollover protective structures (ROPS)** and **overhead protection.** Subpart W contains the following sections:

- 1926.1000 ROPS for material handling equipment
- 1926.1001 Minimum performance criteria for ROPS for designed scrapers, loaders, dozers, graders, and crawler tractors
- 1926.1002 Protective frame test procedures and performance requirements for wheel-type agricultural and industrial tractors used in construction
- 1926.1003 Overhead protection for operators of agricultural and industrial tractors

FIGURE 17–5 Employees working in elevated locations should wear the appropriate personal protective equipment.
Source: Riccardo Arata/Fotolia

ROLLOVER PROTECTIVE STRUCTURES, OVERHEAD PROTECTION, AND TEST PROCEDURES AND CRITERIA

All material handling equipment used in construction must have the proper ROPS when manufactured—if manufactured after September 1, 1972—or by retrofitting, if manufactured before this date.[11] This includes all of the following types of equipment: scrapers, front-end loaders, dozers, wheel-type agricultural and industrial tractors, crawler-type loaders, and motor graders. Figure 17–6 and 17–7 are examples of construction equipment with roller and overhead protection.

The purpose of ROPS is to keep the operator of the equipment from being crushed if the equipment rolls over.

FIGURE 17–6 Piece of construction equipment with rollover protection.
Source: Hoda Bogdan/Fotolia

FIGURE 17–7 Rollover equipment is a must on construction equipment.
Source: Daniel Rönneberg/Fotolia

Because even the best ROPS cannot protect operators who are thrown out of the cab during a rollover, construction professionals should insist on the use of seat belts. In addition, operators should be encouraged to stay with the equipment during a rollover. Their chances of being injured are greater when they try to jump off a piece of equipment that is rolling over. In addition to providing ROPS, contractors should provide overhead protection to prevent materials from falling through the ROPS and injuring the operator. The overhead protection should be installed so that it does not become a hazard itself in the event of a rollover.

Various tests are required by OSHA to ensure acceptable minimum performance by ROPS. Because these tests are typically performed by the manufacturer of the equipment rather than contractors, the tests are not covered in this section. However, because there is a chance, though it is remote, that a contractor might retrofit a piece of equipment with ROPS, readers should be familiar with how to locate the test procedures and criteria; they are contained in 29 CFR 1926.1001 1002 and 1003.

SUBPART X: STAIRWAYS AND LADDERS

Subpart X contains the requirements and safety practices for stairways and ladders used in construction.[12] This subpart contains the following sections:

- 1926.1050 Scope, application, and definitions
- 1926.1051 General requirements
- 1926.1052 Stairways
- 1926.1053 Ladders
- 1926.1054 through 1926.1059 (Reserved)
- 1926.1060 Training

Stairways or ladders must be provided anytime there is a difference of 19 inches or more between levels, unless the difference in elevation is accommodated by a ramp, sloped walkway, runway, or approved personnel hoist. Spiral staircases are not to be used in construction work, unless the staircase is a permanent component in the structure being built. Staircases that are not a permanent part of the structure being built, but will be used by construction personnel, must meet the following requirements:

- *Proper landings.* Landings must be at least 30 inches wide (in the direction of travel) by at least 22 inches at every 12 feet of vertical height in order to be considered proper landings.
- *Slope angle.* Once installed, stairs should run at an angle of 30–50 degrees to the horizon in order to have the proper slope angle.
- *Riser height and tread width.* **Riser height and tread width** must be uniform and shall not vary by more than ¼ inch.
- *Doors and gates.* When a door or gate opens onto a landing or platform, it must not reduce the effective width of the platform or landing to less than 20 inches. This applies to all doors and gates.
- *Metal pan landings and treads.* **Metal pan landings and treads** must be properly secured before being filled with concrete, sand, gravel, or any other material. This type of landing and tread should never be left unfilled. If they are not to be filled until some point in the future, metal pan landings and treads should be covered with temporary treads or landings.
- *Temporary treads.* If permanent treads are to be added later, **temporary treads** must be provided during the interim. Temporary treads may be made of wood or comparable solid materials. They must cover the entire width and depth of the stair tread opening.
- *Protrusions and slippery surfaces.* Stairs must be free at all times of hazardous protrusions such as nails, screws, bolts, or splintered wood. The surfaces of stair treads, landings, and platforms must not be allowed to remain slippery. Protrusions and slippery surfaces are common causes of accidents.
- *Handrails and stair-rails.* **Handrails and stair-rails** are required along all unprotected sides of stairs that rise more than 30 inches vertically or that have four or more risers (whichever is less). Handrails and stair-rails must be able to withstand a force of at least 200 pounds applied within 2 inches of the top rail in any outward or downward direction. This requirement holds true for any point along the rail. When measured from the upper surface of the stair tread, handrails must be at least 30 inches in height but not more than 37 inches. Stair-rails must be at least 36 inches from the upper surface of the tread. In addition, stair-rails must be aligned with the face of the riser at the forward edge of the tread. The ends of handrails and stair-rails must be free of any projection hazards, and the surface of handrails and stair-rails must be free of any jagged areas that might snag clothing or injure a person's hands or arms.

LADDERS AND TRAINING

Ladders used in construction come in several different types, each with its own load capacity.[13] Figure 17–8 shows the various types and grades of ladders and their corresponding load capacities. Most of the requirements for ladders are the responsibility of the ladder manufacturer. However, there are a number of safety practices that should be observed by construction personnel when using ladders. Figure 17–9 contains a checklist that construction personnel can use to inspect ladders—this should be

Ladder Types		
Type	**Grade**	**Load Capacity**
I	Industrial	250 lb
IA	Extra heavy-duty industrial	300 lb
II	Commercial	225 lb
III	Household	200 lb

FIGURE 17–8 Ladder types with corresponding load capacity.

Ladder Inspection Checklist

- Determine whether the ladder has the manufacturer's instruction label on it.
- Determine whether the ladder is strong enough.
- Read the label specifications about weight capacity and applications.
- Look for the following conditions: cracks on side rails; loose rungs, rails, or braces; and damaged connections between rungs and rails.
- Check for heat damage and corrosion.
- Check wooden ladders for moisture that may cause them to conduct electricity.
- Check metal ladders for burrs and sharp edges.
- Check fiberglass ladders for signs of blooming deterioration of exposed fiberglass.

FIGURE 17–9 Checklist for conducting ladder inspections.

Safe Practices for Using Ladders

- Check for slipperiness on shoes and ladder rungs.
- Secure the ladder firmly at the top and bottom.
- Set the ladder's base on a firm, level surface.
- Apply the 4:1 ratio (base should be 1 foot away from the wall for every 4 vertical feet between the base and the support point).
- Face the ladder when climbing up or down.
- Barricade the base of the ladder when working near an entrance.
- Don't lean a ladder against a fragile, slippery, or unstable surface.
- Make sure the ladder extends at least 3 feet above the working surface.
- Don't lean too far to either side while working (stop and move the ladder).
- Don't rig a makeshift ladder; use the real thing.
- Don't allow your waist to go any higher than the last rung when reaching upward on a ladder.
- Don't carry tools in your hands while climbing a ladder.
- Don't place a ladder on a box, table, or bench to make it reach higher.

FIGURE 17–10 These rules will help ensure safe use of ladders.

done before every use. In addition, Figure 17–10 contains a list of "Dos and Don'ts" for safe ladder use.

Employees who use ladders and stairwells in their construction work must be properly trained to recognize hazards and to take the precautions necessary to minimize the hazards. Workers must be trained and retrained as necessary by a competent person. The training must cover all of the following topics:

- Identifying fall hazards related to stairs and ladders.
- Erecting, maintaining, and disassembling fall protection systems.
- Safe construction, placement, use, and care of stairs and ladders.
- Maximum load capacities of ladders.

SUBPART Y: COMMERCIAL DIVING OPERATIONS

Subpart Y contains the requirements and safety practices for **commercial diving operations** undertaken as part of a construction project. The requirements do not apply to instructional diving or search and rescue operations. Subpart Y contains the following sections:

- General
  - 1926.1071 Scope and applications
  - 1926.1072 Definitions
- Personnel Requirements
  - 1926.1076 Qualifications of dive team
- General Operations Procedures
  - 1926.1080 Safe practices manual
  - 1926.1081 Predive procedures
  - 1926.1082 Procedures during dive
  - 1926.1083 Postdive procedures
- Specific Operations Procedures
  - 1926.1084 SCUBA diving
  - 1926.1085 Surface-supplied air diving
  - 1926.1086 Mixed-gas diving
  - 1926.1087 Liveboating
- Equipment Procedures and Requirements
  - 1926.1090 Equipment
- Record keeping
  - 1926.1091 Record-keeping requirements

DIVING REQUIREMENTS AND CORRESPONDING SAFETY PRACTICES

Construction safety professionals should understand **diving requirements** and corresponding safety practices. Diving in conjunction with construction is not uncommon, although most companies that perform this type of work are specialists.[14] However, even construction professionals who do not work with companies that perform commercial diving should know enough about the requirements to be able to contract knowledgeably with these companies. This section explains the requirements and safety practices that every construction professional should know about commercial diving in relation to construction. The requirements are the same as those contained in 29 CFR 1910, Subpart T of OSHA's general industry standard, and are used by ship repairing, shipbuilding, shipbreaking, longshoring, and bridge and dam construction companies.

General Requirements

Any person who performs diving operations in conjunction with a construction project must have an appropriate level of experience or training in the following areas:

1. Proper operation and maintenance of the tools, equipment, and systems used in the specific diving operation.
2. Proper techniques for the diving operation.
3. Emergency procedures related to the diving operation.

In addition to these requirements, there are several requirements that apply to the employer of the divers on the dive team. These requirements are as follows:

1. Employers must ensure that all members of the dive team have been trained in first aid and cardiopulmonary resuscitation by the American Red Cross or an equivalent certified trainer.
2. Employers must ensure that all dive team members who will be exposed to hyperbaric conditions have been properly trained in the physiological aspects of working in such conditions.
3. Employers must ensure that members of the dive team are assigned only those tasks for which they have the proper training or experience.
4. Employers must ensure that members of the dive team are not exposed to hyperbaric conditions against their will (unless it is necessary to do so as part of the decompression process).
5. Employers must ensure that no dive team member with any type of physical impairment that might cause problems is allowed to operate in hyperbaric conditions.
6. Employers must ensure that a competent person (person in charge) is onsite to supervise all aspects of the diving operation for the duration of the operation. This individual must have the appropriate knowledge, skill, training, and experience to carry out the supervision of diving operations.
7. Employers must develop and distribute to all members of the dive team a manual that explains the safe practices required in diving operations. This manual should contain a complete copy of 29 CFR 1926, Subpart Y, and specific policies and procedures for implementing the standard.

Safe Diving Procedures Manual

The company's **safe diving procedures manual** should contain at least the following information *beyond* the actual OSHA standard in 29 CFR 1926, Subpart Y: (1) safety procedures and corresponding checklists for diving operations, (2) assignments and responsibilities of all dive team members, (3) equipment procedures and corresponding checklists, and (4) emergency procedures for all predictable problems, including fire, equipment failure, medical problems and illnesses, injuries, and hazardous environmental conditions.

In addition to the safe-diving manual, a comprehensive list of telephone and emergency numbers must be developed and posted at the dive locations. The list must contain telephone or call numbers for at least the following: (1) the fully operational decompression chamber that will be used by the dive team if necessary, (2) closest, most accessible hospitals to the dive site, (3) physicians who are available to serve the dive team if necessary, (4) all available means of transportation, and (5) the nearest Coordination Center for the U.S. Coast Guard. The dive location must also be equipped with a properly stocked first aid kit that contains all of the materials that might be needed for dive-related emergencies (e.g., American Red Cross first aid manual or equivalent, bag-type manual resuscitator equipped with a transparent mask and tubing).

Planning a Dive Operation

When **planning a dive operation**, a comprehensive safety and health assessment must be conducted. The following areas should be assessed: (1) diving mode; (2) conditions on the surface and underwater; (3) hazards on the surface and underwater; (4) supply of breathing gas; (5) supply of breathing gas reserves; (6) thermal protection; (7) diving equipment, tools, and systems; (8) physical fitness of all dive team members (e.g., fitness level appropriate to the assignment, physical impairments from injuries or illnesses); (9) assignments for each member of the dive team; (10) repetitive dive designation or residual inert gas status for all dive members; (11) decompression and treatment procedures; and (12) applicable emergency procedures.

During Diving Operations

Employers must ensure that the following requirements and safety practices are followed during actual diving operations:

1. A means, such as a ladder, that will support a fully equipped diver must be provided for easing entry into and exit from the water.

2. The means of entry and exit provided must extend far enough below the water's surface to ensure an easy and convenient exit for divers.
3. A means must be available onsite for helping an injured or impaired diver out of the water.
4. An operational two-way voice communication system must be provided between divers and personnel.
5. Decompression, no-decompression, and repetitive-dive tables must be available onsite.
6. Depth-time profiles, breathing gas changes, and decompression information must be maintained on site for every diver.
7. Electrical tools must be de-energized before being placed in the water and before being taken out of the water.
8. Electrical, pneumatic, or hydraulic tools that are powered from the surface must remain de-energized until the diver requests activation.
9. When performing underwater welding operations, a switch that interrupts the flow to the welding unit must be tended continuously by a qualified member of the dive team who is in constant voice communication with the diver doing the welding.
10. In operations that involve using explosives underwater, all involved must comply with 29 CFR 1926.912.
11. A dive must be terminated when a diver requests termination, fails to respond to communication, begins to use reserve breathing gas, or when the two-way voice communication system becomes impaired in any way.
12. Divers must be inspected immediately after completing every dive for physical problems.
13. After dives that are deeper than 100 feet or that use mixed gas for breathing, employers must ensure that the divers stay awake in the vicinity of the decompression chamber for at least one hour.

Record Keeping for Diving Operations

Record keeping for diving operations is an important process. For every dive operation, certain records must be kept by the employer. These records include the following:

1. Full names of every dive team member.
2. Full name of the competent person in charge who will supervise operations.
3. Date, time, location, and diving modes uses.
4. Nature of the work performed (e.g., "repaired piling, made welded connection").
5. Surface and underwater conditions at the dive site.
6. Maximum depth for each diver.
7. Bottom time for each diver.

Each time a diver goes outside the no-decompression limits or uses mixed gas, it is necessary to record additional information. This information includes the following: (1) depth time, (2) breathing gas profiles, (3) decompression table designation, and (4) elapsed time since the last pressure exposure. If symptoms of decompression sickness are observed, yet more information must be maintained, including (1) description of the symptoms observed, (2) time of the onset of the symptoms, (3) description of the treatment provided, and (4) description of the results of the treatment. Any time decompression sickness occurs, the incident should be investigated thoroughly and appropriate steps taken to prevent future occurrences.

Diving hazards are not restricted to decompression sickness. Like other construction workers, divers face a variety of hazards. Consequently, any time a diver is hospitalized for 24 hours or longer, comprehensive records must be maintained regardless of the type of injury. In such instances, the record should show the circumstances surrounding the injury or illness and the extent of the problem.

Retention of Records

The various records that must be maintained by construction companies that perform diving operations have mandated life spans. Retention requirements for the various types of records are as follows:

1. Physician's reports for divers (five years).
2. Depth-time records (one year or through completion of the decompression procedure assessment).
3. Record of all individual dives (one year, unless there has been an incident of decompression sickness within five years).
4. Decompression procedure assessment evaluations (five years).
5. Equipment inspections and testing records for equipment (until the equipment is removed from service).
6. Hospitalization records (five years).

SUBPART Z: TOXIC AND HAZARDOUS SUBSTANCES

Subpart Z contains the requirements and safety practices for a specific set of hazardous substances. OSHA's requirements related to these **toxic and hazardous substances** have to do with proper handling procedures, signs and symptoms of overexposure, and specific regulatory requirements. Subpart Z contains the following sections:

- 1926.1100 (Reserved)
- 1926.1101 Asbestos

- 1926.1102 Coal-tar pitch volatiles
- 1926.1103 13 Carcinogens (4-Nitrobiphenyl)
- 1926.1104 α-Naphthylamine
- 1926.1105 (Reserved)
- 1926.1106 Methyl chloromethyl ether
- 1926.1107 3,3′-Dichlorobenzidine (and its salts)
- 1926.1108 bis-Chloromethyl ether
- 1926.1109 β-Naphthylamine
- 1926.1110 Benzidine
- 1926.1111 4-Aminodiphenyl
- 1926.1112 Ethyleneimine
- 1926.1113 β-Propiolactone
- 1926.1114 2-Acetylaminofluorene
- 1926.1115 4-Dimethylaminoazobenzene
- 1926.1116 N-Nitrosodimethylamine
- 1926.1117 Vinyl chloride
- 1926.1118 Inorganic arsenic
- 1926.1127 Cadmium
- 1926.1128 Benzene
- 1926.1129 Coke oven emissions
- 1926.1144 1,2-Dibromo-3-chloropropane
- 1926.1145 Acrylonitrile
- 1926.1147 Ethylene oxide
- 1926.1148 Formaldehyde
- 1926.1152 Methylene chloride

Dealing with chemicals is a specialized field in construction that requires both unique knowledge and skills. However, every construction professional should have a working knowledge of certain basic information related to chemicals and their safe use. What is presented in this section is an in-depth treatment of asbestos (because it is so commonly confronted in construction renovation and demolition projects) and a general approach to chemical hazards that applies to all of those covered in Subpart Z.

ASBESTOS

Asbestos is a naturally occurring fibrous mineral that has been shown to cause illnesses, such as asbestosis (a debilitating condition similar to emphysema), mesothelioma (a cancerous tumor that can cover the lungs and other vital organs), lung cancer, and gastrointestinal cancer. Symptoms of these diseases may not appear for more than 20 years after the initial exposure to asbestos.

OSHA's asbestos requirements (29 CFR 1926.1101) first took effect in 1970, but they have been revised over the years as more and more has been learned about this hazardous substance.[15] The standard applies to employees involved in new construction, building renovation, remodeling, maintenance, and janitorial work. The latest revision of the standard established four classifications for any type of construction work that has the potential for exposing people to asbestos (Figure 17–11). Class I is the potentially most hazardous type of work; Class IV is the least hazardous. In addition to classifying asbestos work according to the potential risk involved, the standard also establishes the type of work activities that are regulated by OSHA (Figure 17–12).

OSHA's Classifications for Asbestos Exposure in Construction

- **Class I**
  Class I work is the most hazardous and involves the removal of thermal system asbestos that has been sprayed or troweled on.
- **Class II**
  Class II work involves the removal of nonthermal system asbestos.
- **Class III**
  Class III work consists of maintenance and repair tasks in situations where asbestos may be disturbed.
- **Class IV**
  Class IV work consists of janitorial tasks in which asbestos-containing waste is cleaned up.

FIGURE 17–11 OSHA's classification system for asbestos work.
Source: From OSHA's Classification System for Asbestos Work. Published by U.S. Department of Labor.

Checklist of Asbestos-Related Construction Activities

- Demolishing structures where asbestos is present.
- Salvaging structures where asbestos is present.
- Removing materials that contain asbestos.
- Encapsulating materials that contain asbestos.
- Performing any construction operation (e.g., building, renovating, repairing, maintaining) on a structure where asbestos is present.
- Installing materials that contain asbestos.
- Cleaning up after asbestos spills or emergencies.
- Handling, transporting, storing, or using materials that contain asbestos.

FIGURE 17–12 Asbestos-related construction activities regulated by 29 CFR 1926.1101.
Source: From Occupational Safety & Health Administration. Published by U.S Department of Labour.

OSHA's Requirements for Working Around Asbestos

Subpart Z contains specific requirements and safety practices organized in the following categories:

- *Permissible exposure limit (PEL).* The eight-hour time-weighted average (TWA) for exposure to airborne asbestos is 0.1 f/cc. In addition, exposure to airborne asbestos must not exceed 1 f/cc as averaged over a 30-minute sampling period (the short-term exposure limit [STEL]). This is known as the permissible exposure limit.
- *Exposure assessments and monitoring.* All asbestos operations must be assessed to determine their

potential to generate airborne fibers, and monitoring data must be used to assess employee exposure. This is known as conducting exposure assessments and monitoring.

- *Initial exposure assessments.* **Initial exposure assessments** must be conducted by the designated "competent person." They must be completed in time to ensure that all control systems are working properly, and the assessment data are collected soon enough to allow compliance with all OSHA requirements if a problem is detected.
- *Negative exposure assessments.* When a properly trained employee is to perform an asbestos-related task, employers must show that the exposure will be below the established PEL. This is known as negative exposure assessments. Several different methods may be used to confirm the exposure level.
- *Exposure monitoring.* Employers are required to take at least one breathing-zone air sample representing full-shift exposure to determine the eight-hour TWA and one representing a 30-minute exposure (STEL). This is known as **exposure monitoring**. The STEL air sample must be taken from the breathing zone for the most potentially hazardous operations.
- *Periodic monitoring.* Employers must monitor Class I and II jobs daily for every employee working in a regulated area. For Class III and IV jobs, employers must monitor only those work areas where asbestos exposure might exceed the PEL. An exception to the daily monitoring requirement is when all employees working in potentially hazardous areas use supplied-air respirators operated in positive-pressure mode. The process is known as periodic monitoring.
- *Additional monitoring.* Additional monitoring is required when any one of the following conditions exists: (1) a process is changed, (2) the control equipment system is changed, (3) less experienced personnel are introduced into the workplace, or (4) work practices are introduced that could result in levels of exposure that exceed the STEL or PEL.
- *Medical surveillance.* A **medical surveillance** program under the supervision of a licensed physician must be provided for all employees who use negative-pressure respirators, or perform Class I, II, or III jobs for a combined total of 30 or more days per year. Medical surveillance is also required for employees exposed at or above the STEL or PEL. The medical surveillance program must include comprehensive physicals performed at least annually and any other tests recommended by the physician in charge of the worker's case.
- *Record keeping.* The record-keeping requirements of 29 CFR 1926.1101 are extensive. They include the following: (1) objective data that demonstrate asbestos products cannot release fibers at levels exceeding the PEL or STEL, (2) records of employee-exposure monitoring (maintained for a minimum of 30 years), (3) medical surveillance records (maintained for the duration of employment plus an additional 30 years), and (4) training records for all employees (maintained for the duration of employment plus an additional year). When an employer plans to go out of business without a new organization to assume responsibility for all asbestos-related records, that employer must give the director of the National Institute for Occupational Safety and Health (NIOSH) 90 days' notice. NIOSH will provide instructions for the transmittal or disposal of the records.
- *Competent person.* A competent person must be designated at all job sites with asbestos-related operations. This individual must be qualified to ensure the safety and health of workers who might be exposed to asbestos and authorized to inspect job sites, equipment, materials, and processes as frequently as he or she deems necessary. Class I jobs require at least one inspection during every shift and on specific request by a worker. The company's designated competent person is required to complete a special course certified by the Environmental Protection Agency (EPA) or an equivalent course.
- *Regulated areas.* A **regulated area** is any work area where asbestos exposure exceeds or might exceed the PEL. Such areas must be clearly marked so that access can be controlled. All work within a regulated area is under the control of the designated competent person. Signs must be posted containing the information shown in Figure 17–13. Any area where asbestos exposure could exceed the PEL must be marked off and controlled as a regulated area.
- *Notification.* **Notification** requirements apply to both building owners and any contractors that will work on the building. Building owners are required to identify any asbestos hazards installed before 1981 and to notify contractors, tenants, and the owner's own employees of the hazards. Contractors or other employers who identify asbestos hazards on a job site must notify the owner within 24 hours.

DANGER
ASBESTOS
CANCER AND LUNG DISEASE HAZARD
AUTHORIZED PERSONNEL ONLY
RESPIRATORY AND PROTECTIVE CLOTHING
ARE MANDATORY IN THIS AREA

FIGURE 17–13 Example of a warning sign for a regulated area.

DANGER
CONTAINS ASBESTOS FIBERS
AVOID CREATING DUST
CANCER AND LUNG DISEASE HAZARD

FIGURE 17–14 Example of a warning label for a product or container that contains asbestos.

They must also notify owners and workers in the nearby surrounding area of the hazard and the mitigating precautions that will be taken. A contractor has just 10 days after completing a job to notify owners and workers who work in the building of any remaining asbestos hazards and the final monitoring results.

- *Signs.* Any area or room that contains asbestos hazards must be clearly marked with **signs** that contain at least the following information: (1) that hazardous asbestos material is present, (2) the specific location of the material, and (3) appropriate precautions to be taken when entering the area. The employer must also take all feasible steps to ensure comprehension (e.g., foreign language, English notation, graphics).
- *Labels.* All asbestos products and containers holding asbestos must be properly labeled with warnings. **Labels** must be easily seen; printed in large, bold letters against a contrasting background; and carry a statement such as the one shown in Figure 17–14.
- *Employee information and training.* OSHA has both general and specific training requirements related to asbestos. The general requirement is that employers provide training for all workers who perform Class I, II, III, or IV asbestos operations. Figure 17–15 is a checklist of topics that must be covered in general training sessions. In addition to the general training requirements, there are specific requirements for given situations. Class I and II operations that must use isolation techniques, such as critical barriers, increase the training requirements. In addition to the general training required of all workers, training that is equivalent to the model accreditation plan (MAP) for asbestos abatement training provided by the EPA is also required. Class III operations require the general training and the need for the EPA's 16-hour "Operations and Maintenance" course or its equivalent. Class IV operations require the general training and the EPA's two-hour awareness course or its equivalent.

Checklist of General Training Content Requirements for Asbestos Operations

- Recognizing asbestos hazards.
- Health hazards of asbestos exposure.
- Relationship between smoking and asbestos in lung cancer.
- Protective controls for minimizing exposure in selected operations.
- Proper use of respirators.
- Proper work practices for performing asbestos jobs.
- Requirements of the medical surveillance program.
- OSHA's asbestos standard (1926.1101).
- Contact information for public health organizations that provide information or smoking cessation programs.
- Sign and label requirements.

FIGURE 17–15 Employers must provide this training at no cost to employees.

MODEL FOR TAKING APPROPRIATE PRECAUTIONS WITH CHEMICALS

In addition to asbestos, numerous other hazardous substances are listed in Subpart Z.[16] Each has its own route of entry, mode of action, and corresponding requirements for safe use. Construction professionals should consult the specific applicable section in Subpart Z before allowing workers to be exposed to any of the chemicals listed at the beginning of this section. What follows is a generic model that can be used for investigation and preparation before allowing workers to be exposed to any potentially hazardous chemical. The questions asked in the model should be considered the minimum. Refer to Subpart Z for specifics for individual chemicals.

Questions to Ask Before Allowing Workers to Be Exposed to Chemicals

Regardless of the type of chemical or chemicals used on a given job site, construction professionals should answer the following questions and take whatever precautions

SAFETY FACTS & FINES

Removing asbestos is one of the most challenging of renovation, remodeling, or demolition jobs in the construction industry. OSHA's requirements are comprehensive and specific, and it does not pay to neglect any aspect of the standard. When a company in Hastings, New York, cut a few corners during the process of removing asbestos, its owner was sentenced to 37 months in prison and three years of supervised probation. He was fined $5,000 personally, and his company was fined $55,000. The company is now out of business—a heavy price to pay for cutting corners.

their answers indicate are necessary before allowing workers to be exposed:

1. What chemicals are present at the job site?
2. What is the route of entry into the body for this chemical?
3. What are the hazards associated with this chemical?
4. What parts of the body does this chemical attack?
5. How can this chemical be detected?
6. What is the permissible exposure limit (PEL) for the chemical?
7. What is the recommended exposure limit (REL) for the chemical?
8. What are the immediately dangerous to life and health (IDLH) values for the chemical?
9. What is the threshold limit value (TLV) for the chemical?
10. Does a regulated area need to be established to control exposure of workers to the chemical?
11. Do signs need to be posted warning personnel of the presence of the chemical?
12. What types of exposure monitoring equipment and procedures should be used?
13. What type of training should workers have before being exposed to the chemical?
14. What types of personal protective equipment (PPE) should workers use if they might be exposed to the chemical?
15. What types of medical surveillance should be available for workers who might be exposed to the chemical? What types of records should be maintained?

SUBPART AA: CONFINED SPACES IN CONSTRUCTION

For years construction professionals took their guidance on working in confined spaces from OSHA's General Industry Standard (29 CFR 1910.146). But this is no longer the case. In 2015 OSHA adopted 29 CFR 1926.1200, Subpart AA—*Confined Spaces in Construction.* There are two types of confined spaces that construction professionals should be concerned with: (1) confined space, and (2) permit-required confined space. It is important to be able to distinguish between the two because there are different requirements for working in them.

A *confined space* is any working space that meets the following criteria: (1) large enough for a worker to enter, (2) is not suited for continuous occupancy, and (3) has limited accessibility for entry and exit. A *permit-required confined space* meets these same three criteria but also has one or more of the following hazards: (1) hazardous atmosphere (or the potential for one), (2) contains a material that could potentially engulf a worker, (3) internal configuration that could entrap or asphyxiate a worker (sloping floor, converging walls, etc.), or (4) any other recognized serious hazard.

Confined Spaces and Typical Hazards

There are many different situations in construction that might involve workers entering confined spaces. Some of the more common types of confined spaces in construction include the following: ventilation ducts, manholes, tanks, and sumps. Typical hazards confronted in confined spaces in construction include: toxic and/or explosive fumes, vapors, gases. Oxygen deficiency is a common problem in confined spaces.

Subpart AA Coverage

Subpart AA runs from 1926.1200 through 1926.1213 as follows:

- 1926.1200 (Reserved)
- 1926.1201 Scope
- 1926.1202 Definitions
- 1926.1203 General requirements
- 1926.1204 Permit-required controlled space program
- 1926.1205 Permitting process
- 1926.1206 Entry permit
- 1926.1207 Training
- 1926.1208 Duties of authorized entrants
- 1926.1209 Duties of attendants
- 1926.1210 Duties of entry supervisors
- 1926.1211 Rescue and emergency services
- 1926.1212 Employee participation
- 1926.1213 Provision of documents to Secretary

Important Definitions

OSHA provides a list of definitions in 29 CFR 1926.1202, all of which are important and should be known to construction professionals and students. Of these the following are especially important:

- *Acceptable entry conditions.* Conditions that must exist in a permit space before a worker is allowed to enter the space.
- *Attendant.* An individual stationed outside one or more permit spaces who assesses the status of authorized entrants and who is responsible for performing the duties that are specified in 1926.1209.
- *Competent person.* An individual who is capable of identifying existing and predictable hazardous, unsanitary, or dangerous conditions in the work environment and who is authorized to take prompt action to eliminate them.

- *Controlling contractor.* Employer with overall responsibility for construction at the worksite.
- *Entry employer.* The employer that decides if a worker it directs may enter a permit space.
- *Entry permit.* Written or printed document that is provided by the employer who designated the space a permit space to allow and control entry into a permit space (must contain all of the information specified in 1926.1206).
- *Entry supervisor.* Qualified person responsible for determining if conditions are acceptable for entry into a permit space. This individual authorizes entry, supervises operations, and terminates work in the space. This individual may also serve as an attendant or authorized entrant provided he or she is properly trained and equipped for the roles.
- *Host employer.* Employer that owns or manages the property where the construction takes place (this can be the general contractor if the contract transfers overall responsibility from the owner to the general contractor).
- *Monitoring.* On-going observation and identification of hazards once an authorized individual has entered the space.
- *Qualified person.* An individual who by virtue of having a recognized degree, certificate, professional standing, or knowledge, training, and experience has demonstrated the ability to solve problems relating to the work at hand.

Overview of Requirements

The requirements of Subpart AA are extensive and detailed, but in a nutshell they amount to this: construction professionals are responsible for identifying confined spaces, identifying hazards in those spaces, taking appropriate measures for eliminating or mitigating the hazards, following prescribed rules for entry into the spaces, and ensuring that employees who work in confined spaces are properly trained, equipped, monitored, and supervised.

Construction professionals and students are encouraged to download the entire standard and become familiar with all aspects of it before overseeing work in confined spaces. What follows are just a few of the general requirements (1926.1203) of the standard for the purpose of illustration:

- Before work begins employers must ensure that a competent person identifies all confined spaces in which one or more employees might work as well as which space(s) is a permit space.
- When permit spaces are present, the employer who identifies them must: (1) inform exposed employees by posting appropriate danger signs or other equally effective forms of notification, and (2) inform in a timely manner and in a manner other than posting the authorized representatives of its workers and the controlling contractor of their location and the hazards they contain.
- Employers who receive notice of or identify themselves permit spaces must ensure that no worker enters the space who is not authorized.
- Employers must a have a written permit space program that complies with 1926.1204 before authorizing workers to enter permit spaces.

These requirements are just a few of those contained in 1926.1200. Construction professionals and students are encouraged to familiarize themselves with all applicable parts of the standard before overseeing work in confined spaces and permit spaces.

SUBPART CC: CRANES AND DERRICKS

Subpart CC (which used to be part of Subpart N) contains the requirements for safe operation and maintenance of cranes and derricks. Subpart CC contains the following sections:

- 1926.1400 Scope
- 1926.1401 Definitions
- 1926.1402 Ground conditions
- 1926.1403 Assembly/disassembly—selection of manufacturer or employer procedures
- 1926.1404 Assembly/disassembly—general requirements
- 1926.1405 Disassembly—additional requirements for dismantling booms and jobs
- 1926.1406 Assembly/disassembly—employer procedures—general requirements
- 1926.1407 Power line safety (up to 350 KV)—assembly and disassembly
- 1926.1408 Power line safety tips (up to 350 KV)—equipment operations
- 1926.1409 Power line safety (over 350 KV)
- 1926.1410 Power line safety (all voltages)—equipment operations closer than the Table A zone
- 1926.1411 Power line safety while traveling
- 1926.1412 Inspections
- 1926.1413 Wire rope—inspection
- 1926.1414 Wire rope—selection and installation criteria
- 1926.1415 Safety devices
- 1926.1416 Operational aids
- 1926.1417 Operation
- 1926.1418 Authority to stop operation
- 1926.1419 Signals—general requirements

- 1926.1420 Signals—radio, telephone or other electronic transmission of signals
- 1926.1421 Signals—voice signals—additional requirements
- 1926.1422 Signals—hand signals chart
- 1926.1423 Fall protection
- 1926.1424 Work area control
- 1926.1425 Keeping clear of the load
- 1926.1426 Free fall and controlled load lowering
- 1926.1427 Operator qualification and certification
- 1926.1428 Signal person qualifications
- 1926.1429 Qualifications of maintenance and repair employees
- 1926.1430 Training
- 1926.1431 Hoisting personnel
- 1926.1432 Multiple crane/derrick lifts—supplemental requirements
- 1926.1433 Design, construction and testing
- 1926.1434 Equipment modifications
- 1926.1435 Tower cranes
- 1926.1436 Derricks
- 1926.1437 Floating cranes/derricks and land cranes/derricks on barges
- 1926.1438 Overhead and gantry cranes
- 1926.1439 Dedicated pile drivers
- 1926.1440 Sideboom cranes
- 1926.1441 Equipment with a rated hoisting/lifting capacity of 2,000 pounds or less
- 1926.1442 Severability

Overview of the Revisions

Subpart CC was established to prevent the leading causes of fatalities associated with cranes and derricks including electrocution, crushed-by/struck-by hazards during assembly, disassembly, collapse, and overturn. Subpart CC establishes requirements for ground conditions and crane operator assessment and certification, tower crane hazards, and use of synthetic slings for assembly and disassembly work. The primary requirements of Subpart CC fall into the following categories: assembly/disassembly, operator qualification/certification, signal person qualifications, and qualified rigger.

Assembly/Disassembly Requirements

Work must be directed by an Assembly/Disassembly director or A/D who meets the criteria of a "competent person" and a "qualified person" with the exception that the director can be a competent person assisted by a qualified person. The A/D must ensure that all members of the crew know their tasks and the hazards associated with them and must verify all capacities for all equipment used. The A/D must address all hazards associated with the operation including 12 specified areas: site and ground conditions, blocking material, proper location of blocking, verifying assist crane loads, boom and pick points, center of gravity, stability upon pin removal, snagging, struck by counterweights, boom hoist brake failure, loss of backward stability, and wind speed and weather.

Inspection Requirements

Equipment must be inspected after it is assembled but before it is used by a qualified person to ensure that it is assembled according to manufacturer specifications. If manufacturer specifications are not available, the qualified person—with the assistance of a registered professional engineer if necessary—must develop appropriate inspection criteria.

General Requirements

- Before moving to a position out of view of the operator where he or she might be injured by movement of the equipment or its load, a crew member must inform the operator. The operator must not move the equipment until the crew member verifies to the operator that he or she is now in a safe position.
- No person is to be under the boom or jib when pins or similar devices are being removed unless the A/D has taken appropriate precautions to prevent injuries.
- Component weights must be readily available for all components that are going to be assembled and all rigging must be accomplished by a "qualified rigger."
- During disassembly pins may not be removed if the pendants are under tension.
- Booms supported only by cantilevering must not exceed the manufacturer specified limits or RPE limitations—whichever applies.
- Component selection and equipment configuration must be in accordance with manufacturer specifications or RPE limits—whichever applies.

Synthetic Sling Requirements

When using synthetic slings during assembly or disassembly rigging, manufacturer procedures must be followed. This rule holds even when the employer has developed its own alternative A/D procedure. Further, synthetic slings must be protected from any condition that might reduce sling capacity such as abrasion, sharp edges, etc.

Outriggers and Stabilizers Requirements

When the load to be handled and the operating radius require outriggers or stabilizers, the following requirements apply:

- Outriggers and stabilizers must be deployed according to the manufacturer load chart or fully deployed. Further, they must be set in such a way as to remove weight from the wheels except in the case of locomotive cranes.
- When using outrigger floats, the floats must be attached to the outriggers. When using stabilizer floats, the floats must be attached to the stabilizer.
- Each outrigger and float must be visible to the operator or, at a minimum, to a signal person during extension and setting.
- Blocking for outriggers and stabilizers must be placed under the float/pad of the jack. If there is no jack, the blocking should be placed under the outer bearing surface or the outrigger or stabilizer beam. Blocking must be properly placed, and able to sustain the loads in question and to maintain stability.

Tower Crane Requirements

Tower cranes are subject to all requirements set forth above as well as to additional requirements for erecting, climbing, and dismantling. These requirements include a pre-erection inspection (Refer to 29 CFR 1926.1435).

Operator Qualification and Certification Requirements

Any person engaged in a construction activity who operates a crane covered by the crane and derrick rule must be qualified and certified. Exceptions are those who operate sideboom cranes, derricks, and equipment with a rated hoist/lift capacity of 2,000 lbs. or less. Certification has two parts: (1) A written test covering the type of equipment the applicant will operate and the subject matter criteria set forth in 29 CFR 1926.1427(j), and (2) A practical exam in which the applicant must demonstrate his or her ability to safely and properly operate the equipment in question and related skills. There are four ways for an operator to become certified: (1) receiving a certificate from an accredited crane operation testing organization, (2) through an audited program offered by an employer, (3) qualification by one of the armed military services (this does not include private defense contractors), and (4) licensing by a state or local government program that meets all OSHA requirements.

Signal Person Qualification Requirements

A signal person is required at any time the point of operation is not in full view of the operator, the operator's view is obstructed in the direction of travel, or when the operator or person handling the load thinks a signal person is needed because of site-specific concerns. To be considered qualified, a signal person must: (1) know all of the signals used at a worksite and how to properly use them, (2) understand the operation and limitations of the equipment in question, (3) know the signal person qualification requirements set forth in Subpart CC (1926.1419-1422 and 1926.1428), and (4) pass an oral or written test and a practical test. The test may be administered by a third-party evaluator or the employer's evaluator.

Qualified Rigger Requirements

A qualified rigger is required during hoisting operations and assembly/disassembly operations. A qualified rigger is also required when workers are in the fall zone when hooking, unhooking, or guiding a load or doing the initial connection of a load to a component or a structure. A qualified operator is not necessarily considered a qualified rigger. Determining if an individual is a qualified rigger requires consideration of the nature of the specific load, lift, and equipment in question.

Summary

Subpart V covers the requirements and related safety practices for power transmission and distribution. Specific topics include the following: tools and protective equipment; mechanical equipment; material handling; grounding for protection of employees; overhead lines; underground lines; construction in energized substations; external load helicopters; and lineman's body belts, safety straps, and lanyards.

Subpart W covers the requirements and related safety practices for rollover protective structures (ROPS) and for overhead protection in these structures. Specific topics include ROPS for material-handling equipment, minimum performance criteria for ROPS, protective frame test procedures and performance requirements, and overhead protection for operators of agricultural and industrial tractors.

Subpart X covers the requirements and related safety practices for stairways and ladders. Specific topics include stairways, ladders, and training requirements.

Subpart Y covers the requirements and related safety practices for commercial diving operations. Specific topics include personnel requirements, general operations procedures, specific operations procedures, equipment procedures and requirements, and record keeping.

Subpart Z covers the requirements and related safety practices for toxic and hazardous substances. The primary focus of the coverage in this subpart is asbestos. A model for dealing with potentially hazardous substances that may be present at a construction site is also presented.

Key Terms and Concepts

Asbestos
Commercial diving operations
Construction in energized substations
Diving requirements
Doors and gates
Exposure assessments and monitoring
Exposure monitoring
Grounding for protection of employees
Handrails and stair-rails
Initial exposure assessments
Labels
Lineman's body belts, safety straps, and lanyards
Medical surveillance
Metal pan landings and treads
Negative exposure assessments
Notification
Overhead lines
Overhead protection
Periodic monitoring
Permissible exposure limit (PEL)
Planning a dive operation
Proper landings
Protrusions and slippery surfaces
Record keeping for diving operations
Regulated areas
Riser height and tread width
Rollover protective structures (ROPS)
Safe diving procedures manual
Signs
Slope angle
Temporary treads
Tools and protective equipment
Toxic and hazardous substances
Underground lines

Review Questions

1. List six general requirements for power transmission and distribution.
2. What is the minimum working and clear hot stick distance for tools when the voltage range is 46.1 to 72.5 kV?
3. Describe the safe way to transport utility poles.
4. Describe safe practices for grounding for the protection of employees.
5. What precautions should be taken when working on underground lines?
6. What precautions should be taken when work must be performed in energized substations?
7. What is the purpose of ROPS?
8. Is it safer for an operator to jump out of the cab of a piece of equipment that is rolling over or to stay inside? Why?
9. When stairs will be used as part of a construction project, what are the requirements for proper landings, slope angle, doors and gates, and temporary treads?
10. List four unsafe practices that should be avoided when using a ladder on a construction site.
11. What are the required topics for a ladder-training program?
12. What are the three areas in which an individual must have either experience or training before participating in a commercial diving operation?
13. What information must be contained in the safe-diving procedures manual?
14. What are the areas that must be included in the safety and health assessment that is made when planning a commercial dive operation?
15. List the requirements and safety practices that must be followed during a commercial diving operation.
16. What records must be kept by the employer for every commercial diving operation undertaken?
17. How long should decompression procedure assessment evaluations be retained?
18. What is asbestos and why is it so dangerous?
19. Compare and contrast the following classes of work as they relate to asbestos: Class I, II, II, and IV.
20. What is the PEL for asbestos?
21. What is the "initial exposure assessment" as it relates to asbestos work, and what is required concerning the assessment?
22. What is "medical surveillance" as it relates to asbestos work, and what are the requirements for medical surveillance?
23. What is a "regulated area" as it relates to asbestos, and what are the requirements for regulated areas?
24. What are the general training requirements for employees who will perform asbestos work?
25. List at least five questions that construction professionals should ask and answer about any potentially hazardous substance that might be present on a construction site.

Critical Thinking and Discussion Activities

1. Dave Edwards, the supervisor for Jones Electric Company, is feeling a little pushed. His project is behind schedule, and he thinks his employees are "moving too slow." Just this morning before starting work, he held a "chalk talk" and told his team members they needed to step things up a little or the project was going to run over budget. If this happens, late fees will be assessed. Now, two hours later, he is already facing a problem that is slowing things down. Workers rotate in performing certain tasks. One of the tasks requires that workers wear rubber gloves and use rubberized equipment and tools. When Jay Barker took off the rubber gloves and handed them and his tools to his relief, Mike

Clanton, Clanton refused to use the gloves or the tools before they were thoroughly inspected for wear and pin holes. The supervisor, Dave Edwards, is really upset. "Come on Clanton. Jay just wore those gloves for two hours without any problems. They don't need to be tested. You're just wasting time we don't have." "I'm not putting those gloves on until they've been air tested," responded Mike Clanton. Who is right here? Why?

2. "I don't care what you say. If this bulldozer starts to turn over and I can get out, I'm going to," says the operator. "No way. I want you to strap yourself in that cab and stay there. You'll be safer inside," says the supervisor. Debate the pros and cons of this disagreement. Who is right?
3. "Come on Jones," said the carpenter. "I've been using ladders since I was a kid. I don't need training on how to use a ladder." "Yes, you do," responds the supervisor. "You might know how to use a ladder, but you don't know how to use one safely. You're going to attend the training just like the rest of us." Is ladder training really necessary, or is OSHA just being intrusive and overly cautious?
4. "What is this stuff?" asked the mason. "I don't know but it really cleans brick," responded the supervisor. "We used it on our last job, and the brickwork looked great. Using this stuff, you don't have to be so careful to trowel off extra concrete that gets on the brick." "I'm all for that," said the mason. "If you've used it before, it can't be too dangerous." What do you think about this exchange? Would you use this particular brick-cleaning compound without knowing more about it? What would you do if you were this mason?

Application Activities

1. When working at elevated locations, employees are required to wear body belts, safety straps, and lanyards that meet the specifications set forth in ASTM Standard B-117-6. Conduct the research necessary to determine what those specifications are.
2. Visit several job sites where scrapers, loaders, dozers, graders, or crawler tractors are being used. Observe the rollover and overhead protection on the equipment being used. Does it appear to meet OSHA requirements?
3. This activity is for students located in communities in proximity to large bodies of water. Identify a commercial diving company in your community that will cooperate with you in completing this activity. Ask to preview the company's "safe diving procedures manual." Write a critique of the manual. Is it comprehensive? Would you add or change anything?
4. Identify a contractor in your community who performs asbestos abatement work. Ask to review the company's general training program for employees. Write a critique of the training programs. Is it comprehensive? Would you add or change anything?

Endnotes

1. 29 CFR 1926.950
2. 29 CFR 1926.951
3. 29 CFR 1926.952
4. 29 CFR 1926.953
5. 29 CFR 1926.954
6. 29 CFR 1926.955
7. 29 CFR 1926.956
8. 29 CFR 1926.957
9. 29 CFR 1926.958
10. 29 CFR 1926.959
11. 29 CFR 1926.1000–1926.1003
12. 29 CFR 1926.1052
13. 29 CFR 1926.1053 and 1926.1060
14. 29 CFR 1926.1076–1926.1087
15. 29 CFR 1926.1101
16. 29 CFR 1926.1102–1926.1148

Glossary

Accessible Not impeded or guarded by locked gates, doors, elevation, or other means.

Adjustable suspension scaffold A suspension scaffold equipped with a hoist that can be operated by a worker on the scaffold.

Aerated solid powders Any powdered material used as a coating material that is fluidized within a container by passing air uniformly from below.

Alive or live Electrically connected to a source of power (energized or electrically charged) with a potential significantly different from that of the earth.

Anchorage A secure point of attachment for lifelines, lanyards, or deceleration devices.

ANSI American National Standards Institute.

Approved Accepted as satisfactory by a duly constituted and nationally recognized authority or agency.

Approved storage facility A facility that is rated as acceptable for the storage of explosive materials.

Atmospheric pressure The pressure of air at sea level (usually 14.7 psia).

Authorized person A responsible person, approved or assigned by the employer to perform specific duties.

Automatic Self-acting, once activated.

Bare conductor A conductor of electricity that has no covering or electrical insulation.

Barricade A physical impediment, such as a tape, screen, or cone, that limits access to a hazardous area.

Barrier A physical obstruction intended to prevent access to or contact with hazardous conditions.

Bell An enclosed compartment that allows a diver to be transported to and from an underwater work area.

Benching A method of protecting employees from cave-ins by terracing the sides of an excavation to form one or a series of horizontal levels or steps.

Blast area The limited access area in which blasting operations are conducted.

Blaster The individual who is authorized and qualified to use explosives for blasting.

Blasting agent Any material or mixture consisting of a fuel and oxidizer used for blasting, but not classified as an explosive itself.

Blasting cap A metallic tube, closed at one end, that consists of a charge of detonating compounds and is capable of detonation from sparks or flame.

Body belt (safety belt) A strap with a means for securing it around the waist and for attaching it to a lanyard, lifeline, or deceleration device.

Body harness A system of straps that may be secured about the worker so as to distribute the fall arrest forces over the body and that can be attached to other components of a personal fall arrest system.

Boiling point The boiling point of a liquid at a pressure of 14.7 psia.

Bonding Permanent joining of metallic parts to form an electrically conductive path.

Bottom time The total amount of time measured in minutes from the time when the diver leaves the surface in descent to the time that the diver begins ascent.

Brace A rigid connector that holds one member in a fixed position with respect to another member or to a building or structure.

Bricklayers' square scaffold A scaffold composed of framed squares that support a platform.

Buddy system A system of organizing employees into work teams of at least two, so that each member of the team is observed by at least one other member. The purpose of the buddy system is rapid response in the event of an emergency.

Bulkhead An airtight wall separating the working chamber from free air or from another chamber under a lesser pressure than the working pressure.

Caisson An airtight and watertight chamber in which people can work under air pressure greater than atmospheric pressure and below water level.

Catastrophic release A major uncontrolled emission of one or more highly hazardous substances that presents serious danger to people in the vicinity.

Cave-in The sudden movement of soil or rocks into an excavation, either by falling or sliding in sufficient quantity to entrap, bury, or otherwise injure or immobilize a person.

Certified Equipment is certified if it (1) has been tested and found to meet applicable test standards or to be safe for use in a specified manner and (2) is of a kind whose production is periodically inspected by a qualified testing laboratory.

Chemical Any element, compound, or mixture of elements and/or compounds.

Circuit A conductor or system of conductors through which an electric current is intended to flow.

Class A explosives Possessing a detonating hazard and detonating primers.

Class B explosives Possessing a flammable hazard, such as propellant explosives.

Class C explosives Certain types of manufactured products that contain restricted amounts of Class A or Class B explosives, or both.

Clean-up operation An operation in which hazardous materials are processed or handled for the purpose of making the site safer for people and the environment.

Combustible liquid Any liquid having a flashpoint at or above 100°F (37.8°C), but below 200°F (93.3°C).

Combustion Any chemical process that involves oxidation sufficient to produce light or heat.

Competent person A person who is capable of identifying existing and predictable hazards and who has the authority to take prompt corrective measures to eliminate them.

Conductor A material suitable for carrying an electric current.

Connector A device used to couple (connect) parts of a personal fall arrest system and positioning device systems together.

Construction work Work for the purpose of building, altering, or repairing, including painting and decorating.

Contaminant A material likely to cause physical harm to a person.

Controlled access zone (CAZ) An area in which certain work may take place without the use of guardrail systems, personal fall arrest systems, or safety net systems and to which access is controlled.

Coupler A device for locking together the tubes of a tube-and-coupler scaffold.

Cross braces The horizontal members of a shoring system installed perpendicular to the sides of an excavation.

Damp location Locations subject to moderate degrees of moisture, such as some basements and roofed open porches.

Dangerous equipment Equipment that may be hazardous to employees who fall onto or into it.

Dead (deenergized) Free from any electrical connection to a source of power.

Deceleration device Any mechanism that serves to dissipate a substantial amount of energy during a fall arrest or otherwise limit the energy imposed on a person during fall arrest.

Deceleration distance The distance between the location of an employee's body belt or body harness attachment point at the onset of fall arrest forces and the location of that attachment point after the employee comes to a full stop.

Decompression chamber A pressure vessel used to decompress divers and to treat decompression sickness.

Decompression sickness A condition that may result from gas or bubbles in the tissues of divers after pressure reduction.

Decontamination The removal of hazardous substances to the extent necessary to prevent adverse health effects.

Defect Any characteristic or condition that tends to weaken or reduce the strength of a tool, object, or structure.

Designated employee A qualified person assigned to perform specific duties.

Detonating cord A flexible cord containing a center core of high explosives with sufficient strength to detonate other cap-sensitive explosives.

Detonator Blasting caps, electric blasting caps, delay electric blasting caps, and nonelectric delay blasting caps.

Dive team Divers and support workers involved in a diving operation.

Diver A person working underwater using a breathing apparatus that supplies compressed breathing gas at the ambient pressure.

Diving mode A general term used to describe any type of diving requiring specific equipment, procedures, and techniques.

Dry location A location not normally subject to dampness or wetness.

Dry soil Soil that exhibits no visible signs of moisture.

Effectively grounded Intentionally and sufficiently connected to earth to prevent a buildup of voltage that may result in hazards.

Electric blasting cap A blasting cap that detonates by means of an electric current.

Employee Any worker under the OSH Act, regardless of the contractual relationship.

Enclosed Surrounded by a case, housing, fence, or walls that prevent persons from accidentally contacting energized parts or other hazards.

Excavation Any man-made cut, cavity, trench, or depression in the earth formed by earth removal.

Explosive Any chemical compound, mixture, or device intended to function by explosion.

Exposure Subjected to a hazardous substance in the course of work.

Faces or sides The vertical or inclined earth surfaces formed as a result of excavation work.

Fire brigade Organized group of workers who are knowledgeable, trained, and skilled in the safe evacuation of employees during emergency situations and in assisting in firefighting operations.

Fixed ladder A ladder that cannot be readily moved because it is an integral part of a structure.

Flammable liquids Any liquid having a flash point below 140°F and having a vapor pressure not exceeding 40 psia at 100°F.

Flammable Capable of being easily ignited, burning intensely, or having a rapid rate of flame spread.

Flash point The temperature at which a liquid gives off vapor sufficient to form an ignitable mixture with the air near the surface of the liquid.

Foreseeable emergency Any potential occurrence that could result in hazardous conditions in the workplace.

Formwork The total system of support for freshly placed or partially cured concrete.

Free fall The act of falling before a personal fall arrest system begins to apply force to arrest the fall.

Free-fall distance The vertical displacement of the fall arrest attachment point on the worker's body belt or harness between onset of the fall and just before the system begins to apply force to arrest the fall.

Gauge pressure (psig) Pressure measured by a gauge and indicating the air pressure exceeding that of the atmosphere.

Grounded Connected to earth or to some conducting body that serves in place of the earth.

Guarded Protected in accordance with standard barricading techniques to prevent access or potentially hazardous contact by persons or objects.

Guardrail system A vertical barrier, consisting of rails and posts, that is erected to prevent people from falling from a surface to lower levels.

Handrail A rail used to provide employees with a handhold for support.

Hazard warning A message, symbol, or picture, or a combination thereof, appearing on a label to convey the hazard(s) in a container.

Hazardous atmosphere A harmful atmosphere that may cause death, illness, or injury.

Hazardous chemical Any chemical that is a physical hazard or a health hazard.

Hazardous substance A harmful substance that is likely to cause death or injury.

Hazardous waste site Any facility or location at which hazardous waste operations take place.

Health hazard A substance for which there is statistically significant evidence that acute or chronic health effects may occur in exposed people.

Hoistway Any vertical opening or space in which an elevator or dumbwaiter is operated.

Hot work Work involving electric or gas welding, cutting, brazing, or similar flame- or spark-producing operations.

Hyperbaric conditions Pressure conditions in excess of surface atmospheric pressure.

Immediately dangerous to life or health (IDLH) An atmospheric concentration of any toxic, corrosive, or asphyxiant substance that poses an immediate threat to life.

Individual-rung step ladders Ladders without a side rail or center rail support. Individual steps or rungs mounted directly to the side or wall of a structure.

Insulated Separated from other conducting materials by a dielectric substance that offers a high resistance to the passage of current.

Jacking operation An operation that involves lifting a slab or group of slabs vertically from one location to another during the construction of any structure in which the lift-slab process is being used.

Jacob's ladder A marine ladder made of rope or chain with wooden or metal rungs.

Job-made ladder A ladder that is made by employees, typically at the construction site, and is not commercially manufactured.

Kickout The accidental release or failure of a cross brace.

Label Any written, printed, or pictorial material displayed on or affixed to a container.

Landing A platform at the end of a flight of stairs.

Lanyard A flexible line of rope, wire rope, or strap that has a connector at each end for connecting the body belt or body harness to a deceleration device, lifeline, or anchorage.

Leading edge The edge of a floor or roof or formwork for a floor or other walking working surface (such as the deck) that changes location as additional sections are placed, formed, or constructed.

Lifeline A rope, suitable for supporting one person, to which a lanyard or safety belt or harness is attached.

Lift-slab process A method of concrete construction in which floor and roof slabs are cast onsite at ground level and lifted into position using jacks.

Limited access zone An area alongside a masonry wall that is under construction and is clearly marked off to limit access.

Manhole An underground enclosure that people may enter for the purpose of installing, operating, and maintaining equipment or cable.

Material safety data sheet (MSDS) Written or printed material containing important information about a hazardous substance.

Maximum intended load The total load of all employees, equipment, tools, materials, transmitted loads, and other loads anticipated to be applied at any one time.

Misfire An explosive charge that failed to detonate.

Nonelectric delay blasting cap A blasting cap with an integral delay element capable of being detonated by an impulse or signal from miniaturized detonating cord.

Opening A gap of 30 inches (76 cm) or more high and 18 inches (48 cm) or more wide, in a wall or partition, through which employees can fall to a lower level.

Overcurrent Any current in excess of the rated current of equipment or the ampacity of a conductor.

Overhand bricklaying The process of laying bricks and masonry units in which the workload surface is on the opposite side of the wall from the mason, requiring the mason to lean over the wall to complete the work.

Overload Operation of equipment in excess of normal, full-load rating or of a conductor in excess of rated ampacity.

Oxygen deficiency That concentration of oxygen by volume (less than 19.5 percent) below which atmosphere supplying respiratory protection must be provided.

Personal fall arrest system A system used to arrest an employee's fall. It consists of an anchorage, connectors, and a body belt or body harness and may include a lanyard, deceleration device, lifeline, or combinations of these.

Physical hazard A substance for which there is scientifically valid evidence that it is a combustible liquid, a compressed gas, explosive, flammable, an organic peroxide, an oxidizer, pyrophoric, unstable (reactive), or water reactive.

Platform A work surface elevated above lower levels.

Point of access All areas used for work-related passage from one area or level to another (e.g., open areas include doorways, passageways, stairway openings, studded walls).

Portable ladder A ladder that can be readily moved or carried.

Post-emergency response The portion of an emergency response performed after the immediate threat has been stabilized or eliminated and cleanup of the site has begun.

Power-operated hoist A hoist that is powered by other than human energy.

Precast concrete Concrete members (walls, panels, slabs, columns, and beams) that have been formed, cast, and cured before being erected.

Pressure A force acting on a unit of area, typically shown as pounds per square inch (psi).

Primer A cartridge of explosives into which a detonation device is inserted.

Process Use, storage, manufacturing, handling, or the onsite movement of chemicals or a combination of these activities.

psia Pounds per square inch absolute.

psig Pounds per square inch gauge.

Qualified person A person who, through experience, training, or both, is knowledgeable of an operation to be performed and the hazards involved in it.

Radiant energy Energy that travels outward in all directions from its sources.

Ramp An inclined walking or working surface that is used to gain access to one point from another.

Rated load The manufacturer's specified maximum load to be lifted or supported.

Readily accessible Capable of being reached quickly without having to climb over or remove obstacles or resort to portable ladders or chairs.

Safety belt A device usually worn around the waist and attached to a lanyard and lifeline or to a structure to prevent a worker from falling.

Safety fuse A flexible cord containing an internal burning medium that conveys fire for the purpose of igniting blasting caps.

SCUBA diving A diving mode that is independent of surface air supply in which the diver uses a self-contained underwater breathing apparatus.

Secondary blasting The reduction of oversize material to a better size for handling by the use of explosives.

Self-retracting lifeline or lanyard A fall protection or deceleration device containing a drum-wound line that can be slowly let out or wound in slight tension during normal employee movement and that, after onset of a fall, automatically locks the drum and arrests the fall.

Service drop The overhead service conductors from the last pole connecting to the service-entrance conductors at the building or other structure.

Service The conductors and equipment for delivering energy from the electricity supply system to the wiring system of the structure served.

Sheeting The members of a shoring system that hold the earth in position and in turn are supported by other members of the shoring system.

Shield (shield system) A structure that can withstand the forces imposed on it by a cave-in and, thereby, protect workers.

Shore A supporting member that resists a compressive force imposed by a load.

Shoring (shoring system) A structure that supports the sides of an excavation and is designed to prevent cave-ins.

Signals Moving signs carried by workers or by devices to warn of possible or existing hazards.

Single-rail ladder A portable ladder with rungs, cleats, or steps mounted on a single rail instead of the normal two rails used on most other ladders.

Sloping (sloping system) A method of protecting employees from cave-ins by forming sides of an excavation that are inclined away from the trench so as to prevent cave-ins.

Snaphook A connector comprised of a hook-shaped device and a keeper that may be opened to permit the hook to grab an object and, when released, automatically closes to retain the object.

Soil classification system A method of categorizing soil and rock deposits in a hierarchy of stable rock, Type A, Type B, and Type C, in decreasing order of stability. The categories are determined based on an analysis of the properties and performance characteristics of the deposits and the characteristics of the deposits and the environmental conditions of exposure.

Special decompression chamber A chamber to provide greater comfort for workers when total decompression time exceeds 75 minutes.

Specific chemical identity The chemical name, Chemical Abstracts Service (CAS) Registry Number, or any other information that reveals the precise chemical designation of a substance.

Spiral stairway A series of steps attached to a vertical pole and progressing upward in a winding fashion within a cylindrical space.

Spray room A room in which spray-finishing operations not conducted in a spray booth are performed separately from other areas.

Spraying area Any area in which dangerous quantities of flammable vapors or mists, or combustible residues, dusts, or deposits are present due to the operation of spraying processes.

Stable rock Natural solid mineral material that can be excavated with vertical sides and will remain intact without support while exposed.

Stair-rail system A vertical barrier erected along the unprotected sides and edges of a stairway to prevent employees from falling to lower levels.

Stall load The load at which the prime mover of a power-operated hoist stalls or the power to the prime mover is automatically disconnected.

Standby diver A diver at the dive location available to assist a diver in the water.

Stemming An inert, incombustible material or device used to confine or separate explosives in a drill hole or to cover explosives in mud capping.

Stilts A pair of poles or similar supports with raised footrests, used to permit walking above the ground or working surface.

Structural ramp A ramp that is typically used for vehicle access and is constructed of wood, steel, or both.

Surface-supplied air diving A diving mode in which the diver in the water is supplied from the dive location with compressed air for breathing.

Switch A device for opening and closing or changing the connection of a circuit. A switch is understood to be manually operable, unless otherwise stated.

Switching devices (Over 600 volts, nominal) Devices designed to close or open one or more electric circuits. Included in this category are circuit breakers, cutouts, disconnecting (or isolating) switches, disconnecting means, and interrupter switches.

Tag A system or method of identifying circuits, systems, or equipment for the purpose of alerting persons that the circuit, system, or equipment is being worked on; to warn of existing or immediate hazards.

Toeboard A low protective barrier that prevents the fall of materials and equipment to lower levels and provides protection from falls for personnel.

Transportable X-ray X-ray equipment that is installed in a vehicle or that may readily be disassembled for transport.

Tread depth The horizontal distance from front to back of a tread (excluding nosing, if any).

Trench (trench excavation) A narrow excavation (in relation to its length) made below the surface of the ground. In general, the depth is greater than the width, but the width of a trench (measured at the bottom) is not greater than 15 feet (4.6 m; measured at the bottom of the excavation); the excavation is also considered to be a trench.

Uncontrolled hazardous waste site An area identified as an uncontrolled hazardous waste site by a governmental body—federal, state, local, or other—where an accumulation of hazardous substances creates a threat to the health and safety of individuals or the environment, or both.

Unprotected sides and edges Any side or edge (except at entrances to points of access) of a walking or working surface; for example, a floor, roof, ramp, or runway, where there is no wall or guardrail system at least 39 inches (1.0 m) high.

Unstable (reactive) A chemical that will vigorously polymerize, decompose, or condense in the pure state or as produced or transported or will become self-reactive under conditions of shocks, pressure, or temperature.

Unstable material Earth material that, because of its nature or the influence of related conditions, cannot be depended upon to remain in place without extra support, such as would be furnished by a system of shoring.

Unstable objects Items whose strength, configuration, or lack of stability may allow them to become dislocated and shift and, therefore, may not properly support the loads imposed on them.

Uprights The vertical members of a trench shoring system placed in contact with the earth and usually positioned so that individual members do not contact each other.

Vapor pressure The pressure, measured in pounds per square inch (absolute), exerted by a volatile liquid as determined by the standard method of test for vapor pressure of petroleum products (Reid method).

Vault An enclosure above or below ground that personnel may enter and that is used for the purpose of installing, operating, and maintaining equipment and cable.

Ventilated Provided with a means to permit circulation of air sufficient to remove an excess of heat, fumes, or vapors.

Volatile flammable liquid A flammable liquid having a flash point below 38°C (100°F) or whose temperature is above its flash point, or a Class II combustible liquid having a vapor pressure not exceeding 40 psia at 38°C (100°F) with a temperature above its flash point.

Voltage (of a circuit) The greatest root-mean-square (effective) difference of potential between any two conductors of the circuit.

Voltage of a circuit not effectively grounded The voltage between any two conductors if one circuit is directly connected to and supplied from another circuit of higher voltage (as in the case of an autotransformer). Both are considered as of the higher voltage, unless the circuit of lower voltage is effectively grounded, in which case its voltage is not determined by the circuit of higher voltage.

Voltage of an effectively grounded circuit The voltage between any conductor and ground, unless otherwise indicated.

Voltage to ground For grounded circuits, the voltage between the given conductor and that point or conductor of the circuit that is grounded; for ungrounded circuits, the greatest voltage between the given conductor and any other conductor of the circuit.

Voltage, nominal A nominal value assigned to a circuit or system for the purpose of conveniently designating its voltage class (e.g., 120/240, 480Y/277, 600).

Voltage The effective (rms) potential difference between any two conductors or between a conductor and ground.

Volume tank A pressure vessel connected to the outlet of a compressor and used as an air reservoir.

Wales Horizontal members of a shoring system placed parallel to the excavation face whose sides bear against the vertical members of the shoring system or earth.

Walking or working surface Any surface, whether horizontal or vertical, on which workers walk or work, including, but not limited to, floors, roofs, ramps, bridges, runways, formwork, and concrete reinforcing steel, but not including ladders, vehicles, or trailers.

Walkway A portion of a scaffold platform used only for access and not as a work level.

Warning line system A barrier erected on a roof to warn employees that they are approaching an unprotected roof side or edge and which designates an area in which roofing work may take place without the use of guardrail, body belt, or safety net systems to protect workers in the area.

Water gels or slurry explosives A wide variety of materials used for blasting that contain substantial proportions of water and high proportions of ammonium nitrate.

Water-reactive A chemical that reacts with water to release a gas that is either flammable, toxic, or otherwise hazardous.

Watertight Constructed so that moisture will not enter.

Weatherproof Constructed or protected so that exposure to the weather will not interfere with successful operation.

Work area A room or defined space in a workplace where hazards might exist and where workers are present, or that portion of a walking or working surface where job duties are being performed.

Working chamber The working space or compartment under air pressure in which work is being done.

Working load A load imposed by people, materials, and equipment.

Working pressure The maximum pressure to which a pressure containment device may be exposed under standard operating conditions.

Workplace An establishment, job site, or project, at one specific location, containing one or more work areas.

Index

D

E

F

N

O

P